CONNECTING PEOPLE TO SCIENCE: A NATIONAL CONFERENCE ON SCIENCE EDUCATION AND PUBLIC OUTREACH

COVER ILLUSTRATION:

The 2011 ASP meeting was held July 30 through August 3, 2011, in Baltimore, Maryland.

Logo design by Leslie Proudfit, ASP.

ASTRONOMICAL SOCIETY OF THE PACIFIC
CONFERENCE SERIES

A SERIES OF BOOKS ON RECENT DEVELOPMENTS IN ASTRONOMY AND ASTROPHYSICS

Volume 457

EDITORIAL STAFF

Managing Editor: Joseph Jensen
Associate Managing Editor: Jonathan Barnes
Publication Manager: Pepita Ridgeway
Editorial Assistant: Cindy Moody
LATEX Consultant: T. J. Mahoney

MS 179, Utah Valley University, 800 W. University Parkway, Orem, Utah 84058-5999
Phone: 801-863-8804   E-mail: aspcs@aspbooks.org
E-book site: http://www.aspbooks.org

PUBLICATION COMMITTEE

ASPCS volumes may be found online with color images at http://www.aspbooks.org.
ASP Monographs may be found online at http://www.aspmonographs.org.

For a complete list of ASPCS Volumes, ASP Monographs, and
other ASP publications see http://www.astrosociety.org/pubs.html.

All book order and subscription inquiries should be directed to the ASP at
800-335-2626 (toll-free within the USA) or 415-337-2126,
or email service@astrosociety.org

ASTRONOMICAL SOCIETY OF THE PACIFIC
CONFERENCE SERIES

Volume 457

CONNECTING PEOPLE TO SCIENCE: A NATIONAL CONFERENCE ON SCIENCE EDUCATION AND PUBLIC OUTREACH

Proceedings of a conference held at
Baltimore, Maryland, USA
30 July–3 August 2011

Edited by

Joseph B. Jensen
Utah Valley University, Orem, Utah, USA

James G. Manning
Astronomical Society of the Pacific, San Francisco, California, USA

Michael G. Gibbs
Capitol College, Laurel, Maryland, USA

Doris Daou
NASA Lunar Science Institute, Washington D.C., USA

SAN FRANCISCO

ASTRONOMICAL SOCIETY OF THE PACIFIC
390 Ashton Avenue
San Francisco, California, 94112-1722, USA

Phone: 415-337-1100
Fax: 415-337-5205
E-mail: service@astrosociety.org
Web site: www.astrosociety.org
E-books: www.aspbooks.org

First Edition
© 2012 by Astronomical Society of the Pacific
ASP Conference Series
All rights reserved.

ISBN: 978-1-58381-796-4
e-book ISBN: 978-1-58381-797-1

Library of Congress (LOC) Cataloging in Publication (CIP) Data:
Main entry under title
Library of Congress Control Number (LCCN): 2012909200

Printed in the United States of America by Sheridan Books, Ann Arbor, Michigan.
This book is printed on acid-free paper.

Contents

Part I. Plenary Sessions

Part II. Workshops and Special Sessions

Part III. Oral Contributions

Part IV. Poster Contributions

Tremont Grand in Baltimore, site of the conference. Photo by Paul Deans.

Preface

"All things are bound together. All things connect."

Chief Seattle

In the summer of 2011, in Baltimore, Maryland, a national and international community of science education and public outreach (EPO) specialists, scientists, formal and informal educators, science communicators, and others met to share perspectives, learn from each other, seek collaborations, and consider the pressing issues surrounding science, science education, and "Connecting People to Science," which served as the organizing theme for the conference.

In the beautiful setting of the Tremont Grand Historic Venue—the fully restored old Baltimore Masonic Temple—operated by the Tremont Plaza Hotel, conferees enjoyed three and a half days of serious discussion, engaging demonstrations, and good fellowship in what constituted the 123rd annual meeting of the Astronomical Society of the Pacific (ASP), this year in partnership with the American Geophysical Union (AGU), the Space Telescope Science Institute (STScI), our local host, and many co-sponsors who are acknowledged in this volume.

Throughout the conference and its many plenary sessions, presentations, posters, and workshops, the theme of connection was paramount, as the assembled conferees worked to forge stronger bonds with each other that could translate into enhanced connections with their audiences in advancing common goals. This volume serves as a record of the conference, containing transcripts and summaries of many of the presentations given. It is our hope that the contents will provide a useful reference for those who attended and will acquaint others with the good work presented, as we all strive to find new and exciting ways to be the connective tissue that so vitally links the public to both the notion and the practice of science.

Founded in 1889 in San Francisco, the ASP's mission is to increase the understanding and appreciation of astronomy by engaging scientists, educators, enthusiasts and the public to advance science and science literacy. The ASP publishes both scholarly and educational materials, conducts professional development programs for formal and informal educators, and holds conferences, symposia and workshops for astronomers and educators specializing in education and public outreach. The ASP's education programs are funded by corporations, private foundations, the National Science Foundation, NASA, private donors, and its own members. More information about the Society may be obtained at its website: www.astrosociety.org.

James G. Manning
Executive Director
Astronomical Society of the Pacific

Meeting Committees

We are indebted to the following individuals who served tirelessly on the committees that made the Baltimore meeting a success.

Local Organizing Committee

Michael Gibbs, Capitol College
Rev. Frank Haig, S.J. Loyola University Maryland Department of Physics
Wayne (Skip) Bird, Westminster Astronomical Society
Kerri Beisser, The Johns Hopkins University Applied Physics Laboratory
Lou Mayo, NASA Goddard Space Flight Center
Stratis M. Kakadelis, Space Telescope Science Institute - Office of Public outreach
Susana Deustua, Space Telescope Science Institute
Rommel Miranda, Towson University Dept. of Physics, Astronomy & Geosciences
Jim O'Leary, Maryland Science Center
Jordan Raddick, The Johns Hopkins University Institute for Data-Intensive Engineering and Science
Carol Christian, Space Telescope Science Institute
Jennifer Scott
Brooke Hsu
Sten Odenwald

EPO Conference Program Committee:

Greg Schultz, Astronomical Society of the Pacific (committee co-chair)
Suzanne Jacoby, Large Synoptic Survey Telescope (committee co-chair)
Russanne (Rusty) Low, Institute for Global Environmental Strategies (committee co-chair)
Pranoti Asher, American Geophysical Union
Lindsay Bartolone, Adler Planetarium & Astronomy Museum
Lora Bleacher, NASA Goddard Space Flight Center
Bonnie Eisenhamer, Space Telescope Science Institute
Rick Fienberg, American Astronomical Society
Andrew Fraknoi, Foothill College and the ASP
Pamela Gay, Southern Illinois Univ. Edwardsville
Beth Hufnagel, Anne Arundel Community College
Laura Peticolas, Univ. of California, Berkeley
Julia Plummer, Arcadia University
Daniella Scalice, NASA
Theresa Schwerin, Institute for Global Environmental Strategies
Stephanie Shipp, Lunar and Planetary Institute
Christine Shupla, Lunar and Planetary Institute
Denise Smith, Space Telescope Science Institute
Connie Walker, National Optical Astronomy Observatory

Conference Staff:

Meeting website and graphics: Leslie Proudfit
Meeting coordination: Cinndy Hart
Meeting registrar: Albert Silva
Meeting Sponsorship: Michele Pearson

Conference Proceedings Editorial Assistance

S. Blaine Haws, ASP Conference Series
Richard Blake, Utah Valley University
Jonathan Barnes, Salt Lake Community College and ASP Conference Series
Albert Silva, ASP

Support and Sponsorship

The following organizations have contributed significant support to the success of this conference.

American Geophysical Union (partner) • Space Telescope Science Institute (partner)

~

NASA Lunar Science Institute • NASA's Exoplanet Exploration Program (managed by
the Jet Propulsion Laboratory) Infrared Processing and Analysis Center
NASA's Herschel Science Center • Spitzer Science Center

~

Stratospheric Observatory for Infrared Astronomy (SOFIA)
NASA's Chandra X-Ray Observatory

~

The University of Chicago Press • National Radio Astronomy Observatory (NRAO)
Ball Aerospace • Capitol College • Sky-Skan

~

University of Wyoming CAPER Team
The American Astronomical Society (AAS) and the AAS Education Office
Solar Dynamics Observatory (SDO) • Seiler Instrument

~

Explore Scientific • Pratt Street Ale House • MWT Associates, Inc
American Elements • Charlesbridge • Celestron

Other EPO Conference Series Volumes

The ASP Conference Series is pleased to have published several other volumes on the topics of Education and Public Outreach (EPO), including past ASP meeting proceedings, which may be of interest to the readers of the current volume. All are available for sale in print or electronic format at www.astrosociety.org and www.aspbooks.org.

Earth and Space Science: Making Connections in Education and Public Outreach, eds. Joseph B. Jensen, James G. Manning, and Michael G. Gibbs, ASP Conference Series Volume 443 (2011).

The Inspiration of Astronomical Phenomena VI, ed. Enrico Maria Corsini, ASP Conference Series Volume 441 (2011).

Learning from Inquiry in Practice, eds. Lisa Hunter and Anne Metevier, ASP Conference Series Volume 436 (2010).

Science Education and Outreach: Forging a Path to the Future, eds. Jonathan Barnes, Denise A. Smith, Michael G. Gibbs, and James G. Manning, ASP Conference Series Volume 431 (2010).

Cosmology Across Cultures, eds. José Alberto Rubiño-Martín, Juan Antonio Belmonte, Francisco Prada, and Anzton Alberdi, ASP Conference Series Volume 409 (2009).

Preparing for the 2009 International Year of Astronomy: A Hands-on Symposium, eds. Michael G. Gibbs, Jonathan Barnes, James G. Manning, and Bruce Partridge, ASP Conference Series Volume 400 (2008).

EPO and a Changing World: Creating Linkages and Expanding Partnerships, eds. Catharine D. Garmany, Michael G. Gibbs, and J. Ward Moody, ASP Conference Series Volume 389 (2008).

Part I

Plenary Sessions

Connecting People to Science
ASP Conference Series, Vol. 457
Joseph B. Jensen, James G. Manning, Michael G. Gibbs, and Doris Daou, eds.
© *2012 Astronomical Society of the Pacific*

Unscientific America: What's the Problem? What's the Solution?

Chris Mooney

Journalist and Author

Abstract. It's a staggering paradox. Thee United States has the finest universities in the world and invests more money in scientific research than any other nation. Yet we're allowing ourselves to fall behind in science education, and behind other countries like China, in green energy innovation. Meanwhile, most Americans know very little about science, and often don't even understand what they're missing—or why science matters to their lives. No wonder we have unending battles over the science of global warming, the teaching of evolution, and whether or not to vaccinate our children. How could the U.S. become so...unscientific? And what can we do about it? How can we make science popular again, or even...sexy? In this talk, Chris Mooney explains the reasons for the gap between science and the U.S. public, and what we can do to bring these two worlds—both of which need the other—back together again.

Journalist and author **Chris Mooney** is a senior correspondent for *The American Prospect*, a contributing editor for *Science Progress*, and writes an online column named *Doubt and About* for the magazine *Skeptical Inquirer*, where he serves as a contributing editor. His books include *The Republican War on Science*, *Storm World: Hurricanes, Politics, and the Battle Over Global Warming*, and *Unscientific America*. He has been a visiting associate at the *Center for Collaborative History* at Princeton University and a Knight Science Journalism Fellow at the Massachusetts Institute of Technology.

Chris Mooney delivers the keynote presentation. Photo by Paul Deans.

Connecting People to Science
ASP Conference Series, Vol. 457
Joseph B. Jensen, James G. Manning, Michael G. Gibbs, and Doris Daou, eds.
© *2012 Astronomical Society of the Pacific*

Looking Homeward Toward Earth: The Power of Perspective

Waleed Abdalati

NASA Chief Scientist, 300 E St. SW, Washington, D.C. 20546, USA

Abstract.   With the 1968 "Earthrise" image of planet Earth emerging from beyond the lunar horizon, society's view of our celestial home was changed forever. Beautiful and vulnerable, and suspended in dark stillness, this image inspired an appreciation that we are one human race, whose fate hinges delicately on our collective actions. Since that time, space-based observations of the Earth have continued to provide essential insights and information across the full spectrum of human activities and natural processes, and have even become a mainstream part of our daily lives. From documenting disappearing Arctic ice cover, to providing key insights to hurricane evolution, to tracking the amount of movement and cycles of Earth's biomass, these observations allow us to understand how and why our world is changing, and what these changes mean for life on Earth. But beyond their tremendous scientific value, they can be a powerful and inspiring tool for generating a true appreciation of the complexities and beauty of the world in which we live. From that iconic Earthrise photograph to the viral popularity of event-based satellite imagery, the power of the space-based perspective satisfies our need for constant and current information, and fuels our emotional connection to the planet we call home.

Waleed Abdalati was appointed NASA chief scientist on Jan. 3, 2011, serving as the principal adviser to NASA Administrator Charles Bolden on NASA science programs, strategic planning and the evaluation of related investments. He is currently on leave from his position as director of the University of Colorado's Earth Science and Observation Center, which carries out research and education activities on the use of remote sensing observations to understand the Earth. Abdalati is also a fellow of the Cooperative Institute for Research in Environmental Sciences at the University. His research has focused on the use of satellites and aircraft to understand how and why Earth's ice cover is changing, and what those changes mean for life on our planet. He has published more than 50 peer-reviewed papers, book chapters, and NASA-related technical reports, with approximately 1,500 citations in the peer-reviewed literature. He has given featured lectures and keynote addresses to the United Nations, AIAA, SPIE, AGU and various other professional and international organizations, as well as public lectures at The Smithsonian Institution, the American Museum of Natural History, and the Adler Planetarium. Abdalati has received various awards and recognition, most notably the NASA Exceptional Service Medal and The Presidential Early Career Award for Scientists and Engineers from the White House.

Plenary session audience. Photo by Paul Deans.

Connecting People to Science
ASP Conference Series, Vol. 457
Joseph B. Jensen, James G. Manning, Michael G. Gibbs, and Doris Daou, eds.
© *2012 Astronomical Society of the Pacific*

Engaging Girls in STEM: A Discussion of Foundational and Current Research on What Works

Karen Peterson,[1] Jolene Jesse,[2] and Laura Huerta Migus[3]

[1] *EdLab Group / National Girls Collaborative Project*

[2] *National Science Foundation*

[3] *Association of Science-Technology Centers*

Abstract. Diversity in science, technology, engineering, and mathematics (STEM) education and careers occupies center stage in national discussions on U.S. competitiveness in the 21st century. Women constitute roughly half the total workforce in the U.S., but they hold just 25% of mathematical and science jobs and 11% of engineering jobs. Women earn nearly 60% of all bachelors and masters degrees, except in physics, computer science, and engineering, where the percentages are 20–25%. This disparity is even more pronounced at the doctoral level, where women earn fewer than 20% of awarded Ph.D.'s in physics or engineering. However, at the high school level, there is far less gender disparity: both female and male students take comparable advanced physical science and math courses. What, then, accounts for the lack of gender diversity in STEM advanced education and career paths? In fact, there is no consensus even among experts. So, what information and strategies do the EPO community need to know and include as part of designing and implementing programs to encourage more girls and women to engage in STEM for the long term?

The panelists will discuss foundational and current research on pressing questions on why these trends exist and what can be done to change them. They will highlight research and evaluation results from programs that are successfully engaging girls in STEM.

Karen Peterson is the Chief Executive Officer of the EdLab Group and has been active in education for over twenty years as a classroom teacher, university instructor, pre-service and in-service teacher educator, program administrator, and researcher. Currently, she is the Principal Investigator for the National Girls Collaborative Project, SciGirls: a new national TV series, the Computer Science Collaboration Project, and Bio-ITEST: New Frontiers in Bioinformatics and Computational Biology, all of which are funded by the National Science Foundation. These projects all address gender, racial and socioeconomic under-representation in STEM fields. Ms. Peterson serves on local and national boards which develop and administer programs designed to increase under-represented students' interests in STEM. Ms. Peterson has published in *T*he Journal of Women and Minorities in Science and Engineering and has co-authored evaluation reports and promising practices reports in informal information technology education for girls for the National Center for Women & Information Technology and the Girl Scouts of the USA.

Jolene Kay Jesse is a Program Director for the Research on Gender in Science and Engineering program in the Directorate for Education and Human Resources (EHR) at the National Science Foundation. The program funds and promotes research into education and workforce issues aimed at broadening the participation of women and girls across the science and engineering fields. It also funds efforts to diffuse research based innovations in gender equitable teaching, pedagogy, and counseling to practitioner audiences.

Laura Huerta Migus is the Director of Equity and Diversity at the Association of Science-Technology Centers (ASTC). She is responsible for spearheading ASTC's Equity & Diversity Initiative, which seeks to ensure that science centers and museums are capable of effectively serving their diverse audiences. Initiative activities include: ASTC Diversity & Leadership Development Fellows program; gathering and developing resources to support ASTC member institutions in their diversity journeys; and identifying best practices in the field for replication. She works to advance ASTC's equity and diversity agenda through a number of NSF-funded projects, including leadership positions in the Cosmic Serpent project and the Nanoscale Informal Science Education Network (NISE Net). Laura is also the Co-Principal Investigator on the Girls RISE Museum Network.

Connecting People to Science
ASP Conference Series, Vol. 457
Joseph B. Jensen, James G. Manning, Michael G. Gibbs, and Doris Daou, eds.
© *2012 Astronomical Society of the Pacific*

Why Counting Attendees Won't Cut it for Evaluation in the 21st Century: Planning and Evaluating Informal Science Programs

Randi Korn

Randi Korn & Associates, Inc., 118 E. Del Ray Ave., Alexandria, Virginia 22301, USA

Abstract.   Yogi Berra once said, "If you don't know where you are going, how will you know when you get there?" One could ask the same question to those of us who work in science education and outreach—if you haven't articulated clear goals about what you would like your program to achieve, how on Earth will you know whether you have achieved them? While all of us want to do the right thing for our audiences, knowing that we have actually done so is another story. Without sound planning (clarifying outcomes) evaluation is a moot point, and with more and more funding agencies asking their grantees to evaluate their informal science education efforts, planning with the end in mind is becoming a necessity. With budget, staffing, and time limitations, it's easy to lose track of the value of planning and evaluation. And evaluation can seem a daunting task for those who have not done it, especially when an audience is temporary or spread out over a region or the entire nation.

To respond to the demands, many good books are now available with ideas for evaluating projects outside the formal classroom, including several published by the National Research Academies and NSF. In this article, we will discuss the importance of planning and evaluation, no matter what your budget size, we will share examples of how unusual projects have been evaluated, and we will suggest questions you can ask yourself and your audiences that will help you think like an evaluator. To achieve results, program leaders must first clarify what they want to achieve and then align all actions and resources towards achieving those ends. Will your strategic and daily work change as a result? Absolutely! But only if you want your program to make a difference in people's lives.

1. Introduction

Good afternoon. I am delighted to have the opportunity to talk with you today. I have been asked to talk about evaluation, so as I was thinking about what to discuss, it occurred to me that I should probably talk about the prerequisite to evaluation, which is planning. Often misunderstood is the strong relationship between planning and evaluation. Without a sound plan of what you want to achieve, evaluation is a moot point. So this afternoon I will be focusing on planning, and I will be taking a big picture. In a subsequent session, three other evaluators and I will present the nuts and bolts of evaluation.

When I am asked to conduct an evaluation, my first two questions are: what outcomes are you striving to achieve, and who is your primary audience? As an evaluator, I must know a project's intent. It was Yogi Berra who said, "If you don't know where you are going, how will know when you get there?" Without clarifying what results you want to achieve, how will you know that you have achieved them? Without a vision of success, how will you know that your program is successful? If people came, attended, or simply showed up, is that enough? It used to be enough, but it isn't any more. So if numbers are no longer the metrics we live by, how do we even begin to construct metrics to guide and measure our work? That's why I want to talk about planning, because we have before us an evaluation challenge which brings me right back to planning.

My evaluation experience has taught me that without a clear statement of intent and a deliberate plan to achieve that intent, results are not likely to happen in the way you imagined. In this presentation about planning, I will be emphasizing the whole organization, yet each concept is also relevant to program planning.

Today, funders are demanding a demonstration of impact; thus, an important question to ask is, "What impact is your organization hoping to achieve on the audiences you serve?"

2.   Clarifying Intent

Organizations that continually clarify their intent and realign all practices and resources to achieve that intent are addressing the challenge of achieving impact.

It is no small matter for organizations to intentionally plan to achieve a specific, well-defined result. Defining impact is paramount for any organization interested in achieving impact, because knowing what you want to achieve, and continuing to clarify those ideas with utmost precision will help with decision making so every decision can support the results you want to achieve. You must clarify what you want to achieve before you can create activities that support it. And you must clarify what you want to achieve to be able to determine whether you have the resources to support it.

And finally, you must clarify what you want to achieve if you are to determine, through evaluation, whether you have achieved it.

3.   Impact Planning

Impact planning requires that everyone's planning, decisions, and actions are purposeful and focused exclusively on achieving impact.

However, focusing energy and resources towards achieving impact requires enormous discipline because one must say "no" to ideas and programs that move the organization away from its vision of impact.

If organizations are interested in achieving impact, they must exercise discipline, carefully plan their work, and they must clarify what they want to achieve from the start. They must begin planning with a vision of the end in mind.

4.   Cycle of Learning

I would like to introduce this cycle of learning as a visual (Fig. 1) that represents a way of working, both strategically and day to day. This cycle is defined by four actions and corresponding questions.

Impact is in the center of the cycle because impact should be the outcome and driving force for the work of most non-profits, and all actions in the cycle ideally should support an organization's vision of impact. The planning aspect of this cycle is very

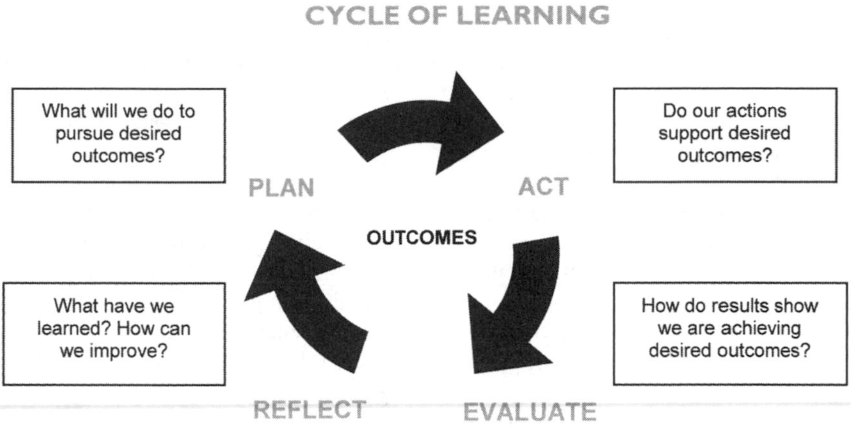

CYCLE OF LEARNING

Figure 1.   Cycle of Learning

much about the process of peeling back layers to achieve clarity so you can know to what end are you doing your work. The first question on the cycle (upper left) embodies the idea of the moment—impact—asking: "What impact do you want to achieve?" The clarity with which you describe the end result is paramount.

Moving around the cycle clockwise, the actions that you take to achieve impact must be aligned with available resources, as resources are not endless. Decisions must be made with a strong understanding of and conviction for what you want to achieve.

The evaluation phase of work asks how well you are doing in your pursuit of impact, and of course the gauge of success is the impact the organization says it wants to achieve. Reflection is about learning and an important result for all of us. Reflective practice requires taking the time to learn from evaluation and daily practice, and consider results compared to aspirations, so organizations can become more intentional as they work towards achieving impact.

5.   Planning for Impact

I will discuss four ideas that constitute planning for impact:

1.   an organization's ability to clarify its intended impact;

2.   the discipline to align practices and resources to support the impact the organization wants to achieve;

3.   conducting impact evaluation as a way to learn about achieving impact; and

4.   taking the time to reflect on evaluation data and/or daily practices to learn how to achieve greater impact.

6. Clarify Intended Impact

If you cannot clarify intended impact, how will you know whether you have achieved it?

Reaching clarity about the impact you want to achieve is about uncovering the meaning and essence of your work.

To What End?

A new awareness is emerging about non-profit organizations' responsibilities to create public value, and many organizations are revitalizing their mission statements to include contemporary thinking about their larger role in society. A museum's mission statement often describes actions like educating and sharing knowledge and as many of you know, it is easy to replace one museum's mission statement with that of another.

If a mission statement describes what a museum *does*, an impact statement can describe the *result* of what a museum does. I believe mission statements need companion statements that focus on the results of a museum's work on the public it serves.

An impact statement can address the vital "To what end?" question. To what greater end do museums carry out their work? For a museum to educate is fine, but how might a museum describe the result of its education work? If all the museums in a community are achieving impact, what do we see in those communities as evidence of their work?

7. Impact Statement

The process of creating an impact statement will help an organization think through what it hopes to achieve, but how might they begin?

I believe an impact statement is a melding of three elements:

1. staff members' deepest passions for their work;

2. the unique characteristics of the organization, and

3. what is relevant to the community.

Personal passion for one's work leads to professional commitment, which in turn builds a sense of common responsibility among colleagues; both personal passion and professional commitment are essential ingredients for doing a job well so one can achieve impact. In our work with museums, we have observed patterns as staff talk about their passions for science, personal learning, teaching others, and in the case of administrators, a passion for organizational effectiveness and efficiency.

Another important ingredient in creating an impact statement is an organization's unique characteristics, qualities, and strengths. Thinking through an organization's greatest assets, including its material and intellectual resources, will help everyone in the organization begin to see its internal capacity to achieve impact externally. Together with passion, the unique value of an organization can power the organization to achieve impact.

Without passion and focused attention on what you do best, or in other words, playing to your strengths, the impact will be subdued, diffuse, and difficult to measure. Items 1 and 2 are about you and your organization. Now it is time to turn attention to the recipients of your work—the *who* part of the conversation. This is where the notion of relevance enters the conversation.

One hundred years ago in Newark New Jersey, John Cotton Dana founded the Newark Museum. He said "a museum is good only insofar as it is of use." His ideas resonate today, as museums pursue creating public value. Relevance is certainly part of the equation and necessary for any organization that desires to make a difference in people's lives. Organizations will need to determine whose lives they hope to affect, as affecting everyone is neither possible nor likely. Organizations cannot be all things to all people and expect to achieve impact—discernable impact that can be measured by an evaluator.

Yet, community relevance alone, without the context of an organization's passions, is akin to pandering. Community relevance must be balanced with what an organization has the capacity to deliver, which is connected to the staff members' passions and the organization's unique value.

We have worked with a several museums to develop an impact statement, and I'd like to share two of them.

- From a museum of natural history and culture: "People value their connection with all life—and act accordingly."

- From an art museum: "Visitors will expand their creative thinking, deepen their understanding of human experiences, and value the museum as a place for personal learning and civic engagement."

8. Cycle of Intentional Practice

With an impact statement in hand, staff will be able to use it as a guidepost moving forward because then the organization will need to align practices and resources with the intent of achieving impact.

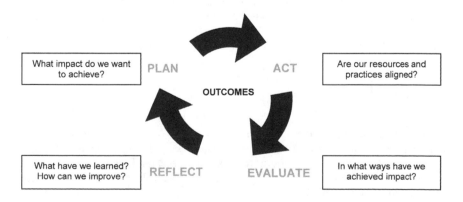

Figure 2.   Cycle of Intetional Practice.

Aligning practices—this is where the programmatic work all of you do feeds into achieving impact. All your programs and projects should support the impact your organization wants to achieve. I use the word "impact" to talk about the results of an entire organization. When talking about results of a program, I use a different word "outcome." The next section describes the relationship bewteen outcomes and impact.

9.   Align Practices and Resources

Aligning practices and resources so they support achieving impact requires thinking about what you should be doing and what you need not do any more. Many non-profits often continue adding more and more onto their workloads while never taking anything away. Such an idea is not sustainable, in terms of staff capacity, but also as a business model. Learning to say "no" is an important survival skill.

The very idea of saying "no" or only focusing on what you need to do to achieve the results you want may mean that you end up doing less, which is a good thing, as so many of us are doing too much already. Sometimes when staff are allowed to do less, they have the time to do a better job and go deeper on the programs they are doing, which would then allow evaluators to detect results.

Use relevance and impact as the lenses through which to explore doing less. By doing so, your decisions will be more deliberate and intentional, strengthening the organization and readying it to achieve impact in its community. I want to share a few ideas on how one might approach doing less. One can consider only doing work that is relevant to the organization's community (and stop doing work that is not). An organization might need to reexamine what it does and its community's needs to determine which programs are no longer *relevant* to its community, and likewise, which programs *do not represent the organization's intrinsic value and greatest strength*. Do away with those programs, thereby freeing up resources and time for programs that *can make a difference*. By examining all the activities an organization does, one might be able to identify those that are relevant to your community in the 21st century and those that were relevant in the last century but are no longer. Such considerations involve knowing well those who comprise your organization and those who comprise the community.

The impact statement that an organization might develop is an internal planning and decision-making tool. While you can share it with your constituents, its real value is as a guidepost to help you know where you should put your time and resources—if impact is what you hope to achieve. An impact statement can also help you to say "no" to initiatives that move you away from achieving your intended impact.

10.   Evaluate Impact

So what does it mean to do less and achieve more? Doing less is about an organization rethinking what it does and for whom, with the goal of being more deliberate and intentional in its daily work. As we move around the cycle to the evaluation quadrant (Fig. 2), it makes sense to talk about metrics of success, as there is a tension between doing less and traditional metrics of success. Success, at least in American terms, is often tangled up in how big and how many.

For museums, high visitation has always been the marker of success; however, one can reasonably argue that numbers no longer suffice as the key success metric because attendance figures only mean that people came—they offer no indication of meaningfulness. At some point someone will ask, "So what difference has the museum made in the quality of people's lives?" In this article, achieving more is associated with quality rather than quantity. That is not to say that quantity is unimportant, and that there haven't been quality programs that also attracted large quantities of people. Numbers have a purpose, but numbers may become more meaningful once the qualitative value of an organization is articulated by those who are affected.

Each organization owes it to itself and its community to define its *unique* brand of quality. Defining success in terms of quantity is easier than defining success in terms of quality. Though difficult, each organization owes it to itself and its community to define its *unique* brand of quality in its vision of impact. I noted earlier that an organization's distinctive characteristics and unique value comprise one of the building blocks for defining impact. They also provide clues for defining the quality of what you offer. Describing what your organization offers qualitatively begins to suggest an individual organization's meaning of quality, which is necessary to help staff think about other potential metrics of success.

Consider the imperative to clarify the impact they would like to achieve and documenting results that focus on quality, meaningfulness, and relevance. Oddly, doing less is a huge undertaking because of the organizational ramifications and its potential interference with tradition and status quo. In this presentation, I associate doing less with prioritizing and focusing the work of an organization according to what it would like to achieve in terms of the public good. Doing less supports organizations saying "no" to resources or projects that prevent them from focusing on their intended results. Nearly all organizations face the pressing reality that it is no longer possible to continue adding more to staff members' at-capacity workloads.

11. Reflective Practice

Continuing around the cycle in Figure 2, we reach reflective practice. As indicated by some of your responses when I asked how you spend your time, it is true that evaluation receives little attention and reflective practice receives almost no attention. Most practitioners have too much on their plate to engage in reflective practice at all, yet there is a strong relationship between *taking the time to think about* your work and *learning from* your work. Learning is not likely to happen if one does not take the time to reflect.

Most of you in this room would agree that you value education and learning and want to create opportunities for others to learn. You probably also identify yourselves as life-long learners. Reflective practice is a process of periodically thinking about an organization's work and questioning how it supports the impact it wants to achieve. Reflective practice can feed practitioners' desire for personal, professional, and organizational learning and provide insight to how an organization can continually improve.

12. Passion Exercise

What about your work are you most passionate about?
Why is that important?
Why is *that* important?
Why is *that* important?

I'd like to leave you with a simple exercise that you can do in your organization to discover what ignites your passion and the passions of those you work with. Repurpose a staff meeting and gather together staff from up and down and across the organization; invite a few board members if you like. Divide people into small groups of about five people each and ask each group to appoint a scribe. Ask them to discuss and respond to the questions above. Then you can then look at those passions against your programs, for example, to see if there is alignment between what people love to do and the work the organization is currently doing. This exercise is a small baby step towards beginning to think about impact and being deliberate and intentional in your work with the goal of your organization achieving the results it wants.

Randi Korn is Founding Director of Randi Korn & Associates, Inc. (RK&A), a company that helps museums plan their strategic and daily work around achieving impact, which often includes conducting impact evaluation, visitor research, and all phases of museum program evaluation. During the company's 23-year history, RK&A has learned a great deal from the hundreds of exhibitions and programs it has evaluated for children's museums, museums of science, art, and history, botanical gardens, zoos, libraries, and aquariums. As an active member in the museum community, Randi enjoys sharing her knowledge, experience, and enthusiasm for evaluation and intentional practice. She presents at conferences, writes for museum journals, and lectures at Johns Hopkins University, The Corcoran College of Art, and The George Washington University, where she was an adjunct instructor in the Museum Education division for 18 years. A recent publication, "The Case for Holistic Intentionality," underscores her commitment to helping museums pursue intentional practice and evaluate the ways in which they are achieving their desired results.

Connecting People to Science
ASP Conference Series, Vol. 457
Joseph B. Jensen, James G. Manning, Michael G. Gibbs, and Doris Daou, eds.
© *2012 Astronomical Society of the Pacific*

New Views of Diverse Worlds

David Blewett

Johns Hopkins University, Applied Physics Laboratory

Abstract.   Spacecraft exploration is in the process of revolutionizing our knowledge of the airless rocky bodies in the inner Solar System. Mercury has long been viewed as a planetary "end-member", but NASA's MESSENGER spacecraft, which flew past the innermost planet three times in 2008–09 and entered orbit in March of 2011, is finding that Mercury is even stranger than we thought. Mercury is weird in essentially all its characteristics: interior structure, surface composition, geology, topography, magnetic field, exosphere, and interaction with the space environment. Closer to home, a flotilla of international probes have targeted the Moon in the past few years. Giving lie to the "been there, done that" attitude held by many toward the Moon, the new missions are making many new discoveries and reminding us that there is much we don't know about our nearest planetary neighbor, and that rich opportunities for exploration are waiting nearby. Finally, I'll present findings from NASA's Dawn spacecraft, which will begin its orbital mission around the asteroid Vesta in mid-July 2011. Vesta is sometimes called "the smallest terrestrial planet" because it has separated into a crust, mantle, and core, and experienced a protracted geological evolution. Vesta is probably the source of a common class of meteorites, so we have abundant samples that help to inform our interpretation of the data to be obtained by Dawn. Mercury, the Moon, and Vesta are worlds who share some characteristics, but have taken radically different evolutionary paths. They provide insight into the most fundamental geological processes that likely affect all rocky planets—around our Sun or beyond.

David T. Blewett joined the Johns Hopkins University Applied Physics Laboratory as a member of the Senior Professional Staff in September 2007. Prior to that he was a Principal Scientist at NovaSol (Innovative Technical Solutions, Inc.), a small employee-owned high-tech company in Honolulu, Hawaii, of which he was a co-founder. His planetary research emphasizes remote sensing, geological analysis, and spectral algorithm development using data from planetary spacecraft including *Mariner 10*, *Clementine*, *Galileo*, *Lunar Prospector*, *MESSENGER*, and *Dawn*. In addition, he has done considerable work in the analysis of Earth-based telescopic spectra of the Moon. He has been a Principal Investigator in the NASA Planetary Geology and Geophysics Program since 2002, was selected as a *MESSENGER* Participating Scientist in 2007, and as a Dawn at Vesta Participating Scientist in 2010. He is presently the Deputy Chair of the *MESSENGER* Geology Discipline Group and a member of the *MESSENGER* Science Steering Committee. He serves as the *MESSENGER* Science Outreach Liaison, in charge of answering questions from the public that come in via the *MESSENGER* website. He was fortunate to spend a field season with the Antarctic Search for Meteorites in the late 1980s.

Neil deGrasse Tyson. Photo by Paul Deans.

Connecting People to Science
ASP Conference Series, Vol. 457
Joseph B. Jensen, James G. Manning, Michael G. Gibbs, and Doris Daou, eds.
© *2012 Astronomical Society of the Pacific*

Tales from the Twitterverse

Neil deGrasse Tyson

American Museum of Natural History

Abstract. "Tales from the Twitterverse" describes Dr. Tyson's running experience communicating science via the medium of Twitter. He now has about 130,000 followers on his twitter handle @neiltyson and was recently selected for Time Magazine's list of the best 140 Twitter feeds. It is perhaps fair to say that, so far, social media and science EPO have not yet been fully introduced to one another. So this plenary talk will be a kind of overview of his successes and failures in the medium, as a way to jumpstart people's interest in what is possible.

Neil deGrasse Tyson is the first occupant of the Frederick P. Rose Directorship of the Hayden Planetarium. In addition to dozens of professional publications, Dr. Tyson has written, and continues to write for the public. He is a monthly essayist for *Natural History* magazine under the title "Universe." And among Tyson's eight books is his memoir *The Sky is Not the Limit: Adventures of an Urban Astrophysicist*, and *Origins: Fourteen Billion Years of Cosmic Evolution*, co-written with Donald Goldsmith. *Origins* is the companion book to the PBS NOVA 4-part mini-series *Origins*, in which Tyson serves as on-camera host. Beginning in the fall of 2006, Tyson has hosted the PBS's NOVA spinoff program *NOVA ScienceNow*, which is an accessible look at the frontier of all the science that shapes the understanding of our place in the universe.

Part II

Workshops and Special Sessions

Connecting People to Science
ASP Conference Series, Vol. 457
Joseph B. Jensen, James G. Manning, Michael G. Gibbs, and Doris Daou, eds.
© 2012 Astronomical Society of the Pacific

Hanny and the Mystery of the Voorwerp: Citizen Science in the Classroom

Kathy Costello, Ellen Reilly, Georgia Bracey, and Pamela Gay

Southern Illinois University Edwardsville, Edwardsville, Illinois 62026, USA

Abstract. The highly engaging graphic comic *Hanny and the Mystery of the Voorwerp* is the focus of an eight-day educational unit geared to middle level students. Activities in the unit link national astronomy standards to the citizen science *Zooniverse* website through tutorials that lead to analysis of real data online. NASA resources are also included in the unit.

The content of the session focused on the terminology and concepts—galaxy formation, types and characteristics of galaxies, use of spectral analysis—needed to classify galaxies. Use of citizen science projects as tools to teach inquiry in the classroom was the primary focus of the workshop.

The session included a hands-on experiment taken from the unit, including a NASA spectral analysis activity called "What's the Frequency, Roy G Biv?" In addition, presenters demonstrated the galaxy classification tools found in the "Galaxy Zoo" project at the *Zooniverse* citizen science website.

1. Purpose

Hanny and the Mystery of the Voorwerp uses a comic book format to introduce teachers and students to the Galaxy Zoo citizen scientist project, which is part of the *Zooniverse*.[1] The *Zooniverse* and the suite of popular and successful citizen science projects it contains, (*Planet Hunters, Moon Zoo, The Milky Way Project,* and more), was developed by the Citizen Science Alliance to use the efforts of volunteers to help researchers deal with the flood of data that confronts them. Galaxy Zoo was the first of these projects and has produced many unique scientific results, ranging from individual discoveries to those using classifications that depend on the input of everyone who has visited the site. *Hanny and the Mystery of the Voorwerp* follows the story of one such individual discovery.

The overarching goal of the instructional unit is to entice students in grades 8 to 12 to participate in the authentic scientific research that is conducted through citizen science projects.

2. The 5E Instructional Model

Each lesson in the *Voorwerp* unit follows the 5E Biological Sciences Curriculum Study (BSCS) 5E instructional model.[2] The format of the presentation also followed this model.

---

[1] www.zooniverse.org

[2] For more information on this lesson format, see http://www.bscs.org/curriculumdevelopment/features/bscs5es.html

3. Engage

Participants were invited to imagine what they would have done if they had been the first to see the voorwerp through a telescope. Discussion questions included: "What do you think this is?" "How would you find out?" and "What would you name it?"

4. Explore

The comic book *Hanny and the Mystery of the Voorwerp* was introduced in the workshop, tracing the development of the mystery as more and more scientists became involved. There were profiles of some of the space- and ground-based telescopes that gathered information about the voorwerp as well as the scientists who helped to unravel the mystery.

   To give attendees a taste of the science involved, they were invited to conduct one of the activities included in the unit. "What's the Frequency, Roy G Biv?" is an exploration of the relationship between wavelengths and their frequencies taken from the NASA website.

5. Explain

This unit includes a website[3] that features an electronic version of the comic book, background information such as astronomy terminology and telescope facts, resources, links to the *Zooniverse* and Hubble websites, and complete lesson plans for the eight-day unit. The presenters gave an overview of the topics covered:

Lesson 1: The History of the Voorwerp

Lesson 2: Electromagnetic Spectrum

Lesson 3: Scientific Method & Problem Solving

Lesson 4: Identification of Galaxies

Lesson 5: Quasars and Black Holes

Lesson 6: Data from Telescopes

Lesson 7: Science, Technology, and Society

Lesson 8: Classification of Galaxies

Participants were also introduced to the Galaxy Zoo websites. There are three Galaxy Zoo websites that may be used by teachers:

- http://zoo1.galaxyzoo.org: Easy galaxy classification

- http://zoo2.galaxyzoo.org: More advanced galaxy classification

---

[3]http://hannysvoorwerp.zooniverse.org/

- `http://hubble.galaxyzoo.org`: Advanced galaxy classification and the current default online

The first two sites are archival in nature; the third is the active website. All three sites guide citizen scientists through basic classification, but progressively more detailed observations are included as you progress to the Hubble website.

6. Elaborate

After the overview of Galaxy Zoo, an introduction to other *Zooniverse* websites was given. Presenters highlighted the importance of citizen science projects that invest in scores of computer-human interactions to help scientists understand the data—and uncover the mysteries—relayed by the many technologies available today.

As a final hands-on activity, UV beads were distributed to the attendees, and suggestions for classroom experiments were shared.

7. Evaluate

The web component of the unit includes evaluative tools for classroom teachers, including lab sheets, quizzes, and tests. All components of the unit address National Standards as listed below.

8. Standards

The instructional unit *Hanny and the Mystery of the Voorwerp* addresses the following national standards:

The National Science Education Standards (NSES)[4]

As a result of their activities in grades 5–8, all students should develop an understanding of:

1. abilities necessary to do scientific research and Understanding about scientific inquiry (Content Standard A, Science as Inquiry);

2. transfer of energy (Content Standard B, Physical Science);

3. earth in the solar system (Content Standard D, Earth and Space Science);

4. understanding about science and technology (Content Standard E, Science and Technology);

5. science and technology in society (Content Standard F, Science in Personal and Social Perspectives); and

6. science as human endeavor, nature of science, and history of science (Content Standard G, History and Nature of Science).

---

[4]`http://www.nap.edu/openbook.php?record_id=4962`

Benchmarks for Science Literacy Project 2061 [5]

By the end of eighth grade, students should know:

- that the universe contains many billions of galaxies, and each galaxy contains many billions of stars (4A/M1bc); and

- that some distant galaxies are so far away that their light takes several billion years to reach the earth. People on earth, therefore, see them as they were that long ago in the past (4A/M3).

---

[5]http://www.project2061.org/publications/bsl/online/index.php?chapter=4#A3

Connecting People to Science
ASP Conference Series, Vol. 457
Joseph B. Jensen, James G. Manning, Michael G. Gibbs, and Doris Daou, eds.
© *2012 Astronomical Society of the Pacific*

Top 10 Things NOT to Do if You Want to Get Published in Astronomy Education Review

Andrew Fraknoi

Foothill College, 12345 El Monte Rd., Los Altos Hills, California 94022, USA; email: fraknoi@fhda.edu

Abstract.   Based on my experience as editor of Astronomy Education Review for its first 10 years of publication, here is a tongue-in-cheek list of things to avoid in publishing papers or articles on astronomy education and outreach.

1.   Introduction

This list was distributed at the ASP meeting as part of a hands-on session on how to publish your work in astronomy education and outreach in a journal such as *Astronomy Education Review*.[1] I would hope some of the negative hints listed are more generally applicable to publishing in any journal. The tips come from many years of discussion with co-editors Sidney Wolff and Thomas Hockey.

2.   Top 10 Things NOT to Do If You Want to Get Published in Astronomy Education Review

1.   Make your sample size too small in numbers or in time. For example, you could survey only a few students, interview only a small number of museum visitors, or do research only over the course of one semester—and then make grand claims for your results anyway.

2.   Use as much educational or astronomical jargon as possible without defining your terms.

3.   Use the abstract to give only a vague introduction and neglect to mention your sample size, your research goal, or any of your conclusions.

4.   Pad the paper with long discursive discussions on many aspects of astronomy or education that have little to do with your research, but that you've always wanted to get off your chest.

5.   Make sure *no one else* has read your paper, either to see if it makes sense to an uninvolved reader or to proof-read it for syntax, grammar, or meaning.

---

[1]http://aer.aas.org

6. When you have described the results of your research, be sure to say absolutely NOTHING about what your investigation means for a practicing educator and how it might affect future educational program planning and execution (even if there are clear connections to be made).

7. If you are describing a public outreach project, make as many sweeping generalizations about its effects as you can, but don't refer to any actual evaluation data. Even better, don't do any evaluation—just rely on a couple of anecdotes from your staff or a few users.

8. In the list of references, list only papers and books you (or your coauthors) wrote. Show no knowledge of the literature or even previous papers in *AER* in the same area of research.

9. If other projects or investigators around the country have done similar work, don't mention anything about them and strongly suggest that the idea is entirely original with you.

10. If you have diagrams or figures, make them as small and complex as possible and don't label the axes clearly.

*Follow all the above rules and we can virtually guarantee that your paper will **not** be published in Astronomy Education Review.*

Connecting People to Science
ASP Conference Series, Vol. 457
Joseph B. Jensen, James G. Manning, Michael G. Gibbs, and Doris Daou, eds.
© *2012 Astronomical Society of the Pacific*

The Sun Funnel: A Do-It-Yourself Projection Device for Safe Solar Viewing by Groups

Richard Tresch Fienberg,[1] Chuck Bueter,[2] and Louis A. Mayo[3]

[1]*American Astronomical Society, 2000 Florida Ave., NW, Suite 400, Washington, D.C. 20009, USA*

[2]*Nightwise.org, 15893 Ashville Lane, Granger, Indiana 46530, USA*

[3]*NASA Goddard Space Flight Center, Mail Code 690.1, Greenbelt, Maryland 20771, USA*

Abstract.   Virtually every commercial telescope comes with a warning not to point it at the Sun, since doing so could not only damage the instrument but also injure the observer. Yet the Sun is typically the only astronomical object visible in the daytime, and even in white light it offers much to see in a telescope: sunspots and limb darkening, the partial phases of solar eclipses, transits of Mercury and Venus, and even transits of the International Space Station. Teachers, planetarians, and other science educators who wish to share these phenomena with their students or visitors face a challenge: how to safely show a magnified image of the Sun to many people at once. Using aperture filters on telescopes is fine if you have lots of telescopes, but if you have only a few telescopes, or perhaps only one, the result is long lines at the eyepiece. One inexpensive solution is solar projection, i.e., projecting an image from the telescope onto a wall or screen. But this technique is fraught with danger, as there is always the possibility that someone will look into the bright beam of sunlight streaming from the eyepiece and risk serious eye injury. Here we describe a novel solar-projection device, the Sun Funnel, that fits in a telescope focuser in lieu of a regular eyepiece. It is quick and easy to build using inexpensive, readily available supplies and simple household tools. The Sun Funnel completely encloses the sunbeam coming from the telescope and forms a clear solar image on a rear-projection screen. With this device, many people can simultaneously and safely enjoy a telescopic view of the Sun.

1.   Observing the Sun

The June 5, 2012, transit of Venus across the Sun's face generated tremendous interest in daytime astronomy. Nobody alive today will ever get another chance to see a transit of Venus, as the next one doesn't happen until December 11, 2117. A transit of Mercury will occur on May 9, 2016, with another one 3.5 years later. Several websites make it easy to determine when the International Space Station will pass in front of the Sun as seen from any location on Earth.[1,2] Meanwhile sunspot cycle 24 is ramping up toward

---

[1]CalSky: http://www.calsky.com

[2]Ed Morana's ISS Transit Predictions:
http://pictures.ed-morana.com/ISSTransits/predictions/index.html

a predicted maximum in 2013. As the number, frequency, and intensity of solar storms are increasing, so too is interest in observing the Sun and sharing the view with anyone who'll take a look.

But our mothers told us never to look at the Sun, and commercial telescopes come with stickers repeating the same warning. So what are we to do?

Experienced skygazers know many ways to observe the Sun safely with a telescope using special-purpose aperture filters and other accessories. But it can be hard to convince a newbie to look through a filtered scope because of Mom's voice still echoing in his or her head: "Didn't I tell you never to look at the Sun?!"

One way around this is to project an image of the Sun onto a card or screen. Now nobody has to look through the telescope (which means no refocusing and no bumping), and many people can view the solar image at the same time. But there's still the risk that someone might accidentally look into the bright beam of sunlight emerging from the eyepiece.

The solution is a rear-screen projection, first popularized (to the best of our knowledge) by Bruce Hegerberg in his aptly named Sun Gun.[3] With this device and others like it, the intense sunlight is fully contained within the projector so that no one is exposed to it.

Our favorite rear-screen solar-projection device is the Sun Funnel,[4] which Gene Zajac and co-author Chuck Bueter adapted from an earlier design and first presented at the 2003 Great Lakes Planetarium Association Annual Conference. The device uses inexpensive, readily available materials and takes just a few minutes to build with very simple tools (Fig. 1).

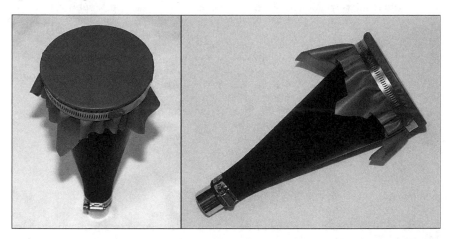

Figure 1.   Oblique (left) and side views of the Sun Funnel. Photos by R. T. Fienberg.

The construction and assembly instructions that follow were successfully field-tested at a workshop during the 2011 annual meeting of the Astronomical Society of the Pacific (ASP), where nearly two-dozen participants built their own Sun Funnels.

---

[3]http://bit.ly/q9WpYU

[4]http://bit.ly/rt16jc

2. What You'll Need

In the following parts list, the URLs point to websites where we purchased materials for our ASP workshop, but you can get the same (or similar) parts from many other suppliers.

- Blitz Super Funnel, #05034, `http://bit.ly/ozgP07`. The dimensions of this round-top black plastic funnel are usually given as $17.75 \times 5 \times 5$ inches. Cost: $2 to $5 at your local hardware or auto-parts store.

- Large hose clamp, e.g., Breeze #62080, `http://bit.ly/qBbvRK`. It needs to open to at least 5 inches. Cost: $1 to $2 at your local hardware store.

- Small hose clamp, e.g., Breeze #62016, `http://bit.ly/oTDjRR`. It needs to open to at least 1.5 inches. Cost: $0.50 to $1 at your local hardware store.

- Da-Lite High-Contrast Da-Tex rear-surface projection screen, #95774, `http://bit.ly/nE0fTU`. You'll need a piece roughly 8×8 inches square. It's usually sold by the square foot, and you'll probably have to buy a minimum of 1 square foot. Cost: a little under $10/square foot.

- Inexpensive (e.g., Huygens, Kellner, Plössl) telescope eyepiece, 1.25-inch barrel. Before you can choose a focal length for the eyepiece, you need to decide which telescope you're going to use with your Sun Funnel. We recommend using a refractor equipped with a star diagonal. We strongly discourage using a reflector or catadioptric (mirror-lens) telescope, as concentrated sunlight can destroy its secondary-mirror assembly. Since the Sun is bright, and since daytime seeing is never all that good anyway, you don't need a big telescope; on the contrary, the smaller the better.

- A 60- to 80-mm refractor with a focal length f_T of 350 to 900 mm works quite nicely. To get a full-disk solar image of about 100-mm diameter on the projection screen, use an eyepiece of focal length $f_E \approx f_T/43$. For example, a 66-mm $f/5.9$ refractor has $f_T = 388$ mm, so $f_E \approx 388/43 \approx 9$ mm, so an old 9-mm Kellner eyepiece works very well with the Sun Funnel on such a telescope.

Don't worry about getting f_E exactly right—any cheap eyepiece you have lying around whose focal length is within $\pm10\%$ of the computed value will work just fine. If you don't have an appropriate eyepiece, try browsing the current offerings at Surplus Shed.[5]

For tools, you'll need a small hacksaw, a 12-inch ruler, a flat-head screwdriver, and a sheet of medium- to fine-grit sandpaper.

3. Construction and Assembly

Step 1. The funnel has a sharp little piece of plastic protruding from the side about halfway down its length. Using the sandpaper, grind it smooth so it doesn't scratch your hands.

---

[5]`http://www.surplusshed.com`

Step 2. Using the hacksaw, cut the little flat tab off the wide end of the funnel. It works best to cut halfway through from one side, then halfway through from the other.

Step 3. Using the hacksaw, cut about 7 inches off the narrow end of the funnel so that what's left measures about 10 inches long (use the ruler). Try to make the cut perpendicular to the axis of the funnel, but don't panic if it ends up slightly tilted. Rotate the funnel as needed to complete the cut. Discard the sawed-off, 7-inch-long piece of plastic.

Step 4. Stand the funnel on its wide end. Using the hacksaw, cut straight down across the middle of the narrow opening, making your cut about 1 to 2 inches deep. The narrow end of the funnel will now have two semicircles of plastic rather than a solid circle.

Step 5. Using the sandpaper, smooth all the cut surfaces on both ends of the funnel.

Step 6. If your eyepiece has a rubber eyecup and/or rubber grip, remove it/them.

Step 7. Insert the eyepiece into the narrow end of the funnel: lens in, chrome barrel out. You may need to pry apart the two semicircular halves of the funnel's opening. If the eyepiece still won't go in, cut away a little more of the funnel to widen the opening, then try again. Aim to get at least a half inch of the length of the eyepiece into the funnel.

Step 8. Place the small hose clamp over the narrow end of the funnel and, using the screwdriver, tighten it around the funnel to securely hold the eyepiece.

Step 9. Turn the funnel wide end up (you might find it easiest to sit down and hold the funnel between your knees). Place the Da-Tex screen over the wide opening; it doesn't matter which side faces down. Lower the large hose clamp over the wide end of the funnel and, using the screwdriver, tighten it around the funnel to securely hold the screen; as the clamp begins to get purchase on the funnel and screen, gently pull down all around the loose edge of the material so that the screen ends up flat and taut over the funnel's wide opening. This is an iterative process; you'll need to pull down on the material after each turn of the screw to keep it taut (not so tight that you risk tearing it, but taut—no wrinkles).

Step 10. Insert the eyepiece barrel into your telescope's 1.25-inch eyepiece holder, secure it with the thumbscrew(s), aim your telescope at the Sun (first taking care to cover or remove your finder scope, if any), focus the image on the screen, and enjoy group viewing of sunspots, partial and annular solar eclipses, and transits of Mercury and/or the International Space Station with your Sun Funnel (Fig. 2)! Always supervise its use, paying particular attention if there are children in the group.

Figure 2.   A 66-mm refractor with a 9-mm eyepiece projects a 100-mm-diameter solar image onto the Sun Funnel's screen. Photo by R. T. Fienberg.

4.   The Galileoscope and the Sun Funnel

With its 50-mm aperture and 500-mm focal length, the Galileoscope build-it-yourself refractor[6] would seem an ideal match to the Sun Funnel. With more than 200,000 of these inexpensive, high-quality telescopes in use around the world, it would be a shame not to use them for public viewing during solar eclipses, sunspot maxima, and other fun-in-the-Sun events. But there are two problems: (1) the supplied eyepiece is plastic and would melt if used for solar projection, and (2) the Galileoscope is designed to be used "straight-through," without a star diagonal, which leaves the Sun Funnel's screen awkwardly pointed at the ground. Fortunately both problems have simple solutions.

First, because the Galileoscope has a standard 1.25-inch eyepiece holder, it can use standard 1.25-inch metal eyepieces with glass lenses. Since $f_T = 500$ mm, you want $f_E \approx 500/43 \approx 11.6$, i.e., an 11- to 12.5-mm eyepiece. If you don't already have one, you can find a cheap one at Surplus Shed. Second, the Galileoscope attaches to a standard photo tripod via the 1/4-20 nut in the bottom of the tube. Simply use the tallest tripod you can find (many extend to a height of more than 60 inches) and ask your Sun-watching audience to sit on the ground around it. Then they can look up at the Sun Funnel and enjoy a convenient view of the projected solar image.

When using the Galileoscope with a Sun Funnel, it's a good idea to use a large piece of cardboard or foam board to shield the eyepiece end of the telescope from a direct view of the Sun. Simply cut a 2.7-inch (69-mm) round hole in the center of the board and press it onto the Galileoscope's dew shield.

---

[6]http://www.galileoscope.org

5.   Pointing a Telescope at the Sun

How do you aim a telescope at the Sun when you're not supposed to look through it, and when you're supposed to remove or cover your finder so that you don't look through that either, and so that bright sunlight doesn't melt its crosshairs? One solution is to watch your telescope's shadow on the ground and adjust the aim until the tube's shadow is as small and as round as you can get it. Another solution is to add a special-purpose Sun finder that projects a shadow or a spot of sunlight onto a target. There are several commercial units available, including these:

- Far Laboratories Helio Pod: http://bit.ly/pcILOW

- Tele Vue Sol-Searcher: http://bit.ly/nkhyOk

- Coronado Sol Ranger: http://bit.ly/phOnds

Yet another solution is to make something yourself based on the design of one of these products. Putting a large cardboard or foam board sunshade on the front of your telescope, as described in the preceding section, may be sufficient—just make sure the board is square to the telescope, then center the shadow of the dew shield on the Sun-facing side of the board.

6.   More Information

A thoroughly illustrated guide to building a Sun Funnel is posted online.[7] That document goes through the math behind solar projection, deriving the formula $f_E \approx f_T/43$ for the focal length of the eyepiece that gives a full-disk solar image in the Sun Funnel when used with a telescope of focal length f_T.

Armed with the mathematical relationships between telescope focal length, eyepiece focal length, projection distance, and projected image size, you can design your own solar-projection device using materials other than those suggested here. We'd be interested in seeing what you come up with, so feel free to contact us by email at rick.fienberg@aas.org, bueter@nightwise.org, and louis.a.mayo@nasa.gov.

---

[7]Illustrated Guide—http://www.transitofvenus.org/docs/Build_a_Sun_Funnel.pdf

Richard Fienberg demonstrates the sun funnel. Photo by Paul Deans.

Connecting People to Science
ASP Conference Series, Vol. 457
Joseph B. Jensen, James G. Manning, Michael G. Gibbs, and Doris Daou, eds.
©*2012 Astronomical Society of the Pacific*

Engaging Girls in STEM: How to Plan or Revamp Your EPO Resources or Activities to be More Effective for Girls

Lora V. Bleacher,[1] Karen A. Peterson,[2] Mangala Sharma,[3] and Denise Smith[3]

[1]*Goddard Space Flight Center, 8800 Greenbelt Road, Code 690, Greenbelt, Maryland 20771, USA*

[2]*EdLab Group, 19020 33rd Avenue W, Suite 210, Lynnwood, Washington 98036, USA*

[3]*Space Telescope Science Institute, 3700 San Martin Drive, Baltimore, Maryland 21218, USA*

Abstract.   This two-hour workshop, which was held as a follow-on to the plenary session "Engaging Girls in STEM: A Discussion of Foundational and Current Research on What Works," offered research-based insights, resources, and tips to help participants plan or revamp programs and resources aimed at encouraging girls in science. Led by Karen Peterson, PI for the National Girls Collaborative Project,[1] the workshop included: a brief discussion about effective strategies recommended for encouraging girls in STEM; hands-on experience, where participants—availing of the expert's guidance—applied the recommended strategies to alter or tailor an existing or planned program/resource to be more girl-friendly; and a sharing out, where the participants reflected on the results of the hands-on exercise and developed action items to continue carrying out the girl-friendly best practices in science, technology, engineering, and math education and public outreach.

1. Introduction

Scientists, teachers, parents, and policy makers all echo the need to get more students involved in science. There is a national impetus to broaden the science participation of girls, in particular. Efforts such as the National Science Foundation-funded National Girls Collaborative Project (NGCP)[2] are creating and fostering new programs and initiatives that use research-based strategies, tools, and collaborations to create gender equity in the areas of science, technology, engineering, and math (STEM). This effort and others like it are providing a large body of best practices that can be utilized by the astronomy, earth, and space science education and public outreach (EPO) communities when developing new programs, or altering existing programs, to be effective at engaging girls and maintaining their interest in STEM.

This two-hour workshop was organized as part of a professional development strand on engaging girls in STEM led by the NASA Science Mission Directorate As-

---

[1]http://www.ngcproject.org

[2]http://www.ngcproject.org

trophysics and Planetary Science EPO Forums.[3] Ms. Karen Peterson, PI for the NGCP, led the workshop. Earlier in the day, meeting participants had the opportunity to attend the related plenary session titled, "Engaging Girls in STEM: A Discussion of Foundational and Current Research on What Works" that was also led by Ms. Peterson (p. 7 in this volume). This article reports on the highlights of the workshop.

After participating in speed networking, where participants had a chance to meet each other and learn about each other's programs, needs, and resources, Ms. Peterson introduced the NGCP and led participants through a review of the research presented during the plenary session. She then continued with an examination of gender bias in our society, a discussion of effective strategies, and provided an opportunity to use the strategies on the development or modification of new or existing programs. After reporting on their efforts and action plans, participants were presented with a variety of printed materials, including copies of the "SciGirls Seven: Proven Strategies for Engaging Girls in STEM," [4] "Summaries of Research Presented by the Social Science Advisory Board" from the National Center for Women & IT meeting [5] held in May 2010, a rubric for evaluating a program's effectiveness for engaging girls, and various flyers that serve as examples of successful programs.

2. National Girls Collaborative Project

The NGCP brings together organizations that are committed to informing and encouraging girls to pursue careers in STEM. Currently about eight thousand programs are part of the NGCP network, serving more than five million girls. The NGCP's project goals are to maximize access to shared resources, strengthen the capacity of existing projects, and collaborate to create the tipping point for gender equity in STEM. NGCP works to meet these goals by fostering collaborations between uncoordinated services to girls interested in STEM careers, reducing competition for scarce resources, strengthening relationships among organizations, gathering information in a centralized location accessible to the general public, and sharing promising practices.

3. Current Status of Gender Equity

Following a topic introduced in the earlier plenary session, workshop participants were presented with a review of current research on gender equity, much of which is covered in the American Association of University Women's report "Why so Few?" [6] and the proceedings from the plenary session associated with this workshop. A short review is presented here. Girls grow up in an environment and context that shapes them via implicit bias, stereotype threat, fixed traits, and self-assessment. The idea of implicit bias suggests that we grow up "knowing" things implicitly as norms of society. For example, math and science are generally considered to be "male" fields and arts and

---

[3]http://smdepo.org

[4]http://www.pbs.org/teachers/includes/content/scigirls/print/SciGirls_Seven.pdf

[5]http://www.ncwit.org/pdf/SSAB_Research_May2010.pdf

[6]http://www.aauw.org/learn/research/upload/whysofew.cfm

humanities are generally viewed as "female" fields. In the case of stereotype threat, girls tend to do worse on a test or project if they perceive a negative stereotype threat, such as having to fill in a bubble on a test indicating that they are female. Fixed traits imply that a person is born with an innate ability to achieve in a particular area, such as playing basketball or doing math, whereas the reality is that with hard work and practice anyone should be able to get better at anything. As for self-assessment, girls tend to self-assess at a lower level than boys. In short, girls often enter into STEM educational settings with the preconceived notions that they will not do as well as boys and that they will not do well if they are not naturally talented in the subject area, both of which are compounded by a tendency to harshly self-assess their performance.

Girls may face additional challenges related to their family and cultural systems when trying to engage in STEM activities. For example, in Native American cultures girls are often caregivers for younger siblings. Including younger siblings in an educational program may allow more girls to participate than would otherwise.

Many of the female workshop participants commented on the "aha" moments that they had during and after the plenary session, as well as the challenges that they had personally faced in pursuing a STEM career. For example, many participants recognized a tendency in themselves to harshly self-assess their performance and/or to be too timid to ask questions in certain situations. Others commented on the differences in attitude and behavior they have observed between their own male and female children.

Ms. Peterson also asked the participants if they had any questions or concerns remaining after the plenary session. One participant mentioned that she was concerned about actually being able to induce gender equity in classrooms, which typically has a rigid structure. This sentiment was echoed among other participants. Karen suggested that we start by working with colleges and universities to train pre-service teachers. Another participant asked the question "What about the boys?" She went on to mention that some schools are banning the advertisement of programs that are for "girls only." The group agreed that it might be better to remove the word "only" when advertising such programs and to simply stress that they are "girl-friendly." Another participant mentioned that their institution had received backlash for hosting "Take Your Daughter to Work Day." They reverted to hosting girls and boys at the same time, which caused the number of attending girls to go down. They recently started separating the girls and boys while still hosting them on the same day, which has caused the number of attending girls to go back up. Another topic that was discussed was the difference between gender equity and gender neutral. Gender equity is parity in areas where girls and women were once underrepresented. Gender-neutral programs make efforts to avoid stereotypes.

4.   Awareness Raising—A Gendered World

In order to demonstrate implicit bias in our society, Ms. Peterson presented workshop participants with an assortment of birthday cards that she picked up at a local grocery store before the conference. Participants were asked to study the cards to look for evidence of gender bias. They noted that cards intended for girls were typically pink and contained images of mythical creatures, such as fairies, and words describing the receiver as "pretty," "sweet," "cute," etc. Cards intended for boys were likely to be in a variety of colors, to contain depictions of boys in career roles, such as a fireman, and to describe the receiver as "strong," "smart," "super," etc. Many participants commented on their own experiences with searching for non-gender biased cards, especially for girls. One participant mentioned that she was recently unable to find an astronaut or space-themed card for a girl.

In addition to raising our awareness of the gendered world in which we live, Ms. Peterson also discussed the need for our awareness of a few key points. First, proven, practical strategies exist for effectively engaging girls in STEM. However, it is also important to remember that not all research applies to everyone. For some girls, hearing that they can't do something is enough to make them want to prove that they can. And last, the involvement of more men is crucial in making a real change in this arena. We need more men to be aware of the implicit bias against girls in our society and to choose to be a part of the solution.

5. Effective Strategies

Ms. Peterson discussed several effective strategies for engaging girls in STEM. These strategies include: 1) making sure that the real-world relevance and meaning of an activity is effectively communicated; 2) providing authentic content and examples; 3) making sure that activities are collaborative; 4) providing opportunities for hands-on, investigative activities; 5) addressing varied learning styles; 6) building participant confidence; 7) providing opportunities for participants to grow and use career-relevant skills; 8) providing opportunities to address stereotype threat; and 9) providing role models and stories from near-peers (in person or via video) for girls to relate to. Adding context to a STEM project helps girls care about why they are doing it. If girls are asked to simply build a robot and make it go in a straight line they will ask "Why?" An example of adding context and real-world meaning would be to tell them that the robot is going to plug an underwater oil leak. One participant asked if it was okay to occasionally separate boys and girls for certain activities. Ms. Peterson indicated that doing so might help girls in some cases, especially for spatial activities, such as those using Legos, where girls usually do not have as much experience in these types of activities as boys. By separating them there is less pressure on girls to complete the activity as quickly as many boys could.

Examples of additional strategies were provided to participants in the form of a program rubric that was developed by the EdLab Group[7] for use with the TechREACH project.[8] The rubric lists elements that should be considered and included when developing an effective STEM program for low-income and underrepresented minority middle school students, including girls. For each element, the rubric helps the user determine how well they are incorporating it on a scale that ranges from "accomplished" to "competent" to "needs improvement" to "limited."

Accomplished programs are inclusive, congenial, supportive, collaborative, and democratic. They focus on STEM, promote student attitudes, and use technology. In addition, they focus on questions/problems/issues that students have identified as interesting and worthwhile, are project-based, make use of adults as mentors or guides, intensely involve students, and involve well-planned and motivating daily activities.

Programs that are rated as limited in their ability to engage underrepresented minority and low-income middle school students, including girls, in STEM allow students

---

[7]http://www.edlabgroup.org/

[8]http://www.techreachclubs.org/

to work alone or in friendship-based cliques, are not inclusive of student decision making, provide activities that are uninteresting or not enjoyable to students, do not include STEM topics by design, do not make use of technology (or in the case of the Internet it is only used for surfing), are classroom-like in their organization, are dominated by teacher-directed learning, and do not make good use of students' time.

6.   Activity Analysis

With these tips and recommendations in hand, participants worked in groups to assess the effectiveness of existing programs or activities, either those that they had brought with them or examples that Ms. Peterson provided during the workshop. Participants found that it was relatively easy to tweak program and activity elements to be more promising for effectively engaging girls now that they knew what to look for and include. For example, encouraging students to work in groups instead of in isolation and including videos or descriptions of female role models associated with the activity topic were mentioned as simple strategies that would make the programs/activities more girl-friendly.

One of the sample activities that was provided for participants to analyze involved having students determine the difference between a pseudo-science article and a real article. The scientists in the pseudo-science article were all female, whereas the scientists in the real article were all male. Although this disparity was probably not intended on the part of the activity's authors, girls who are asked to complete the activity may pick up on it and perceive a negative stereotype threat. By being aware of how we may inadvertently impose such threats and by being armed with the right tools to modify our programs and products accordingly, we can work to reduce the number of these types of incidences.

7.   Next Steps

At the end of the workshop, Ms. Peterson asked participants to write down one or more action items for how they would incorporate some of the information they learned during the workshop into their new or ongoing programs and activities. Ms. Peterson will follow up with participants a few months after the conference to inquire about their progress and to offer advice if needed.

Acknowledgments.   We gratefully acknowledge the Astronomical Society of the Pacific for organizing the conference such that we had an opportunity to plan and implement this workshop. We also appreciate the interest and participation of the workshop participants. The Astrophysics and Planetary Science EPO Forums also gratefully acknowledge the support of NASA's Science Mission Directorate.

Dennis Schatz leads a workshop. Photo by Paul Deans.

Connecting People to Science
ASP Conference Series, Vol. 457
Joseph B. Jensen, James G. Manning, Michael G. Gibbs, and Doris Daou, eds.
©2012 Astronomical Society of the Pacific

Engaging Girls in STEM: A Discussion of Foundational and Current Research on What Works

Mangala Sharma,[1] Karen A. Peterson,[2] Lora V. Bleacher,[3] and
Denise A. Smith[1]

[1] Space Telescope Science Institute, 3700 San Martin Dr., Baltimore, Maryland 21218, USA

[2] EdLab Group, 19020 33rd Avenue W. Suite 210, Lynnwood, Washington 98036, USA

[3] NASA Goddard Space Flight Center, 8800 Greenbelt Rd., Code 690, Greenbelt, Maryland 20771, USA

Abstract.   This article summarizes a panel discussion with Jolene Jesse (Program Director, NSF Research on Gender in Science and Engineering program) and Laura Migus (Director of Equity & Diversity at the Association of Science Technology Centers) on research related to gender in science, technology, engineering and math (STEM). Moderated by Ms. Karen Peterson from the NSF-funded National Girls Collaborative Project, Dr. Jesse and Ms. Migus discussed foundational and current research on pressing questions about the lack of gender diversity in STEM advanced education and careers, and on strategies the EPO community could employ in designing and implementing programs to encourage more girls and women to engage in STEM for the long term.

1.   Introduction: Diversity in STEM Advanced Education and Careers

Diversity in STEM education and careers occupies center stage in national discussions on U.S. competitiveness in the 21st century. Women constitute roughly half the total workforce in the U.S., but they hold just 25% of mathematical and science jobs (U.S. Department of Commerce 2011). Women earn nearly 60% of all bachelor's and master's degrees, except in physics, computer science, and engineering, where the percentages are 20–25% (National Science Foundation 2011). This disparity is even more pronounced at the doctoral level, where women earn less than 20% of awarded Ph.D.'s in physics or engineering. However, at the high school level, there is far less gender disparity: both female and male students take comparable advanced physical science and math courses (AIP Statistical Research Center 2011). What, then, accounts for the persistent lack of gender diversity in STEM advanced education and career paths? Despite the complex nature of the topic and lack of consensus among experts, clear trends and best practices for better engaging girls and women in science and engineering are becoming apparent through the accumulated and forefront research.

This plenary panel session on August 2, 2011 at the Baltimore ASP conference "Connecting People to Science" hosted discussions with Dr. Jolene Jesse (Program Director, National Science Foundation's Research on Gender in Science and Engineering program[1]) and Ms. Laura Migus (Director of Equity & Diversity at the Association of Science-Technology Centers[2]) on research related to gender in STEM. Moderated by Ms. Karen Peterson from the NSF-funded National Girls Collaborative Project,[3] the panelists discussed foundational and current research on why these gender trends exist, and what can be done to change them. The plenary panel was interactive, with opportunity for audience discussion, questions, and feedback to the panelists. The participants also received several handouts with resources and information on gender and STEM including "Why So Few? Women in Science, Technology, Engineering, and Mathematics" from the American Association of University Women (AAUW); "The SciGirls Seven Proven Strategies for Engaging Girls in STEM" from the PBS TV show and website SciGirls;[4] and handouts about the NSF-funded National Girls Collaborative Project. These resources summarize the current research and provide practical applications for EPO professionals.

This plenary session was organized as part of a professional development strand on engaging girls in STEM by the NASA Science Mission Directorate (SMD) Astrophysics and Planetary Science EPO Forums.[5] Meeting participants were encouraged to attend the related, 2-hour, hands-on workshop later the same day titled "Engaging Girls in STEM: How to Plan or Revamp Your EPO Resources or Activities to be More Effective for Girls" led by Karen Peterson.

Here, we summarize key points from the plenary discussions, highlighting the following topics: the research on girls' motivations and participation in STEM, the situation for underrepresented minority groups, and key strategies that help engage and sustain girls' interest and participation in STEM.

2.   Participation of Girls in STEM

Dr. Jesse discussed research detailing girls' actual or perceived abilities regarding, interest toward, and participation in STEM. How and why girls decide on pursuing STEM education and careers depend on their experience in both the academic and informal education environments. Referencing the AAUW publication "Why So Few," she highlighted four major areas that shape the way girls approach science: implicit gender and societal norms, stereotype threat, fixed traits vs. malleable intelligence, and self-assessment. Note, however, that since the study of gender in science is still a developing field as well as a complicated endeavor, various issues remain imprecisely understood, and are the subject of active research.

Implied messages about gender roles that are conveyed by society, or "implicit norms," can have an important influence on girls' interest and participation in STEM. Research using the "Draw a Scientist" test (Chambers 1983) shows that students as early as second grade carry the stereotypical image of scientists as predominantly male, older, and with shaggy hair. Young students may believe science and math are "boy

---

[1]http://www.nsf.gov/funding/pgm_summ.jsp?pims_id=5475

[2]http://www.astc.org/resource/equity/index.htm

[3]http://www.ngcproject.org

[4]http://www.pbs.org/teachers/scigirls/philosophy/

[5]http://smdepo.org

activities" while reading is for girls. Such beliefs and norms (e.g., science and math are "masculine" while humanities are "feminine") can negatively influence girls and keep them from choosing science. Such perspectives can be absorbed from numerous figures in society, including parents and teachers who may sometimes hold unconscious biases about gender roles in science.

Stereotype threat (Steele 1997) refers to poor performance or underachievement due to the anxiety about conforming to a negative stereotype about a group with which the person identifies. It can result from something as subtle as having to fill in your gender in an oval before taking a test, which can function as a reminder that your group is not expected to perform well. Research has shown, for example, that this can hinder the academic performance of females in math, and high-achieving girls may be more vulnerable to stereotype threat.

In response to an audience member's comment that in Astro 101 diagnostic tests, stereotype threat had not been valid, Dr. Jesse commented that the research on this topic still has unresolved issues, and that there also exists "stereotype lift" among non-stereotyped groups, whose performance may improve when they are compared favorably against another group.

Another inhibitor is the notion that intelligence and ability are fixed traits, and that people are born with innate talents. Interventions to reduce the negative effects of such beliefs include teaching students that intelligence is malleable (Aronson et al. 2002), that practice makes us better, and that learning can happen lifelong. For instance, our society tends to regard and extoll the talent of professional athletes as a gift or trait they were born with, while in reality a top basketball player will have logged endless hours of practice. A member of the audience brought up the prominent discussions on Facebook of the "10,000-hour rule" that claims the key to success in any task or field is largely a matter of practice for that long.

The Facebook comment brought up a related example of the ways stereotypes work against STEM fields: Dr. Jesse remarked that while girls spend huge amounts of time with information technology on Facebook and creating websites, they do not see themselves spending their lives working with computers—a pursuit for "geeks."

Even among students with comparable achievement in math, girls assess their ability lower than boys do, holding themselves to higher standards of performance (e.g., Correll 2001). For example, Dr. Jesse said, while boys will continue to feel confident and stick with a field even if they obtain Cs, girls will feel inadequate and depart if they are obtaining Bs. Girls tend to shut down when told they must have an innate talent, and rise to the occasion when told they can perform a task. Students who assess their abilities higher are more likely to pursue advanced courses or degrees in the field. Therefore, helping girls understand and believe that they are equally capable of becoming a scientist or mathematician may increase their interest and participation in STEM fields.

3. Influence of Family and Cultural Values, Especially for Groups Underrepresented in STEM

Ms. Migus discussed the complex influence of socioeconomic, family, and cultural values (e.g., religion, ethnicity, or any group with strongly defined norms) on girls' interest in, access to, and participation in STEM. She recommended reading the National Academy's publication, "Learning Science in Informal Environments," particularly the seventh chapter that addresses diversity and equity. The AAUW report "Why So Few" also provides guidance on this topic.

In general, girls may not be allowed to fulfill their potential in environments that are not gender-neutral and perpetuate gender stereotypes. Given a choice between multiple activities, such as drama, writing, and robotics, girls will avoid activities in which they will be the sole girl in the group. When EPO practitioners remain passive about this situation, they perpetuate the stereotype.

Awareness of cultural values can play an important role in designing an inclusive program. For example, families whose cultural norms are protective of girls may not approve "sleep-aways" in informal learning venues. While this has little to do with actual science content, educators need to assure such parents that the informal science programs provide safe and nurturing spaces for their children. One audience member brought up the topic of working with Native American audiences and the need to engage the entire family in any significant activity. Ms. Migus elaborated on that, and mentioned that some programs take into account that girls are often the caregivers in their families, thus engaging them in informal science program often necessitates engaging their siblings, too.

One cannot assume a homogeneous population or a one-size-fits-all solution. Ms. Migus encouraged educators and EPO practitioners to examine their own assumptions about parental/familial involvement in education. She highlighted the need for better understanding and sensitivity to cultural variations among non-majority racial and ethnic groups. For example, while the majority of Americans assume in-depth parental involvement in their children's education, many Latino parents believe educators are the experts on the topic and may be disinclined to "interfere" with the latter's work. So, educators and EPO providers need to appreciate that the Latino parents' "non-involvement" does not equate to disengagement, and instead need to make explicit that parental involvement is welcome.

In response to an audience comment about the environment of exclusivity that college physics departments tend to cultivate, Ms. Migus exhorted us to think critically about the learning environments we create, and consciously make them more inclusive. Sometimes, cultural norms and stereotypes favor a group in STEM; for instance, an audience member commented that Asian boys are over-represented in the Math Counts program. Dr. Jesse gently cautioned us that the label "Asian" is too general, and ignores the huge differences that exist among Asian cultures. Nevertheless, some of the STEM successes among some Asian communities could be attributed to the cultural belief that intelligence is malleable, and an emphasis on practicing a task to do well on it, e.g., students practicing for a Spelling Bee contest. In response to an audience query about when it is most effective to teach students that intelligence can be developed with effort, Dr. Jesse emphasized "at all ages."

Another thing to keep in mind is that, even among cultures where we assume women do not have equal rights as men such as in Turkey and Egypt, there are significant numbers of women engineers and computer scientists.

4. Barriers Faced in Engaging Girls in STEM

It was evident that the session participants had a variety of experiences in engaging girls in science. Ms. Peterson encouraged us to reflect on the barriers we had encountered in our efforts to broaden the participation of girls in STEM. Several challenges were identified by the EPO community, including reaching non-attentive audiences and reaching families. Panelists and audience members also reflected on the difficulty of getting both boys and girls to contribute equally in group settings, and the advantages or otherwise of working with single-sex groups in educational activities when in the "real world" unisex teams are rare. (Regarding the latter, Dr. Jesse recommended reading the research results on multidimensional teams in the workforce.)

Focusing on process can be an important tool for encouraging learning and participation. Panelists noted, for instance, that process can be very important to girls. In a task to build a robot, girls may focus on the aesthetics of their design, while boys may focus on how fast it can be built. Likewise, girls tend to do all the writing in the task, even if that responsibility is rotated among group members. EPO practitioners can adjust for this by being explicit about the stages of the activity, when discussion will take place, and what defines success. They can consider using multiple milestones for success, in which case the process becomes more important than simply getting to the finish point. This is equally important for boys as well as for girls, as boys also need to learn skill sets to work on teams.

5. Discussion Wrap-Up

During the plenary panel presentations and discussions, it became abundantly clear that multiple, often interrelated factors, contribute to the underrepresentation of girls and women in STEM. Several promising strategies to address gender disparity are beginning to bear fruit.

Nevertheless, the study of gender effects in STEM is an evolving field and much of the research is still not settled because, as Dr. Jesse remarked, we are dealing with complex, nuanced humans. For instance, in some situations, stereotype threat may be fading, or it could be that it is so subtle that we are failing to recognize it. She urged EPO professionals to follow the evolving research and to help identify contributing factors and successful strategies to rectify gender imbalance in STEM.

Stereotypes and cultural norms could be challenges to engaging girls in STEM. But as Ms. Migus urged us, EPO practitioners also need to reflect on our audiences' beliefs and value systems to identify strengths that we could tap into while engaging them in science. When teachers and EPO practitioners encounter a participant who is not engaging in an activity, we should consider what belief or value we are failing to convey.

Ms. Peterson, the moderator, wrapped up the plenary panel with a call to action: to identify one useful piece of knowledge we will take away from the session and use it to encourage more girls to engage in STEM.

Acknowledgments.   The Astrophysics and Planetary Science EPO Forums gratefully acknowledge funding from NASA Science Mission Directorate. We thank the ASP conference organizers and program committee for the opportunity to hold a plenary panel on the topic of girls and STEM, and sincerely thank the more than one hundred session attendees for their active participation through thoughtful questions and comments. The National Girls Collaborative Project acknowledges funding from the National Science Foundation Human Resource Development and Research on Gender in Science and Engineering programs.

References

American Association of University Women 2010, "Why So Few? Women in Science, Technology, Engineering, and Mathematics," (http://www.aauw.org/learn/research/whysofew.cfm)

American Institute of Physics Statistical Research Center (http://www.aip.org/statistics)

Aronson J., Fried C. B., & Good, C. 2002, "Reducing the Effects of Stereotype Threat on African American College Students by Shaping Theories of Intelligence," Journal of Experimental Social Psychology, 38, 2, 113

Chambers, D. W. 1983, "Stereotypic Images of the Scientist: The Draw a Scientist Test," Science Education, 67, 2, 255

Correll, S. J. 2001, "Gender and the Career Choice Process: The Role of Biased Self-assessments," American Journal of Sociology, 106, 6, 1691

National Science Foundation Special Report NSF 11-309, 2011 (http://nsf.gov/statistics/women/)

Steele, C. M. 1997, "A Threat in the Air: How Stereotypes Shape Intellectual Identity and Performance," American Psychologist, 52, 6, 613

U.S. Department of Commerce Economics and Statistics Administration Issue Brief #04–11, Aug 2011, (http://www.esa.doc.gov/Reports/women-stem-gender-gap-innovation)

Connecting People to Science
ASP Conference Series, Vol. 457
Joseph B. Jensen, James G. Manning, Michael G. Gibbs, and Doris Daou, eds.
©2012 *Astronomical Society of the Pacific*

Using the Big Ideas in Cosmology to Teach College Students

Kim Coble,[1] Lynn R. Cominsky,[2] Kevin M. McLin,[2] Anne J. Metevier,[2] and Janelle M. Bailey[3]

[1] *Chicago State University, Department of Chemistry and Physics, 9501 South King Dr., Chicago, Illinois 60628, USA*

[2] *Sonoma State University, 1801 East Cotati Ave., Rohnert Park, California 94928, USA*

[3] *University of Nevada, Las Vegas, Department of Teaching and Learning, 4505 South Maryland Pkwy Box 453005, Las Vegas, Nevada 89154, USA*

Abstract. Recently, powerful new observations and advances in computation and visualization have led to a revolution in our understanding of the structure, composition, and evolution of the universe. These gains have been vast, but their impact on education has been limited. We are bringing these tools and advances to the teaching of cosmology through research on undergraduate learning in cosmology as well as the development of a series of web-based cosmology learning modules for general education undergraduate students. Informed by our research on student learning in cosmology, we are utilizing best pedagogical practices to implement the content in an accessible online student-centered framework. In this workshop, we engaged participants with examples of interactive exercises, illustrations and text from the initial module of the three-module curriculum. We invite interested educators to help us test the materials with their students as the curriculum develops.

1. Introduction

Recent results of cosmology research have revolutionized our understanding of the universe. While we have known for some time that the universe was much hotter and denser in the past, and that it has been expanding and cooling for billions of years, new detailed observations and computer simulations show us directly what our universe was like at various epochs in the past, how it appears today, and how it will evolve in the future. Through observations and measurements of distant supernovae, field galaxies, galaxy clusters, the cosmic microwave background, abundances of light elements, gravitational lensing, and more, we can answer fundamental questions about the composition of the universe, its geometry, its age, how structures are distributed and formed, and even the fate of the universe.

Needless to say, undergraduate astronomy courses have had difficulty staying current with rapidly unfolding cosmological knowledge. Modern topics, such as cosmology, are of primary interest to many educators and students (e.g., Pasachoff 2002), but we are only beginning to understand students' alternative conceptions in this area (Prather et al. 2002). Furthermore, according to a survey by Bruning (2006), only ~20% of a typical introductory astronomy textbook is devoted to cosmological topics. With calls for science education reform at all levels (American Association for the Advancement of Science 1990, 1993; Bransford et al. 1999; Fox & Hackerman 2003; National Research Council 1996, 2003), it is imperative that we design effective instruction to counteract student misconceptions, build upon correct ideas, and provide scaffolds for new understanding (Donovan & Bransford 2005).

We have structured our research and curriculum development around three cosmological themes: (1) structure: the universe is vast in space and time, (2) composition: the universe is composed of not just regular matter, but also dark matter and dark energy, and (3) change: the universe is dynamic and evolving. We also emphasize how this knowledge is supported by observational and experimental evidence and that the processes occur according to the laws of physics.

Our goals for the workshop were to give participants an overview of the curriculum and an opportunity to use some of the interactive exercises. After each topic in the workshop, we presented the relevant results of some of our cosmology education research, which has informed the curriculum. The curriculum and the research are each described further in the sections below.

2. Curriculum

Informed by our research on student understanding of cosmology, we are creating an immersive set of web-based modules that allow students to participate in the process of doing science while learning cosmological concepts. Text, figures, and visualizations are integrated with short and long interactive tasks, which use real cosmological data.

2.1. Organization of the Modules

Table 1 shows the preliminary organization of the curriculum into three modules. Each curriculum module consists of a central theme, with five chapters that can each be pursued over the course of one week. The chapters are further subdivided into sections that focus on a specific topic.

2.2. Pedagogical Approach

Our goal is to move beyond typical curricula, which are predominantly text-based with occasional animations or simulations, toward an environment that realizes a higher level of learner-centered interactions. Our ultimate goal is to create a simulated world in which learners choose their own path to knowledge, much like professional researchers. Such learning experiences provide the most engaging connection to the material, and they go far beyond what has traditionally been possible with textbooks and other passive media such as film.

The pedagogical flow of a section topic is shown in Figure 1. First, student ideas are elicited with a short warm-up question. Insights from cognitive psychology have shown that perceptions are related to prior experiences and current knowledge. Students enter into a course with different mental representations, and these representations can affect their learning. We need to know where the students are in order to take them where we want them to be. We also must explicitly make connections to existing ideas. The warm-up questions target commonly held student ideas as a precursor to learning the chapter material. Next, having thought about the topic themselves, students complete section readings and one or more short, interactive tasks. These tasks have

Table 1.   Preliminary Curriculum Content

Module	Chapter
Our Place in the Universe: Space and Time	1. The Size and Scope of Space 2. Observing the Universe: Light and Telescopes 3. Motion and Time 4. Measuring Distances 5. Special Relativity and Spacetime
The Darker Side of the Universe: Gravity, Black Holes, and Dark Matter	6. Classical Physics: Gravity and Energy 7. General Relativity 8. Black Holes and Spacetime 9. Observing Dark Matter Through the Motions of Objects 10. Observing Dark Matter Through Gravitational Lensing
Our Evolving Universe: Past, Present and Future	11. Expansion and the Hubble Law 12. The Early Universe 13. The CMB and Large Scale Structure Formation 14. Dark Energy and Supernovae 15. Geometry and the Fate of the Universe

styles familiar from education research: ranking, sorting, matching, visual, etc. Finally, students are asked questions to summarize what they have learned by keeping guided notes. By integrating the activities in this manner, students can master a narrow topic before moving on to the next. A student "logbook" provides a framework to tie the work together and allows the instructor to track a student's progress and grade it for effort.

Each chapter features a longer "lab" activity that includes real cosmological data and integrates the concepts learned in the chapter. A "mission report" on the activity can be graded by the instructor. There is also a customizable homework section with problems automatically graded for accuracy. Other features include a scientific calculator, graphing tools, a glossary, and instructor resources.

Any math involved in the activities (high school algebra at the most) is explained conceptually as well as numerically, in order to empower students rather than frustrate them. All examples for the numerical exercises follow a consistent step-by-step approach for problem solving: "Given, Find, Concept(s), Solution, Think About Answer."

Not only have computer simulations and visualizations revolutionized our understanding of cosmology, but students themselves underscore the need for good visuals during interviews. One student explains how a good visualization would help her replace incorrect old ideas:

> QueenB: "I've always heard about the Big Bang Theory and how it was this big explosion and that's how the planets and everything else came about... I don't really remember what the actual Big Bang Theory is now,

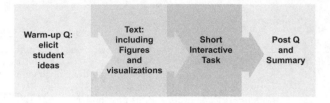

Figure 1.   Pedagogical flow of a section (narrow topic focus) within the curriculum.

but I do know that I was told otherwise in my new class... [but] I need some kind of visual to explain to me how things work in science."

We do not simply show visualizations, but have created exercises to help students understand what they are seeing, because students do not necessarily view them from the same perspective as astronomers. Furthermore, as much as possible we compare models and data and make it clear which is which.

We feel that it is important to include real cosmological data in the curriculum not just because it is a cornerstone of science, but because our research shows that this is important for students. Pre-course surveys show that students are initially skeptical that, for example, the age of the universe is measurable or that the Big Bang model is supported by evidence. In end-of semester interviews, two students describe the value of real data and interactive engagement in understanding the material and in changing their beliefs:

> BigDipper: "Actually, when you told me, I was less inclined to believe it, to tell you the truth... I really didn't understand... At first I thought a lot of scientists kind of sometimes make up facts... I never understood it, so I thought a lot of stuff was made up... Now I do, now that I saw the calculations. And done [sic] it myself."

> Kenya: "We did it during the lab, which was awesome... we took pictures of galaxies and we actually measured them and we recorded them on the graph and we actually came up with real authentic data from real galaxies... it was really real authentic information, so I got a lot from that..."

2.3.   Sample Interactives

The warm-up questions and short interactive tasks are chosen and designed based on our research on student strengths and needs. During the workshop, we presented examples from the first four chapters of the curriculum. Again it should be noted that these activities are all embedded within a full-curriculum and are not intended as stand-alone interactives.

Chapter 1: Starter Question. Warm-up questions such as the example below are typically found at the beginning of a section, are meant to elicit student ideas, and are presented in "student voices." In the workshop, we presented them to participants as clicker questions. In the curriculum, students will choose from a menu of options, and explain their choices in their logbook. The starter question from Chapter 1 involves the hierarchical structure of the universe:

Three students are discussing which objects are in our solar system.

Annie says, "A solar system has different things in it like galaxies and planets and stars and stuff like that. Our solar system has the planets Mercury, Venus, Earth, and so on. The planets have moons so I think moons too."

Brenda says, "I disagree. I think a galaxy has stars inside it. Each one of the stars has planets orbiting around it and that's what a solar system is. So a galaxy has solar systems in it but a solar system doesn't have galaxies in it."

Charles says, "I think that the terms solar system and galaxy mean the same thing."

Do you agree with any of these students, and if so, who? Explain your reasoning in your logbook.

Chapter 1: Structure Hierarchy. The first interactive that participants tested was also from Chapter 1, dealing with the hierarchical structure of the universe (Fig. 2a). In this activity, students must place common objects in the universe in a hierarchy to show how they relate to one another.

Chapter 2: Photon Races. The next warm-up question (Fig. 2b) was again done as a clicker question with workshop participants. It targets student ideas about the speed of light for different colors:

Ladies and gentlemen! Place your bets! Which color of light do you think travels fastest through the vacuum of empty space? In your logbook, rank the following from fastest to slowest:
- Radio ● Optical ● X-ray ● Gamma-ray

In your logbook, describe how the outcome of the race compares with your initial predictions. Explain how the equation that describes the relationship between a wave's wavelength and frequency ($\lambda = c/f$) relates to the outcome of the race.

Chapter 3: Lookback Time. In this interactive activity, students are presented with a star field with an "observer" star in the center (Fig. 2c). At the top of the star field, students are given the order in which three stars go supernova and the time between each event. Students must determine the order in which the light signals from the supernovae will be seen by the observer star. The primary goal for this activity is to confront students with the relationship between distance and lookback time; that it is not just the timing of an event that dictates when it is observed but also its distance from the observer. Several easier activities relating to lookback time precede this one in the curriculum.

Chapter 3: Measuring Redshift. This is an example of an activity that uses real data. Here students use spectra of galaxies (in an intensity vs. wavelength format) and the redshift formula in order to determine the velocities of several galaxies (Fig. 2d). A scientific calculator and answer-checker will eventually be built into the full web-based system. However, for the purposes of the workshop, we programmed the activity to perform some calculations for participants.

Chapter 4: Parallax. In this activity, students must use a conceptual understanding of parallax to rank stars by distance. They are shown a star field and can adjust the position of the Earth around the Sun to observe the stars shifting back and forth.

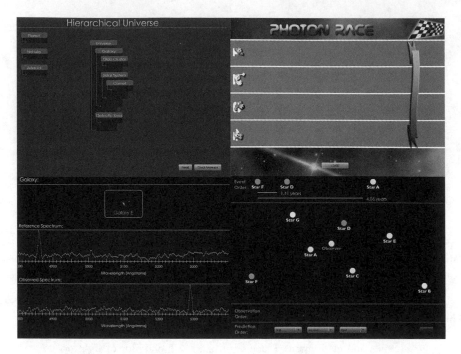

Figure 2.   Sample Interactives. From top left to bottom right: (a) Hierarchical
Universe, (b) Photon Races, (c) Quantitative Redshift, and (d) Lookback Time.

3.   Research

Determining the range and frequency of "alternative conceptions" is an important first
step to improving instructional effectiveness. Through analysis of pre-instructional
open-ended surveys at four different institutions ($N = 703$), follow-up interviews ($N =
14$), and pre-course essays, laboratory assessments, and exams over five semesters at
a single university ($N \approx 60$), our research group is attempting to classify students'
ideas about concepts important to modern cosmology. The areas targeted so far include the universe's structure, age, evolution, and composition (including dark matter
and dark energy), as well as student perceptions of astronomical distances. Survey responses, analyzed through an iterative process of thematic coding, reveal a number of
alternative conceptions. Findings from pre-instructional interviews and homework essays are consistent with results from the multi-institutional surveys. Post-instructional
interviews reveal student progress toward more expert-like ideas as well as areas for
improvement, plus student attitudes toward inquiry-based instruction. In the workshop,
we presented the research results for topics most relevant to Module 1, namely structure
and distances.

3.1.   Structure

Survey results indicate that students frequently conflate structure terms such as solar
system, galaxy, and universe, or do not understand the relationship between the terms
(Table 2). Early interviews confirm results from the written surveys that students do not

have a concrete understanding of the structure of the universe. Students do have some knowledge of the objects in the universe, but their understanding of the structure of the universe is superficial.

Table 2.   Survey results on structure. *"Describe each of the following: galaxy, constellation, solar system, universe. Are any of these related? If so, how?"* Responses are described as "OK" because extremely few were what we would call correct. For example, an OK response might be that a solar system contains stars and planets. The same OK response would also be acceptable for a galaxy.

	OK (%)	Incorrect (%)	No Response
Galaxy	77	12	11
Solar System	87	9	4
Universe	92	1	7
Hierarchical Relationship	60	27	13

Later interviews and exams reveal that students' understanding of structure does increase by the end of the semester. One student described her learning process as follows in an interview:

> Pierce: "Like I kind of thought the galaxy was the universe, and the universe was something they just talked about like when kind of, you know constellation oh that's the universe. That's what I thought the universe was. I didn't know, I just thought it was referring to stars or something."

3.2.   Light and Spectra

Student misconceptions about light and spectra are well-documented in the literature (e.g., Bardar 2006; Comins 2001; Zeilik, Schau, & Mattern 1998). One common source of confusion is the difference between frequency and speed. For example, when questioned deeply, a teacher who participated in NASA's *Multiwavelength Universe* summer program in 2011 (which used activities from Chapter 2 of our curriculum) explains her ideas:

> "Another thing scientists like to know about EM [electromagnetic] waves is how fast they travel. To do this, they calculate how many waves move past in one second. This is called the frequency. . . A wave that travels very fast has a high frequency. A wave that travels slower has a low frequency. EM waves that have a short wavelength move faster than waves that have a long wavelength, so short waves have a higher frequency than long waves . . . The amount of energy contained in an EM wave depends on how fast it moves. That means that shorter wavelength waves, which have a higher frequency, have more energy than longer wavelength waves, which have a lower frequency."

This same teacher had responded to an earlier, lower-level question saying that all light waves travel at the same speed. This kind of varied response demonstrates the need for instructors to probe deeply into their students' ideas.

3.3. Light Travel Time

One of the key concepts that students must learn in order to understand other aspects of cosmology is the relationship among distance, speed, and time, particularly light travel time. Unfortunately this concept is surprisingly difficult for students. Two examples of the pre- and post-test questions that we used to probe student understanding are shown below. The pre-test questions are given after lecture but before lab activities (different than those designed for this curriculum) and the post-test questions are asked on exams. Results are presented in Table 3. Follow-up interviews suggest that students who have difficulty with astronomical sizes and distances have been more strongly influenced by culture and the media, whereas those who had less difficulty expanded on their personal prior experiences.

> Pre-test (N = 57): In the future, when space travel is advanced, you have 3 weeks of vacation time and want to visit the star Sirius, in honor of your favorite Harry Potter character. Sirius is 8.6 light years away. If spaceships in the future could travel at half the speed of light (much faster than current spaceships), would you be able to make the trip to Sirius and back during your vacation? Explain.

> Post-test Q (N = 62): The star Vega is 25 light years away. If you were in a spaceship that could travel at half the speed of light, the amount of time it would take you reach Vega is _____ . (Be specific, use a number.)

Table 3.   Research results for half-speed of light travel time questions. An incomplete response is missing one or more of the identified elements of a correct answer, whereas a "partial" response contains both incorrect and correct elements. On the post-test, the most common wrong response was to divide by 2 instead of multiply by 2.

	Pre (%)	Post (%)
Correct	14	40
Wrong	32	44
Incomplete	23	10
Partial	23	6
True but Irrelevant	7	0
Non-Scientific	0	0
No Response	2	0

3.4. Parallax

In order to assess their understanding of parallax, students were given the question in Figure 3 post-lecture but prior to interactive class activities (different than the ones developed for this curriculum). The question also appeared on exams after activities and homework on the material.

Post-lecture, but prior to the activities on parallax, 28% of students ($N = 36$) answered correctly. On a midterm exam following the activity, 56% of students ($N = 43$)

answered correctly. On the final exam, 58% of students ($N = 45$) answered correctly. The most common incorrect response was that the star that shifts more is farther away.

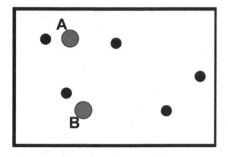 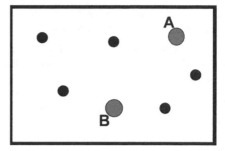

Figure 3.   The following two pictures were taken six months apart. Which star is farther away, A or B? How do you know?

4.   Evaluation and Open Invitation

Throughout the workshop and at the end, participants were invited to fill out a survey, providing feedback on the workshop materials. We are grateful to the 16 participants who did so; we will be incorporating this feedback into our curriculum development. Participants offered particularly useful specific critiques and suggestions in the free response section.

When asked about the instructions for the interactives, responding participants generally were able to understand the instructions well enough to do the activities and, similarly, found the instructions easy to follow. Only two or fewer participants disagreed with either statement among three of the activities. However, about a third (6) of the respondents had trouble with instructions for the *Hierarchical Universe* activity. While participants could follow and execute the current instructions, many of them nevertheless also suggested that some changes be made to the instructions.

Participants overwhelming felt that their students would respond positively to the interactives, with only one or two respondents disagreeing with any of the four relevant items in this question set, for any of the four activities. More specifically, participants felt that the activities would attract their students' use (by being interesting and engaging), and that students would learn from the activities, particularly by addressing misconceptions. For the most part, respondents strongly agreed with the value of the activities for addressing misconceptions, and all 16 respondents "strongly agreed" that *Hierarchical Universe* would address some misconceptions. The ratings for *Quantitative Redshift* were slightly less positive than for the other activities.

This workshop was the first time that any of our curriculum materials have been tested outside of our development group. We are eagerly seeking instructors to pilot-test Module 1 in their classes starting in January 2012. We feel that it is important to develop and pilot-test the curriculum with a diverse range of institutions and students from the onset of the process. Physics and astronomy education research has had a major impact on how we view student learning and many materials developed as a result have been shown to be effective. However, materials have often been developed at large R1 research universities with traditional college students. The institutions that have already signed letters of commitment for pilot-testing include community colleges, undergraduate institutions, research universities, and one online university; some of these are minority-serving institutions. We are also seeking cosmologists with data or simulations that can be incorporated into the curriculum. If you would like to sign up for pilot-testing or to contribute cosmological data, please contact the authors.

Acknowledgments. This three-year project is being funded by the Education and Public Outreach program for NASA's Fermi Gamma-ray Space Telescope, grant NNX10AC89G from NASA's EPOESS program, and the Illinois Space Grant Consortium. Background research was also funded in part by National Science Foundation CCLI Grant #0632563 at Chicago State University. We gratefully acknowledge the artistic and technical expertise of the Education and Public Outreach group at Sonoma State University, especially Kevin John and Aurore Simonnet, and the help of our current and former research students on the project: Melissa Nickerson, Carmen Camarillo, Virginia Hayes, Donna Larrieu, K'Maja Bell at Chicago State University; Roxanne Sanchez at UNLV; and Geraldine Cochran at Florida International University. We thank Ted Britton and his staff at WestEd for developing and analyzing the evaluation. Kendall-Hunt will be publishing the curriculum modules for national distribution.

References

American Association for the Advancement of Science (AAAS) 1990, Science for all Americans (New York: Oxford University Press)
AAAS. 1993, Benchmarks for science literacy (New York: Oxford University Press)
Bardar, E. M. 2006, Ph.D. dissertation, Boston University
Bransford, J. D., Brown, A. L., & Cocking, R. R. (eds.) 1999, How people learn: Brain, mind, experience, and school (Washington, D.C.: National Academy of Sciences)
Bruning, D. 2006, Astron. Ed. Rev., 5(2), 182
Comins, N. 2001, Heavenly errors: Misconceptions about the real nature of the universe (New York: Columbia University Press)
Donovan, M. S., & Bransford, J. D. (eds.) 2005, How students learn: Science in the classroom (Washington, D.C.: National Academies Press)
Fox & Hackerman 2003, Evaluating and improving undergraduate teaching in science, technology, engineering, and mathematics (Washington D.C: National Academy Press)
National Research Council (NRC) 1996, National science education standards (Washington, DC: National Academy Press)
National Research Council (NRC) 2003, Learning and understanding: Improving advanced study of mathematics and science in U.S. high schools (Washington, D.C.: National Academy Press)
Pasachoff J. M. 2002, Astron. Ed. Rev., 1(1), 124
Prather E. E., Slater T. F., & Offerdahl, E. G. 2002, Astron. Ed. Rev. 1(2), 28
Zeilik, M., Schau, C., & Mattern, N. 1998, The Phys. Teacher, 36, 104

Connecting People to Science
ASP Conference Series, Vol. 457
Joseph B. Jensen, James G. Manning, Michael G. Gibbs, and Doris Daou, eds.
© *2012 Astronomical Society of the Pacific*

Exploring Transiting Extrasolar Planets in your Astronomy Lab, Classroom, or Public Presentation

Peter Newbury[1,2]

[1]*Department of Physics and Astronomy, University of British Columbia, 6224 Agricultural Road, Vancouver, Canada V6T 1Z1*

[2]*Carl Wieman Science Education Initiative, 358A-6174 University Blvd, Vancouver, Canada V6T 1Z3*

Abstract. The search for life beyond our Solar System is topic that appears in almost every introductory, general-education astronomy course, typically referred to as "Astro 101." School teachers and science museum presenters might cover this topic, too. This article is our instructor's manual for facilitating a 50-minute, hands-on activity that explores the light curves produced by transiting extrasolar planets: how they form and how to extract characteristics about the extrasolar planet from the shape of the curve. Students apply their skills to the system HD 209458 by examining data collected by the Microvariability Of STars (MOST) telescope.

1.   Introduction

Before we find life beyond our Solar System, we must find places to look: extrasolar planets, that is, planets beyond ("extra-") our own Solar System. We have already found hundreds of them. A growing number of discoveries are being made by NASA's Kepler mission which uses the transit method to detect extrasolar planets. When a planet passes directly between us and its sun, the planet transits the star, and there is a periodic dip in the brightness of star as the planet blocks some starlight from reaching us. By decoding the star's light curve, we can uncover some of the characteristics of the planet: its orbital period and diameter and, if we know the mass of the star, the extrasolar planet's orbital radius. From these, we can determine if the extrasolar planet is in the habitable zone, the "Goldilocks" region around the star that is not too cold (and water is frozen) or too hot (and water is vaporized.) Around distant stars, just as it is on Mars, Europa, and Enceladus and here on Earth, our search for life is really a search for liquid water.

The search for life beyond our Solar System is topic that appears in almost every introductory, general-education astronomy course, typically referred to as "Astro 101." At the University of British Columbia, the 100 to 200 students in our Astro 101 class attend bi-weekly, 50-minute, hands-on activities in groups of 30 to 40. With the support of the Carl Wieman Science Education Initiative, we created activities for these sessions which (i) explore those course learning goals best-suited for hands-on experience (ii) employ, as much as possible, research-based instructional strategies. We also take the opportunity to teach and train our teaching assistants (TAs) about astronomy education by providing them with a detailed "TA Guide" for each activity. These guides are not simply a list of equipment and materials and a recipe of steps for running the activity.

Rather, we justify why each step in the activity occurs, alert the TAs about what to expect, and give samples of dialogue the TAs can use to drive the activity forward.

In Section 2, we reproduce the "TA Guide" for the extrasolar planets activity. In Section 3, we give some suggestions for presenting this concept in other settings: in the lecture hall as an interactive demo and in a science centre, museum or other public presentation. All materials mentioned here, including all the hand-outs, the poster of the HD 209458 lightcurve, and the LoggerPro file "LightCurve.cmbl" are available online.[1]

2. TA Guide

This section reproduces the TA guide for the transiting extrasolar planets activity. The guide is written in 2nd-person ("Place your hand in front of the light sensor...") to better connect with the TA preparing to facilitate the activity. Our suggestions for dialogue the TAs can use to lead the activity *are written in italics*.

Description
_____

In this 50-minute tutorial, students discover how to interpret the light curves of stars with transiting extrasolar planets and how to extract characteristics of the planet. They apply these techniques to the real planet HD 209458b. In order to get there, students first explore how light curves form (Part 1) and then what a light curve reveals qualitatively (Part 2) and quantitatively (Part 3) about extrasolar planets.

Learning Goals
_____

By the end of the tutorial, a student should be able to

- illustrate how extrasolar planets are detected and extract properties of the planets and stars from the observations, and

- compare extrasolar planets to our own.

Preparation
_____

There is a lot of equipment to get set up for this tutorial.

1. Computer (hooked up to a digital projector) with

   - LoggerPro program
   - LoggerPro file LightCurve.cmbl. When you close the program at the end of the tutorial, it will likely ask you if you want to save the changes. Click No. If you accidentally save a changed version of the file, there is a backup copy on the computer's desktop.
   - PDF of hand-outs open in a PDF reader like Acrobat

2. Light sensor (Vernier LS-BTA photometer) plugged into Go!Link plugged into a USB port on the computer. The sensor is inside a white tube of paper, coloured black inside, which reduces the amount of scattered light hitting the sensor.

_____

[1]PDF and source files are available at blogs.ubc.ca/polarisdotca/astrolabs.

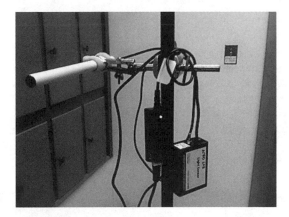

Figure 1. The Vernier light sensor is inside a tube of paper to block scattered light. The sensor connects through a Go!Link interface to the computer USB port.

3. Equipment stand with clamp for holding the light meter.

4. One small and one large styrofoam ball planets. Ideally, the large one has exactly twice the diameter of the small one.

5. White globe lamp.

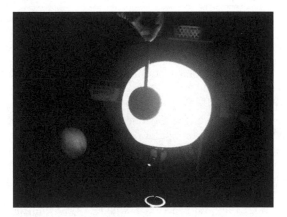

Figure 2. A globe lamp is mounted on a wooden block. Demonstrators pass styrofoam "planets," held at the ends of pencils, in front of the "star" to simulate a transit.

6. Poster of HD 209458 light curve taped to the wall. Data were collected by the Microvariability and Oscillations of STars (MOST) telescope, a Canadian space telescope operated from the University of British Columbia.

7. Hand-outs for the students:

   • Pages 1–3, single-sided, one for each group.

• Page 4 (Questions), one for each student.

Set up the computer cart at the front of the room. Clamp the light sensor into the equipment stand at the height of the middle of the globe lamp. Place it about 1 metre from the globe lamp and aim it at the center of the lamp. Set up the projector (on the overhead cart) so it shines on the screen next to the globe lamp. The goal is to make it possible for the students to simultaneously see the globe lamp, the transiting planets and the LoggerPro graph projected onto the screen, without the light sensor picking up too much light reflected off the projector screen. The diagram below shows one possible configuration.

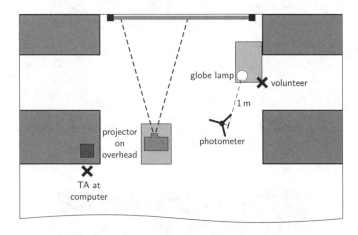

Figure 3.   Configure the room so students can simultaneously see the transiting planet and the projected light curve.

Boot the computer, plug in the Go!Link USB cable and double-click the LoggerPro file "LightCurve" on the computer desktop. After it initializes, the LoggerPro window should recognize the light sensor and begin collecting measurements (look for the lux readings at the top-left of the LoggerPro window). Click the green Start (and red Stop) button to start graphing the brightness of the globe. The graph cycles every 20 seconds, so once you've pressed Start, you can just let it continue.

The readings might be "ragged." If it's bad, try moving the light sensor so it receives less light scattered off the screen from the projector. The paper tube, coloured black inside, reduces the scattered light, though it has to be more carefully aimed at the globe light. If there are fluorescent lights in the room, those can also mess up the readings if the data is collected at 50–60 Hz. It's surprisingly sensitive. Experiment with the sampling rate (Ctrl-D in LoggerPro) to find a rate that reduces that interference. In the end, it's okay if the readings are a bit ragged: it makes it look more like the HD 209458 observations the students will see shortly.

Don't start collecting data and plotting the curve, though. That will be too distracting for your introduction. Instead, project the top of Page 1 of the hand-outs. It will encourage the students to read Page 1 when you give it to each group. We want the students to discover for themselves what characteristics of the transiting planet are important, so *don't* hand-out Pages 2 and 3 of the worksheets (listing characteristics that matter) since that will short-circuit the discovery stage.

Part 1a: Introduction to Extrasolar Planets (5 minutes)

This section provides some background and get the students "up to speed" in case the instructor hasn't covered extrasolar planets yet. With all the new Kepler results, check `planetquest.jpl.nasa.gov` for the latest number of confirmed extrasolar planets. You can use that number in the intro:

> *Astronomers have discovered hundreds (or use actual `planetquest` number) of planets orbiting other stars. These are planets and stars outside our Solar System, so we call them extrasolar planets.*
>
> *There are 3 main methods for finding extrasolar planets:*

> **Radial velocity (Doppler) method:** *When a planet orbits a star, the star wobbles back and forth a small amount as it (and the planet) orbit around the center of mass. [Pick up the globe lamp and swing it around in a small circle towards and away from the students.] The star light is redshifted and blueshifted as the star moves away or towards us. By detecting these periodic Doppler shifts, we know a planet is there.*

> **Direct detection:** *A very small number of extrasolar planets have been observed directly. This is very difficult because planets are small and stars are so bright. If the globe lamp is our Sun, then Earth is about 1 mm in diameter, 10 metres away. That is very hard to see.*

> **Transit method:** *When an extrasolar planet crosses ("transits") directly between us and the star, it temporarily blocks some of the star light and the star's brightness dips. In this tutorial, we'll explore this transit method. [Be sure to define and emphasize the word "transit" so that when the students see "transit method" in the future, they know it's about dips in the light curves.]*

Part 1b: Observations (5–10 minutes)

Ask for a volunteer ("John") to come to the front to move the planets—warn John to be careful with all the cords. Explain we're pretending the globe lamp is a star that we're observing with our telescope, the light sensor. Switch the projection to the LoggerPro window and click Start to begin plotting the light curve. Reading and interpreting graphs is another skill we're teaching—they are not experts—so be sure to orient the students to what the graph shows:

> *The horizontal axis is time. The vertical axis is intensity or brightness of the star. The graph, called a light curve, it tracks how much light the sensor is receiving.*

Put your hand in front of the light sensor to demonstrate how the curve dips when you block some light.

Ask John to use the small planet first, moving the planet horizontally around the globe lamp like a planet orbiting its star. Encourage the rest of the students to try to record what they observe in the Observations table on Page 1. Encourage the students to ask the volunteer try things. If they're hesitant at first, you can suggest the volunteer do a slow transit followed by a fast one so that both dips are visible on the graph. You might want to "freeze" the light curve by clicking Stop in the LoggerPro window. Ask the students what the difference is in the shapes of the curves and what that tells us about the planet. Then click Start to continue watching for patterns.

Ask John to switch to the big planet ("John, could you hold up the big planet for the class to see?"). Before he does the transit, get the students to make a prediction about what the dip will look like:

> *John is about to use a planet that's twice as big. Do you think the dip will be deeper or shallower?* "Deeper!"
> *Deeper, good. How much deeper?* "2 times deeper!"
> *[This is common mistake. Don't let them know it's wrong, though.]*
> *Good prediction, let's try. John, go ahead...*

Certainly the dip is deeper, but how much? Ask John to use both planets: two transits with the big planet followed by two transits with the small planet. Get him to start when the graph cycles so you can get all four dips on the same plot. Press STOP to freeze the plot after the four dips.

> *You can see the dips are deeper for the big planet. How much deeper? Let's measure...*

Use a ruler to measure the depth of the dip. It's quite easy to measure the curve right on the projector screen. The "100% illumination" line might not be horizontal in the light curve so measure the depth from just before it starts to drop or at the middle of the dip. Measure both deep dips and both shallow dips—they should be in a 4:1 ratio!

> *Hmm, it's not two times deeper, it's four times deeper. Why is that? Right, because the big planet's AREA is four times bigger.*

Most students incorrectly predict the dip will be two times deeper but once they see the demonstration and recognize that it depends on area, they usually have no trouble adjusting. However, it's important to emphasize the "squared" relationship between the depth of the dip and the diameter of the extrasolar planet because it comes up again in the activity and in the Questions at the end.

Part 2: Characteristics of Light Curves (10 minutes)

The next Part of the tutorial is for the students to make the links between the changes in the light curve and the characteristics of the transiting planet. Hand-out Page 2. Invite the students to complete the sheet:

> *Exactly how planets in a solar system orbit the star and block its light can be very complicated. A very good approximation, though, depends on only **two characteristics**: the diameter of the planet relative to the star and the length of time it takes the planet to orbit the star. Look at the light curves on Page 2 and figure out what they tell you about the planets.*

After about 5 minutes, or sooner if their attention starts to drift, project Page 2 of the hand-outs. Go over the answers stressing first, how the light curve changed and second, what that tells us about the planet:

> **Planets A and B:** *The dip is deeper in B indicating the planet has a larger diameter. How much larger? Two times because the dip is four times deeper.*

> **Planets C and D:** *The dips in D occur twice as far apart compared to C. Planet D takes twice as long as Planet C to orbit the star.*

Part 3: Decoding the Light Curve (15 minutes)

Hand out Part 3 and project it on the screen. Explain that we want to get *quantitative* values for a transiting planet's period and size, not just *qualitative* (or relative) values. Briefly go over the "tools" to find the period and diameter, reinforcing that the depth of the dip depends on the *area* of the planet compared to the area of the star, and those areas depend on the *squares of the diameters*.

The students will use these tools to learn about an actual extrasolar planet. Ask the students to take their worksheet and a pencil to look at the light curve for HD 209458. The big horizontal tick marks on the lightcurve are days; the small crosses are in 1-hour intervals. The large vertical ticks are 1% drops in intensity. They must *measure two quantities*: the period P and the depth of the dip ΔI. Ask them to be careful not to write on the poster.

Wander around the room and help with the calculations. If they don't have a calculator or phone capable of doing the calculations, they can always use google. The results are something like this:

Orbital period	Planet diameter
Measure the time between dips:	Measure the depth of dip:
P = 3 days, 12 hours	ΔI = 2%
Write the orbital period in days:	Convert % drop to a decimal (for example, 1% = 0.01)
P = 3.5 days	ΔI = 0.02
and years	Find the ratio of diameters:
P = 0.0096 years	$\dfrac{d}{D} = \sqrt{\Delta I} = \sqrt{0.02} = 0.14$
	The star HD 209458 has diameter D = 1,400,000 km, the same size as our Sun. Find diameter d in km:
	$d = (0.14)(1,400,000) = 196,000$ km

Part 4: Questions (Remainder of tutorial)

As students finish the calculations in Part 3, give them the Questions sheet, one per student. They can collaborate but we want the students to hand in their own papers. Here are some answers and comments:

1. We can use the simplified $a^3 = P^2$ version of Kepler's Law because the star HD 209458 has the same mass as our Sun ($M = 1$ solar mass). With $P = 3.5$ days or 0.0096 years, $a = \sqrt[3]{0.0096^2} = 0.045$ AU.

    The characteristics typically extracted by the students are remarkably close to those published for HD 209458b (Miller-Ricci et al. 2008):

period	$P = 3.52474832 \pm 0.00000029$ days
diameter	$D = 191,500 \pm 5700$ km
semi-major axis	$a = 0.045$ AU

2. The extrasolar planet has a size like Jupiter but an orbit far inside Mercury's.

3. Planets in the habitable zone of stars like our Sun take one year to orbit. It takes at least two to three years for the planet to make three transits. We need three dips to make sure: the first could be a random dip. The second dip suggests there a planet in orbit though it could be two planets of similar size. The third dip is strong evidence there is one extrasolar planet in orbit.

    If students are curious, ask them to imagine what the light curve of Earth transiting the Sun would look like to astronomers on HD 209458b: The Earth's diameter is about 100 times smaller than the Sun's so the dip in the intensity would be about $(0.01)^2 = 0.0001$ or 0.01%. That tiny dip is 20 times smaller than the "noise" in the HD 209458 lightcurve (about 0.2%) and occurs only once per year. Imagine trying to find a planet like Earth on the poster!

4. In the past, students have quite easily recognized the 4-day and 7-day periods (notice there are multiple choice answers with 2-day and 3-day periods: that's the number of dips, not the periods.) The key to choosing the right choice is remembering the dip is proportional to the square of the diameters. The 7-day dips are four times deeper which means Planet 2's diameter is two times bigger, not four times bigger.

Clean-up

The computer and other equipment must be locked up in the storage room. Return the digital projector to the Main Office.

3. In-class demonstrations and science center presentations

For Astro 101 courses without a lab, we also have a version of this extrasolar planets activity that runs as an in-class demonstration. It could easily be adapted to a presentation at a science center or a public outreach event.

We set up the equipment at the front of the room. The LoggerPro output is projected onto the screen so the audience can simultaneously see both the equipment and the light curve. Each participant receives a handout.

A volunteer comes up and makes a few orbits of the star using the big planet. With this practice, we get the volunteer ("John") to make a few nice, steady orbits. Then we ask the participants to make a prediction:

> *Have a look at your worksheet. Suppose the orbits John just made are represented by the light curve at the top. Now John is going to make the planet orbit twice as fast, that is, with half the period. What will the new curve look like? Pick one of the graphs A–D in Question 1.*

Students make predictions with clickers, ABCD-cards or a show-of-hands (or fingers). Students typically have no problem recognizing there would be twice as many dips in the light curve but many do not anticipate the dip being narrower like it is in the correct choice A because the planet transits the star in only half the time. With the prediction made, John goes ahead and makes the quick orbits. Having made predictions, the students are more likely to focus on the critical features of the graph: the period of the dips and their width.

Next, we switch to the small planet:

> *(John, would you grab the small planet, please? Thanks.) John is going to try this small planet. Its diameter is one half of the first planet. He's going back to the original, slower orbit. Which light curve in Question 2, A, B, C or D, do you think it will make?*

After making their predictions, John goes ahead and makes the orbits. The dips are definitely shallower. How much shallower, though? With the LoggerPro software, the instructor can measure the depth of the dips. Sure enough, the big dip was four times deeper.

There are lots of directions to go from there, including analyzing the MOST data for HD 209458, deducing orbital radius via Kepler's Laws and exploring the habitable zone.

4. Conclusions

The discovery of extrasolar planets is a story that appears daily in the media. It has grabbed the public's attention. Astro 101 students and science center attendees familiar with extrasolar planets, particularly the ones discovered by the transit method used by the Kepler mission, will have an opportunity to share their new-found knowledge with their community. More importantly, these people will contribute to building a more scientifically-literate society.

Acknowledgments. This work is supported by the Carl Wieman Science Education Initiative.

References

Miller-Ricci, E., Rowe, J. F., Sasselov, D., Matthews, J. M., Guenther, D. B., Kuschnig, R., Joffat, A. F. J., Rucinski, S. M., Walker, G. A. H., & Weiss, W. W. 2008, ApJ, 682, 586

Participants in "Exploring Transiting Extrasolar Planet" workshop. Photo by Paul Deans.

Connecting People to Science
ASP Conference Series, Vol. 457
Joseph B. Jensen, James G. Manning, Michael G. Gibbs, and Doris Daou, eds.
© *2012 Astronomical Society of the Pacific*

STOP for Science! A School-Wide Science Enrichment Program

Patrick Slane,[1] Robert Slane,[2] Kimberly Kowal Arcand,[1] Kathleen Lestition,[1] and Megan Watzke[1]

[1] *Chandra X-ray Center, 60 Garden Street, Cambridge, Massachusetts 02138, USA; slane@cfa.harvard.edu*

[2] *Section Elementary School, W 318 S 8430 County Road Ee, Mukwonago, Wisconsin 53149, USA; slanero@mukwonago.k12.wi.us*

Abstract.   Young students are often natural scientists. They love to poke and prod, and they live to compare and contrast. What is the fastest animal? Where is the tallest mountain on Earth (or in the Solar System)? Where do the colors in a rainbow come from? And why do baseball players choke up on their bats? Educators work hard to harness this energy and enthusiasm in the classroom but, particularly at an early age, science enrichment—exposure outside the formal classroom—is crucial to help expand science awareness and hone science skills. Developed under a grant from NASA's Chandra X ray Center, "STOP for Science!" is a simple but effective (and extensible) school-wide science enrichment program aimed at raising questions about science topics chosen to capture student interest. Created through the combined efforts of an astrophysicist and an elementary school principal, and strongly recommended by NASA's Earth & Space Science product review, "STOP for Science" combines aesthetic displays of science topics accompanied by level-selected questions and extensive facilitator resources to provide broad exposure to familiar, yet intriguing, science themes.

1. Introduction

By capturing the attention of students at certain ages and stages of development, and focusing their thoughts on diverse science topics, we can aim to further develop critical thinking skills that foster a broad appreciation and understanding of science. Imagine a lunchroom debate, not about which new video game is the most awesome, but about whether or not the Sun will ever explode ("Will too." "Will NOT! There's not enough mass!"). That's an argument worth having.

This is the goal of the "STOP for Science" (STOP) project: to promote access to science concepts and spark discussion of these topics among students, teachers, parents, and others. It does this by putting engaging science content in accessible locations tied to the school environment where social experiences are likely to occur, such as hallways, auditoriums, after-school programs, etc. Unlike traditional classroom activities, the STOP project aims to attract individuals who are curious about the content and participate with it voluntarily during their own free time.

STOP consists of a series of five posters and accompanying questions designed to pique student interest in science concepts and their application to the world in which we live. Accompanying each poster is a series of question sheets of increasing difficulty levels that students answer and submit at a designated location (collection box, office, etc.). Random prize drawings can be used to recognize/celebrate student participation. The purpose is to expose students to and create school-wide interest about science so students want to "STOP for Science" as displays are changed throughout the year. Although the focus is building-wide, content can be linked to classrooms through use of accompanying teacher/facilitator resource guides.

A central idea of STOP is not to burden teachers with more curricula, but to foster continuous exposure to science, offering learner-driven participation (Bell et al. 2009) and also neutral territory, where visitors don't have to actively seek out science activities (Riise 2008). The STOP program is intended to be facilitated by parents (e.g. PTA), school administrators, teachers, or other educators or volunteers. Buy-in from the principal or other administrator is key, since space for materials and communication about the program is required.

2.   STOP Program Description

The STOP materials are part of a science enrichment program intended for Grades K–6. As described above, the program is designed for implementation outside of the formal classroom curriculum, but is carefully designed with materials to promote integration of the topics into the formal classroom as desired, and with links to appropriate national standards for Science, Math, and English Language content.

In pilot testing, a series of five different science topics were introduced over the course of a year. All of the topics were introduced in the initial pilot study at Fiske Elementary School in Lexington, Massachusetts, USA. Additionally, two of the posters were used with students in pilot tests, and the other three reviewed for content with elementary schools in Mukwonago, Wisconsin, USA. Because the STOP program is intended to be school-wide, additional topics are being developed for use in subsequent years.

Each STOP poster stands alone and the concepts covered are not directly tied to those in the others. Therefore, the STOP posters may be displayed in any sequence or combination desired. STOP exposes students to a variety of familiar, yet challenging, science concepts utilizing simple explanations and illustrations as applied to real-life events. The materials in the STOP project are designed to serve as a precursor to in-depth studies included in upper grade level curriculum. As in any informal science learning setting, the desire to educate and inspire needs to be balanced with both the background and the interest level of the participant (Smith & Smith 2006; Smith, 2008). The STOP materials are designed to engage students in topics that are potentially relevant and applicable to their interests or experiences.

2.1.   STOP Program Features

The STOP project seeks to promote access to science concepts, encourage critical thinking, and increase scientific awareness, by positioning the science displays as an accessible, social experience (Bell et al. 2009). The STOP project takes advantage of out-of-class time and the voluntary nature of participation in such informal learning settings (Falk 2005). Visitors will choose to be there, remaining at the displays that are interesting to them for however long that interest lasts (Rounds 2004; Smith et al. 2010).

There are several key features of the STOP project that are essential to its potential widespread use and ultimate impact on student interest in science. The first feature we discuss is the ease of administration of the STOP material. The program is designed to be manageable so that one individual (building principal, teacher, librarian, or parent volunteer) can coordinate the program. The involvement of more people in the community can lead to more robust levels of scientific engagement, as described below. However, even a single individual may have the important positive effect of exposing students to science concepts and enrichment beyond articulated classroom curriculum, and his or her responsibilities can also be expanded to support classroom activities— even by simply hanging the posters in useful locations.

To support the possible availability of teacher involvement, a teacher/facilitator resource guide is provided for each of the current five topics. These resource guides are designed to fit into classroom learning settings and are tied to U.S. national standards for science, math, and reading. The target for the STOP resource guides is to allow for independent use by students in Grades 3–6. Children in grades K–2 could participate in the classroom activities but would require adult or upper-grade level assistance. Details on the resource guide are found in section 2.3.

Beyond the print products in the STOP project, interested students and adults may find more information on the STOP website. This website features multimedia and other resources related to the concepts found in the posters and other print material. The STOP website will evolve over time as implementation tips and other ideas contributed by users are posted as they are sent.

2.2. Recommendations for Implementing STOP

While STOP is designed to give any local facilitator the independence to implement the project the best way he or she sees fit, there are several recommendations included in the project materials. These suggestions are intended to help expose the maximum number of students to the content in the most engaging manner possible.

For example, we recommend that posters be displayed in a high-traffic area, where students have ample opportunity to observe, discuss and reflect on their content. School lobbies or walls near lunch lines are examples of these areas. It would be ideal if these posters were located away from any designated "science" areas to ensure that the students of all interests see them.

We also provide recommendations about how long particular posters should be available for student viewing. In particular, we suggest that each poster and accompanying question set be displayed for a sufficient amount of time to allow students time to explore, gain interest, and answer questions, but not so long as to promote complacency or boredom. In testing, periods of four to six weeks provided an appropriate timeframe. It is also advisable to establish a timescale for the poster series (e.g., perhaps five posters over the course of one academic year.)

Other steps for implementation may include "advertising" the STOP materials through an introduction to the program from an educator, principal or other official. A preview or summary of the materials to educators and other staff in regular meetings may also create more comfort and excitement to talk about the material. The inclusion in newsletters that are distributed to parents and other care-givers may spark questions in the home about the materials.

We recommend additional student engagement through scheduled opportunities for student viewing times, contests (with small prizes) for participation, or even "Family

Figure 1.   Example photo of the STOP display in pilot testing.

Night" events. Accompanying each poster is a series of question sheets of increasing difficulty levels that students answer and submit at a designated location (collection box, office, etc.). Random prize drawings can be used to recognize/celebrate student participation.

2.3.   STOP Teacher/Facilitator Resource Guides

The STOP classroom materials are classified into three different levels. They are:

- Level One: K–2 (requires adult or upper-grade level assistance)

- Level Two: Grade 3–4

- Level Three: Grade 5–6

Question sheets are written with increasing difficultly. Grade ranges identify approximate ranges, however, and shouldn't restrict use of any set by interested participants.

The authors used the results from initial testing (Section Elementary and Clarendon Elementary Schools, Mukwonago, Wisconsin, USA in 2009–10 and 2010–11 school years) to refine and expand the associated support materials for the program, and to update the description of the program implementation. The facilitator resource guide provides detailed information about each poster, including: summary of main points, background science, fun facts, common questions and misconceptions, demonstrations and activities, and suggested resources.

The resource guide information is organized such that it makes this program adaptable to use in classrooms, after-school programs, science clubs, libraries, and other informal science education venues.

3. Science Topics

The following five topics are included in the STOP program.

3.1. Size

Young students often make comparisons, and one of the earliest quantitative concepts they encounter is that of size. Some things are small and others are BIG! And there is nothing quite like the engaged argument about which is biggest. But the question of exactly how one measures size can actually be complicated. The *How Tall is Tall?* exercise introduces a rather simple question: What is the tallest mountain on Earth?

Figure 2.   *How Tall is Tall?* poster.

How Tall is Tall? **Key points:**

- Having students make comparisons is a useful tool, and one of the earliest quantitative concepts they encounter is that of size.

- Despite hosting very tall mountains, the Earth is actually very smooth, though slightly oblate.

- Assessing the relative heights of mountains is tricky business, particularly when consideration is given to the starting point for the measurements.

- Olympus Mons is the tallest mountain in the solar system.

3.2. Stars

Most students have some awareness of the nighttime stars, and many are aware of the fact that the Sun is one such star. Because we do not see obvious changes in stars on timescales of days or even years, most students don't give much thought to the question of how long stars last, and what happens to them when they stop shining. *When Stars Go Boom!* introduces several key ideas about stars, and focuses in particular on the supernova explosions that mark the ends of the lives of the most massive stars. The poster provides a particular example of one such supernova that occurred in 1987, explaining briefly how the explosion was discovered (by a student) and how the most advanced telescopes in the world are currently used to study the effects of the explosion. The idea is extended to a stellar explosion that occurred in the year 1054, as recorded by ancient astronomers, and whose remnants can still be seen in a telescope today.

Figure 3.   *When Stars Go Boom* poster.

When Stars Go Boom Key points:

- Our Sun is a star.

- Stars don't live forever.

- Massive stars end their lives in huge supernova explosions. Our Sun is not sufficiently massive to explode.

- Supernova explosions are so bright they can easily be seen in other galaxies.

- We can observe the remains of supernova explosions, even from thousands of years ago.

3.3. Force & Rotational Inertia

It's a familiar baseball refrain. The pitcher winds up and delivers the ball, and the batter swings, but too slowly. Strike one. "Choke up on that bat!" says the coach. Most players know that this means they should slide their hands up higher along the bat handle, leaving a gap between their grip and the bottom end of the bat. But why? *Choke Up on That Bat!* explores the science behind this technique by introducing the concepts of inertia, force, and torque, though in somewhat more conceptual terminology. The poster starts with a discussion of what is required to take an object at rest and put it into motion. This is extended to investigate what is required to rotate an object. A simple example is given to illustrate how the motion resulting from a twisting action depends on where the action is applied. This is then used to connect to the concept of "choking up" with reducing the rotational inertia of the bat, thus increasing the bat speed.

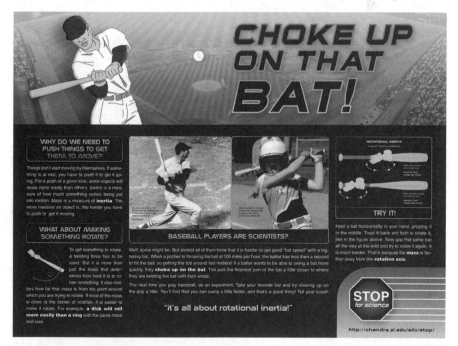

Figure 4.   *Choke Up On That Bat!* poster.

Choke Up On That Bat! Key points:

- A force is required to put a body into motion. Its resistance to this is called inertia, and is related to the mass.

- A twisting force (torque) is required to for rotation. Resistance to this is called rotational inertia.

- Rotational inertia depends on how far the mass is located from the rotation point. The farther the mass is from the rotation point, the larger the rotational inertia.

- Choking up on a bat reduces the rotational inertia, and increases bat speed.

3.4. Speed

Whether from traveling in a vehicle or from racing across the playground at recess, the concept of speed is one that students of nearly all ages have encountered. While most can express what the concept of "fast" means, many do not understand how speed is actually defined. Moreover, many do not have a good understanding of the vast range of speeds that we encounter, or know about. Since "fast" is a relative concept, it is instructive (and fun) to investigate the speeds of different things students know about. *That's Fast!* introduces the basic definition of speed, giving simple examples in units that the students can relate to. It then provides examples of things that we think of as being "fast," and provides other examples of things that are much faster. In this way, students are introduced to the importance of comparison. Using the natural interest that students have in ranking things, a graphical illustration of objects that cover a very large range of speeds is presented.

Figure 5.   *That's Fast!* poster.

That's Fast! **Key points:**

- Students know what "fast" means, but not necessarily how to measure it.

- Speed is the distance an object will travel in a fixed amount of time.

- We encounter a huge range of speeds in everyday life. A cheetah is the fastest land animal, but travels much more slowly than sound, for example.

- No object can be accelerated to a speed greater than the speed of light in a vacuum.

3.5. Light

Many students have encountered rainbows, either spotting them directly on those special days when the raindrops fall while the Sun still finds cloudless regions to peek through, or at least seeing pictures of them in books or movies. Many have also seen "rainbows" formed in the spray of water from a garden hose on a sunny day. Although most students thus make the connection between water droplets and sunlight in the formation of these colorful arcs, few think about how this actually happens, or about the special geometry required for a rainbow to be visible. *Somewhere Over The Rainbow* introduces key ideas about light—and sunlight in particular—to explain its behavior when it travels through a raindrop, and how this results in the formation of rainbows. The concepts of refraction and dispersion are introduced, with an example to explain how the direction of a wave is changed when it enters a medium in which its speed is reduced. This is extended to the case where white light enters a drop of water, and the geometry of the resulting light path is illustrated to explain how the colors in a rainbow are formed, and why they come from the directions we observe to form these familiar arcs.

Somewhere Over the Rainbow **Key points:**

- Light is a form of energy and behaves like a wave.

- Different colors correspond to light of different wavelengths. Sunlight is made of all different colors.

- When light travels from one medium into another, its path can be bent, or refracted. Different colors are refracted by different amounts, allowing them to separate.

- A rainbow is formed by sunlight entering raindrops, refracting, and reflecting back to the observer.

4.   Conclusion

Science is all around us, but sometimes it escapes our notice. With young students, whose interests in science topics seem particularly strong, efforts to introduce science in multiple settings, both formal and informal, should help maintain and foster that interest. We need to keep engaging students in science concepts and investigations.

Figure 6.   *Somewhere Over the Rainbow* poster.

STOP for Science! approaches this goal through the introduction of a science enrichment program aimed at stimulating interest in science outside the formal classroom and by capitalizing on the process of learning as participation in a social world (Lave & Wenger 1993). Based on science topics developed to spark student interest, and supplemented by well-tested resource materials to extend the program by including demonstrations and other activities, the program engages students through aesthetic displays of science topics accompanied by level-selected questions to provide broad exposure to familiar yet intriguing themes.

Acknowledgments.   This material is based upon work supported by the National Aeronautics and Space Administration under NASA contract NAS8-03060.

References

Bell, P., Lewenstein, B., Shouse, A. W., & Feder, M. A., eds. 2009, "Learning science in informal environments: People, places and pursuits," (Washington, D.C.: The National Academies Press)

Falk, J. H. 2005, "Free-choice environmental learning: Framing the discussion," Environmental Education Research, 11, 265

Rounds, J. 2004, "Strategies for the curiosity-driven museum visitor," Curator, 47, 389

Smith, J. K. 2008, "Learning in informal settings, Psychology of Classroom Learning: An Encyclopedia," (Farmington Hills, MI: Cengage Learning)

slane5 Smith, L. S., Smith, J. K., Arcand K. K., Smith, R. K., Bookbinder, J., & Keach, K. 2010, "Aesthetics and Astronomy: Studying the public's perception and understanding of imagery from space," Science Communication, 33, 201

Smith, L. F., & Smith, J. K. 2006, "The nature and growth of aesthetic fluency," P. Locher, C. Martindale, L. Dorfman, V. Petrov, & D. Leontiev, eds., New directions in aesthetics, creativity, and the psychology of art, (Amityville NY: Baywood), 47

Lave, J. & Wenger, E. 1993, "Situated learning: Legitimate peripheral participation," (New York: Cambridge University Press)

Riise, J. 2008, "Bringing Science to the Public," D. Cheng, M. Claessens, T. Gascoigne, J. Metcalfe, B. Schiele, & S. Shi, eds., Communicating Science in Social Contexts, (Brussels: Springer), 301

Capitol College exhibit. Photo by Paul Deans.

Connecting People to Science
ASP Conference Series, Vol. 457
Joseph B. Jensen, James G. Manning, Michael G. Gibbs, and Doris Daou, eds.
© *2012 Astronomical Society of the Pacific*

Barriers, Lessons Learned, and Best Practices in Engaging Scientists in Education and Public Outreach

Sanlyn R. Buxner,[1] Mangala Sharma,[2] Brooke Hsu,[3] Laura Peticolas,[4] M. Alexandra Matiella Nova,[5] and Emily CoBabe-Ammann[6]

[1] *Planetary Science Institute, 1700 E Fort Lowell Rd., Suite 106, Tucson, Arizona 85719, USA*

[2] *Space Telescope Science Institute, 3700 San Martin Dr., Baltimore, Maryland 21218, USA*

[3] *LPI at Goddard Space Flight Center, 8800 Greenbelt Rd., Code 690, Greenbelt, Maryland 20771, USA*

[4] *Space Science Lab, University of California at Berkeley, Berkeley, California 94720, USA*

[5] *Johns Hopkins University Applied Physics Laboratory, 11100 Johns Hopkins Road, Laurel, Maryland 20723, USA*

[6] *Emily CoBabe & Associates, Inc., 2410 Vassar Drive, Boulder, Colorado 80305, USA*

Abstract.   This Astronomical Society of the Pacific conference brought together a group of specialists interested in education and public outreach (EPO) from a wide variety of contexts including NASA centers, non-profits, museums, and universities. Active engagement of scientists in EPO activities results in benefits for both the audience and the scientists. Despite this, education research has shown that many barriers exist that keep scientists from engaging in EPO activities. To counter these barriers, many stakeholders in this community are working to bridge the gap and help scientists make a meaningful contribution to these efforts. There are many documented roles for scientists including giving public talks, classroom visits, large outreach events, radio broadcasts, engaging in curriculum development and teacher workshops. Over the past year, members of the NASA science mission directorate forums have been actively working with their community members to understand the reasons that scientists in our community do and do not participate in EPO activities. This session expanded the discussion to the larger community of stakeholders across science, education, and outreach contexts. It was an open forum for discussion of ideas about barriers and lessons learned regarding engaging scientists in education and public outreach.

1. Introduction

This special interest group session brought together scientists, EPO professionals, and program officers to discuss barriers, successes, and opportunities to engage scientists in education and public outreach. The session was structured so that participants could respond to previously documented barriers and resources, leading to a larger discussion so that everyone could express their opinions. Participant feedback was collected in both written and spoken words. This document synthesizes feedback from the thirty-five people who participated in the one-hour discussion.

A list of documented barriers that prevent scientists from participating, or participating more fully, in EPO activities was compiled from education research, evaluation, and surveys conducted by members of NASA's Science Mission Directorate over the last ten years (Andrews et al. 2005; Crettaz von Roten 2010; Poliakoff & Webb 2007; Thiry, Laursen, & Hunter 2008; Word Craft 2008). The working list of documented barriers presented during the workshop can be found in Appendix A.

In the next section we present the challenges and successes of doing EPO discussed by participants in the session. We have sorted comments by topic and included representative feedback (written and verbal) collected during the discussion. We have organized the feedback we collected into four themes: i) support for EPO involvement, ii) considerations of cost versus benefit, iii) connecting with the EPO community, and iv) skills to engage in EPO.

2. Support for Scientist Engagement in EPO

2.1. Challenges

Participants discussed how a lack of support from their supervisors coupled with the pressure to publish and a lack of release time posed barriers to doing EPO for scientists. Specific comments included:

"[I] do not feel there is enough incentive or recognition [for doing EPO]."
"[There is] no support from management for spending 'work' time on EPO."
"The lack of support is real and affects many researchers that try EPO."
"[Support] needs to come from the top down on the org chart."

2.2. Successes

In contrast to those who expressed a lack of support to do EPO, several participants were able to share success stories. Each of the successful stories documented support from managers and/or supervisors to do EPO so that it was not done during "off" time but rather was part of normal activities. Specific comments included:

"At [our lab], the lab director has dictated that scientists and engineers should be spending 1% of their overall effort on EPO; this makes management much more supportive and scientists and engineers more involved."

"We have been successful in getting a ROSES supplement grant that provides a week of funding for a group of scientists to do EPO work each year. Being able to pay them makes it more likely that they would be engaged and lets them know that we value their time."

3. Cost versus Benefit for Scientist Engagement in EPO

3.1. Challenges

Participants expressed that doing EPO activities took away too much time from doing science and that overall it was too time consuming to do if there was no funding for the activities, which was the experience of at least some of the people in the room. Specific issues related to NASA education proposals raised concerns, reflected in the literature and previous surveys of NASA scientists, which included the difficulty of obtaining EPO grant funding, too little money for the effort spent, and damaging one's career. Specific comments included:

"[Writing a] SMD supplement grant is too much effort for too little money (too many hoops to jump through)."

"The biggest problem — [EPO] conflict with personal time; EPO takes away from this and scientists aren't willing to give that up."

"For post docs and grad students, danger of doing EPO is that it could derail their science career."

"Women scientists sometimes are particularly sensitive to being "assigned" to EPO; feel it damages their credibility."

This last point raises a concern well-known to the community and is worth further investigation. Several issues are worth further discussion including possible gender bias in the community with regards to who does EPO as well as promoting a balanced number of role models engaged in this work.

3.2. Successes

In contrast to those who expressed barriers, some scientists in the room shared that they had had been successful in crafting winning EPO proposals funded by NASA. Thus the sentiment that the "effort was not worth the time to do it" was not shared by all scientists in the room. Specific comments included:

"I disagree with 'too hard to get funding' [for EPO work]."

"For post docs and grad students [doing EPO] can benefit their career toward teaching."

4. Connection to Education and Public Outreach Community

4.1. Challenges

Scientists in the session expressed that they were in need of better ways to connect to teachers and classrooms. Additionally, that they needed more ways to find opportunities to do EPO. Specific comments included:

"I need better EPO contacts."

"I need better ways to get my name out to EPO community."

"[We have] pedagogical barriers—translating content to teaching practices."

"Don't know (or have time to figure out) how to get a relevant education professional partner."

"Many scientists feel that the public could/would not understand what they do."

4.2. Successes

In contrast, other participants were able to point to successful practices that they had used to make connections to classrooms and EPO opportunities. Specific comments included:

"We have had success using master teachers to make connections to skills and develop pedagogically appropriate materials."

"At our institution, we have hired a retired teacher who not only helps us make connections back to the school district, but helps scientists understand the standards."

5.   Skills to Engage in EPO

5.1.   Challenges

Some participants shared that they felt that they and some of their peers lacked the opportunities, either due to skills or complexity of their science content area, to engage in education and public outreach. Specific comments included:

"My science is too esoteric to explain."

"Many scientists feel intimidated by children; they feel they don't know what is age-appropriate (terminology, concepts, developmental understanding)."

"Lack of communication skills in scientists."

5.2.   Successes

In contrast, some participants expressed positive contributions that scientists bring to EPO. Specific comments included:

"Scientists may not be aware/convinced that they bring a unique and valuable experience to EPO that others (ed specialists/former researchers) cannot provide alone."

"[Our program] has an "Ask an Astronomer" service in which they ask for volunteers; they have 12 scientists who participate and in one week it takes care of all the questions."

6.   Exemplar Programs to Look Toward for Best Practices

Project ASTRO: The program provides a manual for both teachers and astronomer partners to work through together, thus sets parameters and gives strategies for meaningful partnerships.

NSF GK-12: Graduate students partnered with classroom teachers for content support, and teacher give grad students strategies to work with students.

6.1.   Best Practices

In addition to discussing systematic programs that help connect scientists to education and outreach opportunities, participants discussed best practices from programs that they had facilitated and participated in. Specific comments included:

"Encourage grad students to be involved. It helps if you have incentives, a recognition program for doing EPO work, so that they feel that it is valued and worth their time, and that there is support from program management."

"At Princeton University Astrophysics department, a professor has organized some grad students to volunteer to teach math and science in a youth correctional facility. This program has helped both grad students and those who made a positive connection to the community."

7. Resources to Support Scientists Engagement in EPO

Throughout the discussion, participants asked for and provided resources to connect scientists with EPO opportunities. These resources are used both by scientists who wanted to find avenues to become more involved and from educational professionals who wanted to find a way to connect with scientists for various projects. Below is a list of resources discussed during the session.

NASA Speakers Bureau: A resource for both scientists and the public.
http://www.nasa.gov/about/speakers/

The Higher Education Clearinghouse: A One-Stop Shop for information and resources for undergraduate education in planetary science and solar and space physics. The site hosts the latest news, funding opportunities, and educational research for undergraduate faculty. Additionally, the clearinghouse is a place where faculty can submit and find materials for their classroom, including lectures, activities, homework, and other assets. http://www.lpi.usra.edu/hecl/

NASA's Science Mission Directorate Forum Community Site: The site provides resources to assist individuals funded by NASA's Science Mission Directorate (SMD) in carrying out their SMD education and public outreach programs. Information and documents that may be of interest to all, such as tips for getting involved in SMD education and public outreach and meeting notes. The site may be accessed by everyone. http://smdepo.org/

National Lab Network: The National Lab Network is a nationwide initiative to build local communities of support that will foster ongoing collaborations among volunteers, students and educators. http://www.nationallabnetwork.org/

8. Issues Raised and Opportunities for Growth

There were many issues raised during the session related to barriers and successes of engaging scientists in education and public outreach. One issue revolved around communicating the importance of engaging in EPO activities to scientists. A program officer from the National Science Foundation pointed out that putting EPO into an NSF research grant was considered a positive attribute of the research grant.

There was a general consensus in the room that that message was not getting out to faculty. This was an issue even brought up in the session by a scientist participant.

"As a soft-money funded research scientist, I can't ask for any support money for me in an EPO supplement."

Although program officers from several organizations discussed the importance of supporting scientists to conduct EPO, there persists an idea that scientists must do the work unfunded. This is an ongoing issue that we as a community need to help mediate to encourage more participation. Several other questions were raised and requests made for improving our work in engaging scientists in these activities. Specific comments included:

"How do we address different levels of scientist engagement, viz., (i) engaging more scientists in proposing for funded EPO projects vs. (ii) engaging and enabling more scientists in volunteer EPO activities?"

"I would love to see a rational online database of scientists interested in pub-lic/classroom speaking, that educators could search for a match with; specifically by broad category (i.e. astronomy, planetary science, environmental science, etc)."

"Be careful what you ask for...what to do about scientists eager to do EPO but are bad at it?"

"We need resources for knowing what is age appropriate, what types of things to use with students at different stages in development."

Each of these concerns provides an opportunity for partnership and development of new tools to assist and engage more scientists in meaningful EPO opportunities. Additionally, there is a need for better dissemination of resources to scientists who would like to use them to support their EPO efforts.

9. Future Work

This discussion demonstrates that as a community we have many opportunities to sup-port scientists in doing education and public opportunities. Over the past year, NASA SMD Forum teams leading efforts for scientist engagement in EPO have been com-piling resources to help EPO-engaged scientists make their EPO effort more effective. These resources will be made available on the SMD EPO community workspace[1] in the near future.

Acknowledgments.   This work is supported by funding for the Planetary Science Education and Public Outreach Forum and the Astrophysics Science Education and Public Outreach Forum, both provided by NASA's Science Mission Directorate.

References

Andrews, E., Hanley, D., Hovermill J., Weaver, A., & Melton, G. 2005, Journal of Geoscience
        Education, 281
Crettaz von Roten, F. 2011, Science Communication, 52
Poliakoff, E., & Webb, T. L. 2007, Science Communication, 242
Thiry, H., Laursen, S. L., & Hunter, A. B. 2008, Journal of Geoscience Education, 23
Word Craft 2008, COSEE NOW with ASLO Annual Scientist Survey (New Brunswick, NJ:
        Rutgers University)

Appendix A

Documented barriers that prevent scientists from participating, or participating more fully, in Education and Public Outreach (EPO) activities:

- Lack of support

- Little support from supervisors

- No funding (charge accounts for civil servants, grants for soft money scientists, and funding for university scientists)

---

[1]SMD EPO Community Workspace – http://smdepo.org

- Has to be done on their "own" time so it does not conflict with their paid employment
- Not worth it
- Writing NASA EPO proposals is too cumbersome
- The amount of work outweighs the amount of money offered
- Takes too much time away from doing science
- Too time consuming to do without funding
- Too difficult to get EPO funding
- NASA education proposal standards are too restrictive
- Lack of support on crafting EPO proposals that match expectations
- Proposals often rejected
- Too hard to compete with programs that are staffed with education professionals
- Do not know how to get started
- Lack of support with EPO resources
- Need help connecting to teachers and classrooms
- Need knowledge of available opportunities and how to plug into them
- Other professional/systemic barriers
- Feel marginalized by the professionalization of EPO efforts
- Other scientists see them as less serious scientists if they engage in EPO too much

ASP Baltimore, 2011 SMD Forum Working Document

John Grunsfeld and ASP Board member Schyleen Qualls. Photo by Andrew Fraknoi.

Connecting People to Science
ASP Conference Series, Vol. 457
Joseph B. Jensen, James G. Manning, Michael G. Gibbs, and Doris Daou, eds.
© *2012 Astronomical Society of the Pacific*

MY NASA DATA: An Earth Science Data Visualization Tool for the Classroom

Preston Lewis,[1] Daniel Oostra,[1] Sarah Crecelius,[1] and Lin H. Chambers[2]

[1]*Science Systems and Applications Inc., 1 Enterprise Parkway Suite 200, Hampton, Virginia 23666, USA*

[2]*NASA Langley Research Center Mail Stop 420, Hampton, Virginia 23681, USA*

Abstract.   Have you needed a source of up-to-date authentic data to use in the classroom? Do your students wonder why they need to know how to use data? This workshop shows how to engage your science students using authentic NASA satellite data! Explore the MY NASA DATA Live Access Server along with classroom-ready lessons using real satellite data. These data can be visualized in a number of ways to suit your established curriculum while grabbing the attention of your students. A focus on the implementation and use of Earth Systems data sets, developed for student researchers in grades K–12, will allow you to better make use of this wonderful tool. All of the data sets are derived from an archive of remotely sensed data retrieved from the myriad of NASA's Earth Observing System Satellites. The data that you and your students will be using and manipulating is the same data, formatted for educational use, that NASA scientists rely on every day to better understand our Earth. Regardless of what subject you teach, you will find multiple lesson plans that will fit right into your class. For more information on the project, go to the MY NASA DATA webpage at http://mynasadata.larc.nasa.gov/.

Outcome

Participants explored topics in Earth and atmospheric science and educational application of data sets, and gained the knowledge to use the Live Access Server to access authentic data. Participants learned how data visualization can be used to enhance their curricula and how students can utilize real NASA data for inquiry and problem-based learning activities/research. By learning background information related to atmosphere, radiation budget, clouds, and other Earth Science topics, particants can walk away with a classroom-ready set of skills to engage their students.

Space Telescope Science Institute exhibit. Photo by Joseph Miller.

Connecting People to Science
ASP Conference Series, Vol. 457
Joseph B. Jensen, James G. Manning, Michael G. Gibbs, and Doris Daou, eds.
© *2012 Astronomical Society of the Pacific*

Bring NASA's Year of the Solar System into Your Programs

Christine Shupla,[1] Stephanie Shipp,[1] Keliann LaConte,[1] Heather Dalton,[1] Sanlyn Buxner,[2] Don Boonstra,[3] John Ristvey,[4] Alice Wessen,[5] Rachel Zimmerman-Brachman,[5] and Emily CoBabe-Ammann[6]

[1]*Lunar and Planetary Institute, 3600 Bay Area Blvd., Houston, Texas 77058, USA*

[2]*Planetary Science Institute, 1700 East Fort Lowell, Suite 106, Tucson, Arizona 85719, USA*

[3]*Sustainability Schools Consulting, LLC, Tempe, Arizona 85280, USA*

[4]*McREL, 4601 DTC Boulevard, Suite 500, Denver, Colorado 80237, USA*

[5]*Jet Propulsion Lab, 4800 Oak Grove Dr., Pasadena, California 91109, USA*

[6]*Emily A. CoBabe & Associates, Inc., 2410 Vassar Dr., Boulder, Colorado 80305, USA*

Abstract. NASA's Year of the Solar System (http://solarsystem.nasa.gov/yss) is a celebration of our exploration of the solar system, which began in October 2010 and continues for one Martian year (687 Earth days) ending in late summer 2012. The diverse planetary missions in this period create a rare opportunity to engage students and the public, using NASA missions to reveal new worlds and new discoveries. Each month focuses on a particular topic, such as the scale of the solar system, its formation, water in the solar system, volcanism, atmospheres, and more! All educators are invited to join the celebration; indeed, the EPO community is needed in order for this event to be successful! Participants at the 2011 ASP Conference surveyed a variety of thematic activities, received resources and implementation ideas, and were invited to share their own experiences and upcoming events!

1. Introduction to the Year of the Solar System

Spanning a Martian Year—23 months—the Year of the Solar System celebrates the amazing discoveries of numerous NASA missions as they explore our near and distant neighbors and probe the very outer edges of our Solar System. It is an unprecedented time in planetary sciences as we learn about new worlds and make new discoveries!

To engage a variety of audiences, each month in the Year of the Solar System has a specific thematic topic, ranging from the formation of the solar system to the search for life. Materials are organized within each month on the website, with background information on the topic, activities for formal and informal learning environments, special events that month, online resources such as interactives, podcasts, and videos related to the topic, and more.[1]

All are invited to join the celebration, using the various materials and suggesting additional resources, registering their events, sharing their stories, and more!

2. Featured Activities

During the workshop, the participants experimented or modeled a variety of original activities selected to demonstrate the range of topics within the Year of the Solar System, and to meet the needs of the variety of educators that attend ASP, from elementary teachers to college faculty, amateur astronomers and outreach professionals.

Activities featured included:

- Vegetable Light Curves, an innovative activity for high school and college students who model the reflection of light off of a rotating potato.[2]

- Goldilocks and the Three Planets, a data-analysis activity for middle and high school students who determine what some of Earth, Venus, and Mars' atmospheres are composed of and then mathematically compare the amount of the greenhouse gas, CO_2, on the planets Venus, Earth, and Mars.[3]

- The Pull of the Planets, where children model the gravitational fields of planets on a flexible surface by placing and moving balls of different sizes and densities on a plastic sheet to develop a mental picture of how the mass of an object influences how much affect it has on the surrounding space.[4]

- Impact Paintings, a hands-on activity for the informal education setting, in which children use cotton balls dipped in paint to model how scientists use craters to determine the ages of lunar surface.[5]

- Mars Uncovered, which guides students through an inquiry-based, critical thinking approach to studying the surface of Mars; students create a geologic map of part of Mars and use relative age dating techniques of craters to analyze the information and interpret the geologic history of the region.[6]

- Cake Batter Lava, where middle school students use cake batter to understand how different lavas flow and the structures that are created.[7]

---

[1]http://solarsystem.nasa.gov/yss

[2]http://dawn.jpl.nasa.gov/DawnClassrooms/light_curves/index.asp

[3]http://lasp.colorado.edu/education/spectra/lessons.htm

[4]http://www.lpi.usra.edu/education/explore/solar_system/activities/bigKid/planetPull/

[5]http://www.lpi.usra.edu/education/explore/marvelMoon/activities/familyNight/splat/index.shtml

[6]http://marsed.mars.asu.edu/node/1000891

[7]http://www.spacegrant.hawaii.edu/class_acts/CakeLavaTe.html

Connecting People to Science
ASP Conference Series, Vol. 457
Joseph B. Jensen, James G. Manning, Michael G. Gibbs, and Doris Daou, eds.
© *2012 Astronomical Society of the Pacific*

Citizen Science: Mapping the Moon and Mercury

Georgia Bracey, Kathy Costello, Pamela Gay, and Ellen Reilly

Southern Illinois University Edwardsville, Center for Science, Technology, Engineering and Mathematics Research, Education and Outreach, Edwardsville, Illinois 62026, USA

Abstract.   The familiar face of our Moon is brought even closer to home by experiencing "Moon Zoo," an engaging online citizen science project from the creators of Galaxy Zoo. Using high-resolution images from the Lunar Reconnaissance Orbiter, Moon Zoo lets the public explore the lunar surface in breathtaking detail, mapping craters and discovering new features as they go. The maps that they generate will be used by scientists to understand solar system ages and to comparatively study geology across worlds. The less-familiar face of Mercury is also being explored and mapped through Mercury Zoo, thanks to images from MESSENGER, the first spacecraft to orbit Mercury. As citizen science projects, both of these Zoos let the public participate in authentic scientific research. This workshop offers participants the opportunity to make new and stronger connections to both of these solar system objects while getting a glimpse of the process and nature of science.

1.   Brief History of the Zooniverse

In July of 2007, a citizen science project called "Galaxy Zoo" made its debut on the Internet. Galaxy Zoo asked volunteers to view pictures of galaxies and classify them according to shape (spiral, elliptical, or irregular) and orientation (clockwise or counterclockwise) if spiral. Astronomers used this data to study the distribution of different types of galaxies in the universe. The project became extremely popular, and in less than a month about 80,000 volunteers had classified over 10 million images of galaxies. Since then, Galaxy Zoo continues to generate science and inspire the public. Other online projects have joined Galaxy Zoo to form the Zooniverse, which includes a variety of online projects such as Galaxy Zoo Hubble, Solar Storm Watch, Old Weather, and Planet Hunters. Each project incorporates a different set of tasks for its volunteers and focuses on different scientific questions.

Two Zooniverse projects focus on the surface geology of solar system bodies: Moon Zoo and Mercury Zoo. Both projects ask volunteers to look at high-resolution images of the surface in order to map craters and a variety of other features. Data from these projects is used to create maps that will be used by scientists to understand solar system ages and to comparatively study geology across worlds.

This workshop described the main tasks involved in Moon Zoo and presented a hands-on activity that connects the science of Moon Zoo with classroom content.

2. Why Citizen Science?

Citizen Science is a wonderful way to bring real science data and research into the classroom. There are a variety of citizen science projects available covering most areas of science, and many—like the Zooniverse—are online and require only an internet connection and a computer. These projects do not require advanced knowledge or skills, so students and teachers can be involved at many levels, from simply collecting data, to making observations and classifications, to helping with the analysis of data and possibly discovering something new. Citizen science lets teachers bring authentic scientific research into the classroom, and by participating in the project tasks—as well as interacting with the accompanying forums and blogs—students get a glimpse into the nature of science and scientists.

3. Citizen Science and the Moon

Moon Zoo is one of the newest projects in the Zooniverse. Launched in 2010, Moon Zoo[1] asks participants to look at high-resolution pictures of the Moon's surface taken by the Lunar Reconnaissance Orbiter in order to count craters, look for boulders, and find new objects. After watching a short tutorial, participants may choose between two main tasks: Crater Survey and Boulder Wars.

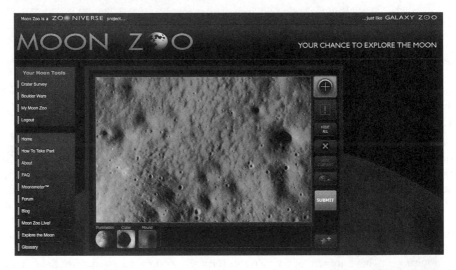

Figure 1.   Moon Zoo's "Crater Survey"

3.1. Crater Survey

In the Crater Survey task, Moon Zoo asks its volunteers to look at an image of the lunar surface (see Fig. 1 below) and mark and measure craters. Using the crater tool, users can mark the center of each crater they see, then draw and adjust an oval to match

---

[1]http://www.moonzoo.org

the crater's size and shape. Users are also asked to look for boulders and white areas around the craters—signs of a newer impact. Finally, as with many of these citizen science projects, users are asked to keep their eyes open for anything new.

3.2. Boulder Wars

In Boulder Wars, participants are presented with two images of the lunar surface as seen in Figure 2 below. The participant must decide which picture has more boulders and indicate that by clicking on the button "Most Boulders" under the chosen image. If neither image contains boulders, the user chooses "No Boulders." As with the "Crater Survey" task, the user should also be on the lookout for anything unusual or interesting.

4. Exploring the Lunar Surface in the Classroom: Making Regolith

In order to have a more tangible experience with lunar surface geology, teachers may guide their students through one of the many NASA classroom activities available that focus on various aspects of lunar science content. In this workshop we presented NASA's "Making Regolith" activity. This activity may be found on the NASA website.[2] Using easily available everyday materials, students model the process of lunar regolith formation by "bombarding" a layer of graham crackers with dried, powdered sugar mini-donuts. The resulting layer of crumbs and powdered sugar represents the fine layer of regolith found over much of the lunar surface.

After creating the regolith, students can use magnifiers to examine its properties and compare and contrast it with the larger pieces of graham crackers ("bedrock") still left underneath.

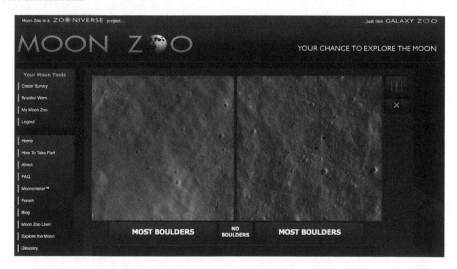

Figure 2.   Moon Zoo's "Boulder Wars"

---

[2]http://www.nasa.gov/audience/foreducators/topnav/materials/listbytype/Making_Regolith_Activity.html

5. Citizen Science and Mercury

Mercury Zoo (launched in late 2011) is the newest of the Zooniverse citizen science projects. Like Moon Zoo, Mercury Zoo will focus on surface geology and will ask participants to count and measure craters, mark crater positions, trace linear features, and indicate the locations of unusual or interesting features. Images from the MESSENGER probe will be used in this project, and as with all Zooniverse projects, a forum and blog will be created to facilitate communication and community-building between the participants and the scientists.

6. Summary

Citizen science projects like Moon Zoo and Mercury Zoo bring authentic scientific research into the classroom. Requiring only a computer and an internet connection, these online projects are flexible enough to be used as large-group classroom projects or independent study, and work well with students of many ages and at many levels of scientific expertise. By incorporating hands-on activities with related content, teachers can guide students in making strong connections between these fascinating solar system objects and the ongoing scientific research that helps us learn more about them every day.

Connecting People to Science
ASP Conference Series, Vol. 457
Joseph B. Jensen, James G. Manning, Michael G. Gibbs, and Doris Daou, eds.
© *2012 Astronomical Society of the Pacific*

Science Standards: The Next Generation; Minimum Astronomy and Space Science Concepts in a K–12 Curriculum

Lee Ann A. Hennig,[1] Jeanne Bishop,[2] Harold Geller,[3] Alan Gould,[4] and Dennis Schatz[5]

[1]*Middle Atlantic Planetarium Society: lahennig@earthlink.net*

[2]*Great Lakes Planetarium Association: jeanneebishop@wowway.com*

[3]*George Mason University: hgeller@gmu.edus*

[4]*Lawrence Hall of Science, University of California, Berkeley: agould@berkeley.edu*

[5]*National Science Foundation: dschatz@nsf.govs*

Abstract.   Educators from all U.S. planetarium associations have been working for months to provide input into the construction of the National Research Council's (NRC) new framework for K–12 science education. Simultaneously, the planetarium associations produced a document useful to the entire K–12 community and the future authors of the Next Generation Science Education Standards. "Astronomy Literacy: Essential Concepts for a K–12 Curriculum" includes both big picture concepts and grade specific concepts. This document has an extensive grade-by-grade list of astronomical concepts, making it a simple task to construct a school-wide astronomy curriculum. After an introduction to the curriculum document, stakeholders may explore different parts of the document in depth and provide feedback that can be used to refine the document and provide additional input to the Next Generation Science Education Standards. Furthermore, the workshop allowed participants an opportunity to provide more detailed input into the development of the new science education standards to be developed by the organization Achieve. Stakeholders may now provide input regarding the planetarium societies' list of grade appropriate astronomy concepts that should be included in the Next Generation Science Education Standards, provide input that will be shared with the developers of the new science education standards, and know how they can continue to be involved in the standards development process.

1. Overview

The session preceding this Special Interest Group (SIG) session was devoted to a discussion of the framework for K–12 science education, where they are and where they are going. We incorporated some of the information from that session into this article, in terms of drawing connections and addressing issues raised during that presentation. We were fortunate to have in attendance the presenters from that SIG and appreciated their interaction in the discussion. Our facilitator, Dennis Schatz, was also the facilitator for the preceding Framework/Standards SIG, and the two sessions were well blended.

Jeanne Bishop started the session with an overview of the Great Lakes Planetarium Association document "Astronomy Literacy: Essential Concepts for a K–12 Curriculum," which covers the history of the document, the value of the planetarium, the importance of astronomy in education, and the teaching concepts within astronomy. She expressed the opinion that the document could have the greatest value in guiding classroom teachers who do not know much astronomy as well as providing a structure for teachers and teacher candidates (teacher training, education journals, conferences, etc.) after becoming familiar with the more general requirements put forth in official standards.

Harold Geller followed up with a discussion of the collegiate point of view, the importance of including math throughout the standards (not isolated in one area), teacher training, his experience with evaluating planetariums, and the interdisciplinary nature of astronomy. Harold noted that the recently released "Framework" document covers a lot of ground in terms of the cross-cutting concepts that it defines and encompasses. Perhaps in the next edition of the "Literacy" document the authors may want to point out how many of the disciplinary core concepts included in the document are indeed themselves cross-cutting concepts.

Alan Gould stressed the importance of Astronomy outreach in the curriculum and the standards in general. He expressed a concern that although the "Framework" document will likely lead to Next Generation Science Standards that are better than the old National Science Education Standards (NSES), there is a distinct possibility that the Next Generation Standards will still contain more material than can comfortably fit into the total time that can be reasonably allocated through the entire K–12 time span. He put forth the view that it would be better to err on the side of having fewer content topics/concepts and give teachers more "wiggle room," or freedom to include topics of their own as special interests in their curricula. He holds a similar view of the "Literacy" document in that although each concept is fine, the sum total of concepts might not quite fit into time available for science in K–12. In addition, it will be very important for planetarium personnel to identify which of the essential astronomy concepts are most appropriate for treating in a planetarium, and which are best suited for teaching in a classroom setting. If it is not overtly stated in the document, it should be understood that the entire set of essential concepts are not a recipe for designing a repertoire of planetarium presentations, only a subset of concepts that are truly ideal for planetarium teaching.

Lee Ann Hennig closed the session with acknowledgement of the contributions and comments of the previous speakers' and noted how the "Framework" document related closely to the GLPA "Literacy" document (as Jeanne had pointed out earlier). We recognized that the "Framework" document had not been published at the time the "Literacy" document was formulated, and a subsequent review and evaluation now would be important. She also pointed out that the "Framework" document addresses two other topics that are important to Astronomy in particular and science education in general: it emphasizes "Scientific & Engineering Practices" by encouraging research by high school students (a chance to do real astronomy with actual data and with astronomers); and it acknowledges the importance of teacher training and interaction with scientists in their fields of expertise. Her closing statement was a reminder that the planetarium experience serves as a venue for addressing the standards, demonstrating the interdisciplinary nature of astronomy, and encouraging students to explore, investigate, ask questions, inspiring them to consider their place in the universe.

Following the prepared remarks from the presenters, participants engaged in discussion and were urged to pass on comments concerning astronomy standards to Achieve.org. To that end, the session included about 20 minutes of questions, comments, and suggestions, some of which are listed below:

- Knowledge of structure of the Universe is very weak (now)—it needs emphasis in K–12.

- Some states have a small number of teachers prepared to teach Earth and Space Science and Astronomy at the 9–12 level now—how will that be impacted by the new standards?

- Regarding the writing of the "Framework" document—it was difficult to write to a level that would reach everyone—it was a tricky process that required judgment.

- The "Framework" document is a vision now—it represents some progress.

- Some states have adopted astronomy curricula which seem inadequate—Minnesota is one (the documents are too late to make a difference).

- Reorganize concepts into groups different from just grade-level and developmental ability.

- Planetariums are an avenue for sharing recent discoveries and concepts in the astronomical community.

- If the new "standards" are "jam-packed" into the curriculum, it leaves no room for other topics which are valid in classroom learning.

- We can go beyond the "Framework"—it does not say what the content is, but it should serve as the "floor" of the Standards.

- The work in the Achieve segment of developing the new standards is very important in giving direction to how disciplinary concepts will be integrated in other subjects.

1.1. Conclusion

The entire astronomy community, especially the planetarium community, is encouraged to maintain active participation in the process in place to develop the next generation of national science education standards. Currently, the best way to do this is to monitor, and to participate with state education partners, the ongoing process now in the hands of the organization called Achieve. The online web presence of this process can be found at `http://www.achieve.org/next-generation-science-standards`. We cannot afford to have astronomy and its underpinning concepts overlooked in the education of those who will lead us into the next century.

Acknowledgments. The authors gratefully acknowledge the participation of all those present at both special interest group sessions.

References

National Research Council 2011, "A Framework for K–12 Science Education: Practices, Cross-cutting Concepts, and Core Ideas," Committee on a Conceptual Framework for New K–12 Science Education Standards, Board on Science Education, Division of Behavioral and Social Sciences and Education (Washington, D.C.: The National Academies Press)

Tomlinson, G. 2011, "Keeping astronomy in science education: The United States planetariums' regional response to the current science standards revision," Planetarian, Journal of the International Planetarium Society, 40, 3, 10

United States Planetarium Affiliates 2011, "Astronomy Literacy: Essential Concepts for a K–12 curriculum," Planetarian, Journal of the International Planetarium Society, 40, 3, 13

Connecting People to Science
ASP Conference Series, Vol. 457
Joseph B. Jensen, James G. Manning, Michael G. Gibbs, and Doris Daou, eds.
© *2012 Astronomical Society of the Pacific*

Collaborating with Public Libraries: Successes, Challenges, and Thoughts for the Future

Denise A. Smith,[1] Bonnie Eisenhamer,[1] Mangala Sharma,[1] Susan Brandehoff,[2] Jennifer Dominiak,[2] Stephanie Shipp,[3] and Keliann LaConte[3]

[1]*Space Telescope Science Institute, 3700 San Martin Dr., Baltimore, Maryland 21218, USA*

[2]*American Library Association, 50 E. Huron, Chicago, Illinois 60611, USA*

[3]*Lunar and Planetary Institute, 3600 Bay Area Blvd., Houston, Texas 77058, USA*

Abstract. Public libraries serve learners of all ages and backgrounds, provide free and convenient access to resources, and have strong ties to local schools and community-based organizations. Libraries recognize the importance of science literacy to our culture and strive increasingly to include science in their programming portfolio. What are our shared goals in communicating science to the public? What resources, events, or programs are available through your local public library? How can we work with public libraries to connect people to science? This interactive Special Interest Group discussion facilitated by representatives from the American Library Association's (ALA) Public Programs Office, the Lunar and Planetary Institute's *Explore!* program, and the Space Telescope Science Institute's *Visions of the Universe: Four Centuries of Discovery* library exhibit explored successes, lessons learned, and future opportunities for incorporating science programming into public library settings.

1. Introduction

The American Library Association (ALA) Public Programs Office provides a variety of programming resources and opportunities that support libraries in their role as a place for patrons to discover and reflect on the wealth of information available to them in our modern world. Library programs that present current and accurate scientific information are needed and desired by this vibrant community, as exemplified through the success of the Space Telescope Science Institute (STScI) *Visions of the Universe* traveling exhibit and the Lunar and Planetary Institute's (LPI) *Explore!* program. Created for the 2009 International Year of Astronomy, *Visions of the Universe* remains on tour and interest remains strong. The *Explore!* program has expanded from its roots in a collaboration involving the State Library of Louisiana to a community involving hundreds of librarians in more than 20 states. During this one-hour Special Interest Group discussion, participants discussed the needs of libraries, goals we have as a community for engaging library audiences in science, the successes and challenges that we have encountered in developing or implementing our programs, and how we might move forward as a community.

2. Needs, Challenges, and Successes: Perspective from the American Library Association

As part of the session, Susan Brandehoff, ALA Director of Programming Development and Partnerships, outlined results from an ALA survey of libraries concerning their use of and need for traveling exhibitions. Of the responses received, 40% were from rural libraries, 35% were from suburban libraries, and 25% were from urban libraries. Findings included: 82% had presented exhibits on history; 88% on art; 38% on science; 18% on technology; 6% on mathematics; 4% on engineering; 91% would like to host science exhibits; and 77% would like to host technology exhibits. Libraries that responded to the survey suggested the following potential exhibition topics: women in science, climate change, sustainability, nanotechnology, robotics, DNA demystified, development of energy sources, the U.S. space program, water, and animals that live in the dark. Using *Visions of the Universe* as an example, Ms. Brandehoff also illustrated the wide range of programming that libraries are able to implement to engage their local schools and community members in a science-themed traveling exhibit.

Based on the ALA's experience in working with public libraries to implement science programming, including the *Visions of the Universe* traveling exhibit, examples of challenges facing libraries include: locating authoritative speakers and resources on STEM topics in their own region; feeling unfamiliar with science/technology program formats that would be successful; feeling unprepared to present science/technology programs and to select science/technology materials for collections because they are unfamiliar with science subjects; and creating a demand for science/technology programs with limited resources.

From the ALA's perspective and final reports from participating libraries, examples of successes encountered by libraries participating in *Visions of the Universe* include: creating community interest in STEM topics and receiving requests for more programs; making connections with scientists and with regional and national science/technology organizations; having the library recognized as a community center for science/technology programming for all ages; and becoming more comfortable with developing and presenting science/technology programming.

Establishing a community of practice to support libraries and those working with libraries was also discussed as a strategy to address challenges and share successes. The ALA established a Google group to support *Visions of the Universe* participants; libraries continue to use this communication tool even after the exhibit has left their institution. A new online community of practice for librarians and professionals interested in working with libraries[1] is being established as part of the STAR (Science-Technology Activities and Resources) Library Education Network (STAR_Net) program; STAR_Net is a partnership between the National Center for Interactive Learning at Space Science Institute, the American Library Association, the Lunar and Planetary Institute and the National Girls Collaborative Project, and is funded by the National Science Foundation.

---

[1]http://community.discoverexhibits.org/

3. Example: Challenges and Successes from the *Explore!* Program

LPI began the *Explore!* program in 1998 in recognition of the potential role that libraries have in engaging children in space and planetary science through youth programs. Typically, LPI conducts two-day training sessions state-by-state or within a region to build a partnership with librarians to bring earth and space science to their community. During these trainings, librarians are provided with content background, explore access to scientists and local partners, undertake program activities, and examine paths of engagement and ways that the activities/programs could best fit in the individual library. The *Explore!* program offers eight free, flexible, online, field-tested earth and space science modules of multiple activities (standards-aligned) for 8 to 13 year olds that are available for use in the library learning environment.[2]

Overarching challenges that program leaders hear from librarians in pre-program surveys to which the *Explore!* program is designed to respond include: knowing *how* to engage patrons in science and having or locating the resources to help engage patrons in science. The top reasons preventing librarians from implementing *Explore!* programs include time for programming, resources to offer programming, and support by administrators for offering programming in response to the first two challenges. These are not exclusive to the *Explore!* program, and have impacts across other library programs. Challenges that LPI faces in offering the *Explore!* program include recruiting in rural areas, probably in response to library staffs being smaller and resources possibly being more restricted. Sustaining a conversation with the librarians using the *Explore!* materials (building a community of practice) is another challenge which is currently being evaluated.

Successes that LPI *Explore!* team members hear from partnering librarians about the *Explore!* program in follow-up surveys include that librarians have the understanding and confidence to offer science programming to their patrons (the vast majority leave the workshop stating that they have the materials to lead science programs and that they have a specific plan for what programs they will implement and when). In addition, librarians observe that patrons are participating in science programs and utilizing more of the library's science resources. LPI as program leader has succeeded in partnering with over 600 librarians in more than 25 states to bring current earth and space science into their library programs. LPI has started to build collaborations with other partners—and scientists—in the project regions to continue to support libraries in their science programming (noting that many libraries already are developing these strategic partnerships).

4. Example: Challenges and Successes from *Visions of the Universe: Four Centuries of Discovery*

STScI partnered with the ALA and the Smithsonian Astrophysical Observatory to create *Visions of the Universe: Four Centuries of Discovery* to help advance international, U.S., and NASA goals for the 2009 International Year of Astronomy (IYA). These included helping citizens of the world to rediscover their place in the universe, and within the U.S. and NASA, helping to offer an engaging astronomy experience to every person

---

[2]http://www.lpi.usra.edu/education/explore/

in the country, to nurture existing partnerships, and to build new connections to sustain public interest in astronomy. *Visions of the Universe* is a traveling exhibit for public libraries that portrays how our knowledge of the universe and the objects within it has changed over the past 400 years. The exhibit features stunning imagery from multiple NASA missions, including the *Hubble Space Telescope*, along with historic astronomical sketches and photographs. The project includes six two-sided free-standing exhibit panels, supporting educational materials, a scientist speaker's bureau, and training and support for participating libraries and scientist speakers. In addition, each exhibit panel is available online as a downloadable, poster-size file. *Visions of the Universe* will travel to 64 libraries during 2009–2012 that serve rural towns and large cities with limited access to NASA resources; these institutions were selected from over 120 peer-reviewed applications from public libraries located across the United States.[3]

Challenges encountered while working with *Visions* libraries have included a need for resources for elementary-aged audiences and a lack of background in science amongst librarians. Librarians were hesitant and needed both information and encouragement to incorporate science topics into their programming. Possible strategies to address these challenges include developing story-time activities for elementary aged audiences to accompany science-themed traveling exhibits and planning and building science background into the program.

Successes emerging from STScI follow-on interviews with participating libraries include the fact that new partnerships remain intact; the exhibit has a long-term effect in the community. The exhibit also has been successful in reaching and involving underrepresented populations and in filling a void in local communities—providing their first exposure to science programming.

5. Compiling a Community Summary of Library Needs

As part of an effort to support education and public outreach professionals who are interested in partnering with public libraries, the NASA Science Mission Directorate (SMD) Astrophysics Science Education and Public Outreach Forum is compiling a summary of library needs based on the ALA experiences and surveys discussed above, additional conversations with and surveys of librarians attending ALA conference sessions about *Visions of the Universe* and NASA resources, and the experiences of other NASA SMD-funded education and public outreach professionals working with libraries. Librarians indicate that they provide after-school activities, training in using library resources, and specific programs, e.g., reading/book clubs, story times, art, music, dance, and science/astronomy activities. For afterschool activities, most offer handouts, printed materials, and books. Some also have DVDs, audio books, and webpages with downloadable files and links. They proactively work with a variety of local community partners and audiences, including 4-H clubs, local astronomy clubs, schools, and families, and actively engage underserved populations.

As cited by the ALA, libraries need programs that present accurate, authoritative, and up-to-date scientific information. Overall, library programs are gender-neutral, but may be geared for specific audiences such as young/beginning readers, girls, 4-H clubs,

---

[3]http://www.ala.org/visionsoftheuniverse/ and http://amazing-space.stsci.edu/visions/

families, etc. Themed events, e.g. "Space Day" or "Family Astronomy Evening" help attract schools and community members to library programs; afterschool programs are generally timed for the summer, Saturdays, or evenings during the school year. In addition to the items outlined earlier in this paper, librarians cite the following science programming needs: "programs in a box" that include hands-on activities and demos, age-appropriate reading lists and bibliographies, material or templates that libraries can use to promote programs to local schools and the community; engaging speakers (especially women, bilingual/Spanish language speakers, scientists, astronauts; preferably, in-person speakers, although *Visions of the Universe* has had success in using remote speakers via Skype); assistance with stargazing/telescope nights or locating and using a mobile planetarium; age-appropriate science and cultural educational resources that are available in quantity and are easily accessible (e.g. hardcopy handouts, lesson plans, CD-ROMS, DVDs, etc.); resources that combine science with content or skills from other curriculum subject areas (science across the curriculum); bilingual resources/resources for English language learners; focused webpages with links. Some libraries may also need assistance forming partnerships with local institutions that can provide access to science expertise and science speakers.

Professional development needs include information on science and science education resources that are available for libraries and how to find them; current science discoveries and science background, including explanations of the science in the newspapers; and focus on science and math process skills (especially for school librarians). Librarians have expressed interest in professional development in conjunction with library conferences as well as online/remote professional development.

6. Next Steps and Outstanding Questions: Perspective of Session Participants

Session feedback forms indicated that participants agreed or strongly agreed that the information discussed in this session will be useful for their future work. Participants indicated that new and useful information included: information and survey results on what libraries are looking for; contacts made with the American Library Association; the existing programs and community that participants can tap into, learn, and share with in the future; and the idea of partnering with libraries and establishing a community of practice as a resource for libraries. As a result of the special interest group discussion, participants stated that they intend to learn more about library needs and existing educational materials for libraries; to follow-up with session leaders and the emerging community of practice; to include library programming in future EPO discussions and program planning; and to investigate collaborating with local libraries. Multiple participants indicated that they had remaining questions about how libraries and EPO partners can connect with each other and how to let libraries know about resources and programs. Other remaining questions were related to how to best feature an online resource at a library, the amount of space libraries have for exhibits, the preferred form of display for library exhibits, the potential for collaboration with existing library-based projects, and what people are learning in the programs that libraries conduct.

Acknowledgments. *Visions of the Universe: Four Centuries of Discovery* is supported by the National Aeronautics and Space Administration (NASA) under Grant NNX08AG33G issued through the Science Mission Directorate's Education and Public Outreach for Earth and Space Science (EPOESS) program. The program is a partnership between the Space Telescope Science Institute, Smithsonian Astrophysical Observatory, and the American Library Association. The Astrophysics Science Education and Public Outreach Forum is supported by NASA under Cooperative Agreement NNX09AQ11A issued through the NASA Science Mission Directorate.

Connecting People to Science
ASP Conference Series, Vol. 457
Joseph B. Jensen, James G. Manning, Michael G. Gibbs, and Doris Daou, eds.
© *2012 Astronomical Society of the Pacific*

Incorporating the Performing Arts and Museum Exhibit Development in a Multidisciplinary Approach to Science Learning for Teenage Youth

Irene Porro,[1] Mary Dussault,[2] Ross Barros-Smith,[1] Debra Wise,[3] and Danielle LeBlanc[4]

[1]*MIT Kavli Institute for Astrophysics and Space Research, 77 Massachusetts Avenue, Cambridge, Massachusetts 02139, USA*

[2]*Harvard-Smithsonian Center for Astrophysics, 60 Garden Street, Cambridge, Massachusetts 02138, USA*

[3]*Underground Railway Theater, 450 Massachusetts Avenue, Cambridge, Massachusetts 02139, USA*

[4]*Museum of Science, Science Park, Boston, Massachusetts 02114, USA*

Abstract.   It is not unusual for science educators to experience frustration in implementing learning initiatives for teenage youth who are not already hooked with science. Such frustration may lead them to focus their attention on different audiences, missing an opportunity to break the chain of science apathy among these youth. Youth's apparent lack of interest in science is associated with behavior typical of adolescence and the inadequacy of many science programs to adapt to meet the need of this audience. Teenage youth identify effective programs as those that engage them in challenging but fun activities and that contribute to their social development. Youth are looking for opportunities for skills and knowledge development that are otherwise unavailable to them in or out of school, and for positive relationships with adults with unique expertise in science and other fields. The Youth Astronomy Apprenticeship (YAA) has been successful in reaching out to teenage youth through the implementation of a model that incorporates principles of positive youth development in a multidisciplinary approach to science education. The project-based outcome of YAA participation is the creation and implementation of artistic performances, planetarium shows, museum exhibits, and even entertaining PowerPoint presentations.

1.   Introduction: The Youth Astronomy Apprenticeship Program

The Youth Astronomy Apprenticeship (YAA) is a yearlong, out-of-school initiative that connects urban teenage youth with astronomy in order to promote scientific literacy and overall positive youth development. The program employs the strategies of a traditional apprenticeship model common in crafts and trades guilds as well as in higher education. During the apprenticeship, youth develop knowledge and skills to create informal science education projects; through these projects they demonstrate their understanding of astronomy and use their communication skills to connect to general audiences. For some youth, participation extends across multiple years and their responsibilities for program implementation become multifaceted. YAA progressively develops a youth's science knowledge and 21st century employable skills through several stages (Fig.1):

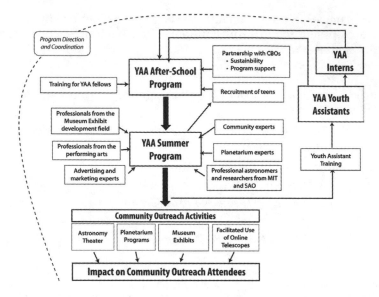

Figure 1.   Figure 1: YAA Program Model as implemented in 2009

Afterschool Program. Youth engage in astronomy investigations, take astronomical images using robotic telescopes they can operate via the Internet, learn to use software tools to process astronomical images, and produce reports and presentations about their investigations. Typically, afterschool sessions take place across three months, meeting two times per week, at local community-based organizations.

Summer Apprenticeship Program and YAA Outreach. With the support of many local professionals, YAA apprentices write, produce and perform science/astronomy plays, design and facilitate activities to introduce a lay audience to the use of the telescope, create components for professional museum exhibits, and create and run planetarium shows that they perform at various venues using a portable planetarium. Throughout the program year that follows, youth present their science/astronomy projects at community outreach events in communities across the city. Youth engage attendees in enthusiastic dialogue about astronomy, science in their everyday lives, and the overall importance of science literacy to the community.

YAA Internship. A small number of youth assistants remain with the program, earning the opportunity for a paid internship. Interns are an integral part of the program staff, participating in planning, curriculum development and review, teaching the sessions for the afterschool program, guiding the summer youth apprenticeship program, and supporting community outreach events.

2. The Apprenticeship Model

The pedagogical importance of *apprenticeship*, which is receiving increasing attention in schools, youth-serving organizations, and arts, civic, and other cultural institutions, is thoroughly documented in Robert Halpern's book *The Means to Grow Up* (2009). The apprenticeship movement aims to re-engage older youth through in-depth learning and unique experiences under the guidance of professionals that "share their disciplinary knowledge and skills with youth." Through his research, Halpern describes how youth strengthen skills, dispositions, and self-knowledge that is critical to future schooling and work, and renew their sense of vitality. He also argues that the attributes of apprenticeship-like experiences fit well with the tasks and tensions of adolescence and that, though not perfect, they are a better fit than school to provide means to foster skills thought to be critical to the 21st century workforce.

The Youth Astronomy Apprenticeship implements an apprenticeship model where multiple levels of skilled youth and adults help and mentor those at lower skill levels and, through guided, hands-on learning, use new acquired skills to create a quality end product (an outreach project). Of the various participants in the YAA program, youth apprentices, youth assistants, and youth interns represent various levels of the apprenticeship program. Afterschool and summer adult facilitators are the journeymen (the skilled workers who practice the trade) and the content experts (scientists, educators, artists, museum developers) are the masters. Many of the results of the program's summative evaluation reinforce the effectiveness of this model by showing a progression of scores on outcome measures, from a beginning apprentice through the youth assistant, to the youth intern.

In the YAA model, the profession that the apprentices are learning is that of an informal science educator. For an apprentice it is not enough to learn science and astronomy: the goal is to create a project that meaningfully illustrates the acquired knowledge and skills and then to share it with peers and public audiences. Non-science content and processes, which we refer to as integrating, play a critical role as the vehicle for project creation. These non-science disciplines offer the content and processes that provide a set of authentic experiences and alternative meanings to science learning. The integrating content for YAA includes theatre (scriptwriting and performance), museum exhibit (design and creation), planetarium shows (scriptwriting, performance, and maintenance of a portable planetarium), use and maintenance of optical devices (telescopes), astrophotography, communication and marketing.

In the selection of the integrating content for the YAA apprenticeships we aimed to match our available resources with the needs of our target audience. For example, we understand that many of today's youth and adults view contemporary science with a lack of relevance if not with distrust. They often feel mystified and even threatened by it. To provide youth with an opportunity to connect with scientists and experience science in a way that will help them break through these preconceptions, we considered a science theater experience as a promising alternative to engage the youth's imagination and intellectual curiosity. In science theater, the subject is science, but the process is art: this can open up new ways of thinking about science and especially captivate young people's attention. Research suggests that through the use of the arts, youth become more engaged in learning and more motivated to explore new ideas, are encouraged to work in groups, and to communicate effectively (Rooney 2004). The creation of theater performances and, with a few different technical requirements, planetarium shows, gives young people opportunities to learn through multiple intelligences (kinetic, verbal, visual) and to engage multiple habits of mind essential for both scientific understanding and personal development (careful observation, the taking of multiple perspectives, problem-finding, and exploration).

The idea of introducing the development of a science exhibit as integrating content was also motivated by the needs expressed by adolescents to cultivate new skills, learn about possible employment opportunities, and build up their resumes with concrete work experiences. When developing professional science museum exhibits, interdisciplinary teams with a wide range of expertise (including design, science content, graphic arts, education, engineering, evaluation, computer/media skills, writing, and project management) work together to create learning experiences that will engage, inform, and delight their target audiences. Exhibit development is an intensely creative and challenging problem-solving activity with tangible outcomes and deliverables; the road to success is filled with moments of inspiration and imagination as well as tough decision-making, compromise, and frustrating setbacks. These features make an exhibit development apprenticeship an ideal opportunity for youth to learn and apply science content knowledge, to learn from adult professionals, and to develop a wide range of practical 21st century skills (Partnership for 21st Century Skills 2004).

The introduction of distinct integrating contents and processes in the apprenticeship occurs through the participation of adults who share their expertise with the young apprentices. Through the YAA experience we identified three areas of adult expertise that are required to support a successful apprenticeship: science content (the subject matter around which the project will be built); skills of the craft ("technical know-how" of what it takes to design and produce a specific project); and positive youth development (attention to the particular developmental and social needs of adolescent youth). These areas of expertise should come from different individuals, so that one adult is not responsible for overseeing, or even advising, all aspects of the projects' development. To illustrate some of the integrating content and processes implemented in YAA, we briefly describe three apprenticeships adopted by the program.

2.1. Science Theater Apprenticeship

In the YAA science theater apprenticeship, youth with widely varying levels of science and verbal literacy collaborate with professional theater artists (actors, dancers, and playwrights) and scientists to create and perform original plays exploring topics in astronomy. The topics range from the life cycle of stars, to exoplanets, to telescopes, to a number of other sub-topics that reflect youth interest.

Integral to the process of creating the plays and performances are opportunities for deepening youth and audience understanding about scientific inquiry, principles and phenomena. Using theater, youth discover new possibilities to explore science, to create personal connections to science, to generate rich questions about science, and to share information about science with targeted audiences they themselves selected (e.g. peers, families, adults or children). Instructors, mentors, and youth typically start the creative process by asking themselves what scientific topics are most powerful for public exploration and engagement through theater. Among the criteria for topic selection that were identified over the years, some are shared by all apprenticeship projects, including topics that offer opportunities for deepening understanding about the nature of science, or connect with pressing questions of contemporary society. Other criteria suggest unique opportunities for investigation through theater and interaction with a theater audience, consequently providing experiences and ways of knowing and understanding not readily available in other forms of dialogue about science. These include the possibility to provide insight into the culture of science and the impact of that culture on our society, and to invite deep reflection about potential personal engagement, as young people and world citizens, with the science topic presented.

Because of the highly interdisciplinary work required for this apprenticeship, some of the most valuable outcomes youth are likely to gain through the science theater experience include both practical skills and new understanding: youth develop public speaking skills, which they consciously transfer to presentations about science content designed for other settings, but also an appreciation of what can be defined as authentic exploration, in both science and the arts.

2.2. Planetarium Show Apprenticeship

In the planetarium apprenticeship, youth write and produce original planetarium shows that they perform with the use of a portable planetarium (comprised of an inflatable dome and a projector), an educational tool that is becoming relatively common in both informal and formal education settings. In their execution, each show consists of an introduction to the planetarium and night sky, a presentation on the selected topic, and an opportunity for the audience to pose questions. A finished show includes a written script that contains dialog and dome cues, visual media prepared in a PowerPoint file, and digital sound files for audio cues.

In the initial phase of developing shows, youth new to the project are presented with a live performance of a planetarium show developed on a previous year. New apprentices are then introduced to the operation of the dome and are tasked with giving a very short presentation in which they outline and describe a constellation in the sky. This serves as a confidence building and familiarization exercise. From here, apprentices begin to develop a complete script while engaging in further research directed by staff and science experts so as to become fully competent in the area of their show's content. Simultaneously, they become more comfortable with the use of the dome and in leading others in navigating the planetarium's night sky.

Opportunities for constructive feedback are built into the apprenticeship itself: every other week apprentices provide a report of the work in progress to staff and peers engaged in other projects. In addition, halfway through the apprenticeship, planetarium shows in their developing state are shared with an outside audience, giving the apprentices an opportunity for real world presentation experience and preparing them for a larger public event at the conclusion of the apprenticeship. Opportunities for feedback and revisions continue throughout the school year following the summer project: theater performances and planetarium shows are often presented multiple times at various sites (schools and after-school sites) and events (local science festivals) and they are subjected to ongoing improvements and updates.

As such, there is considerable latitude in the possibilities for professional collaborations and many local resources can be adapted to excellent use. In our applications, members of the theater and planetarium apprenticeship have enjoyed interaction and input from scientists at local universities and observatories to assist with the development of science content knowledge. Theater professionals engaged in the YAA theater apprenticeship have shared their talent in ensuring that planetarium apprentices use strong, crisp voices in the dark environment of the dome where body cues are not seen. Observing sessions shared with amateur astronomers, a ubiquitous resource, give apprentices a chance to apply their use of the planetarium to navigating the real sky above them.

112   *Porro et al.*

Dedicated planetarium professionals, while rare, contribute their own unique combinations of skill sets, and they model and impart immediately applicable knowledge and presentation skills.

2.3. Museum Exhibit Apprenticeship

Between 2007 and 2010, YAA apprentices pursued three museum exhibit apprenticeships. Our initial exhibit apprenticeship was a unique partnership with a professional exhibit development team for a traveling museum exhibition on black holes. The youth astronomy apprentices were asked to be "co-developers," in charge of creating a subset of prototype exhibit components that would be tested with audiences at Boston's Museum of Science for inclusion in the final exhibition. In the summer of 2009, a new group of apprentices was tasked with developing hands-on experiences for groups coming to visit MIT to tour an on-campus telescope control facility. The youth designed, built, and tested a number of working prototypes out of arts and crafts supplies and then "contracted" a local exhibit design firm to create more robust versions of these activities. The third summer was entirely driven by the YAA program—the topic was telescopes, the intended audience was the apprentices' peers and relatives, and the final exhibits were primarily fabricated from simple arts and crafts supplies.

The particular experience with the exhibit development apprenticeship, and the marked difference of each implementation, helped the whole YAA program define a set of project elements characteristic of a successful apprenticeship, independently of the specific nature of the project. We illustrate here two of these elements that are especially important for the implementation of an apprenticeship model with older youth.

One of the most valuable—and difficult—set of skills for youth to learn is that of planning and time management. The exhibit development apprentices were provided with a project-planning framework that professionals use, and this framework allowed them to sequence their action items, interim milestones, interactions with experts, and deliverable deadlines according to well identified project phases: concept development (research and planning); design development (planning and testing); prototyping (testing and building); production (refining and building). The sequence of project phases and planning strategies implemented by the museum exhibit apprentices ultimately became a model for all the other projects across the YAA program.

The exhibit development apprenticeship also offered a clear illustration of the range of roles and relationships that might be established between the youth participants and their adult mentors, independently of the specific apprenticeship being implemented. Key tensions typically arise from the nature of the youth-adult relationships, often relating to issues of ownership, authority and control over project ideas. Youth programs with adult mentors span a wide range of working relationships, from youth- or adult-driven to co-collaborative, and previous research has found that success does not depend on the program's location on the youth-adult role spectrum, but rather, on the agreement between youth and adult perception of this location (Jones & Perkins 2005; Londhe, Houseman & Goodman 2009). Based on our experience, we developed the following set of questions to guide adult facilitators of the program in negotiating this agreement:

1. Who is responsible for the youth's learning and development?

2. Who defines the limitations of the apprenticeship and the parameters for success? (Is any project idea appropriate, or do experts get to weigh in?)

3. How will each party benefit from the collaboration?

4. Which areas of expertise are most useful at the various stages of the project?

Engaging in the process to answer these questions also guides the facilitators of the program to set clear measures of success for each project. As it should be expected, in each one of the of the four YAA summer experiences, the measures of success for the apprentices' projects differed by the project scope and the intended audience. In regards to the collaboration between youth and adult mentors, however, a broader perspective emerged. Success became clearly defined in terms of an adult-youth relationship that, over the course of the apprenticeship, evolved from expert-novice to co-collaborators. The apprenticeship begins with consultations between youth and experts to share ideas and basic knowledge, but quickly moves into youth-driven project development, filled with research and experimentation. At this stage, youth can receive feedback from advisors and review sessions, but true mentorship develops when the youth are driven to enlist the adults as trusted co-conspirators in solving real-world problems that have arisen from their own development process. With the proper structure, the relationship becomes one of collaboration and shared professionalism.

3. Conclusions

The experience we have gathered over the years and the results from the summative evaluation of the YAA program confirm some important understandings about the implementation of an apprenticeship model to engage older youth in science learning experiences. We list them below, organized in terms of youth outcomes and programmatic elements.

3.1. Youth Outcomes

Working with older youth in out-of-school time presents unique opportunities not available with other age groups. Adolescents integrate an understanding of content knowledge and a new science identity with the process of personal identity development that carries on into adulthood. Focusing on older youth also allows for applying the development of real-world professional skills as a platform for creating an understanding of content knowledge in a manner that emphasizes both these skills and knowledge.

A key to developing successful projects with adolescents is giving youth opportunities to test and get feedback on their ideas early and often, from front-end focus group testing during a project concept development to frequent formative evaluation during the remaining stages of the project. By undertaking multiple reviews in the course of the apprenticeship, youth gradually develop an understanding that to fail is not only OK but it is part of the learning experience, and that evaluation is an important tool for project improvement.

Data from the summative evaluation demonstrate that confidence in advocacy for science is connected to time spent in the program. This underscores the need for a program environment that sustains youth participation over time and continuously builds upon their past positive outcomes. Multiple years of involvement, with opportunity for increased personal responsibility and leadership engagement, also produce the best results in terms of youth development, strengthening of content knowledge and mastery of professional skills. A program design that is adolescent-centered and encourages long-term engagement provides youth with multiple entry points to reflect youth personal growth and change of interests. This approach leads to deeper engagement and meaningful science learning and produces results that implicitly support the idea that "it is never too late" to develop an interest in STEM disciplines (the authors would argue that this statement is true for other disciplines as well). Finally, because it takes time for youth to trust—and sometimes confide in—the adults who work with them, interaction with professionals is most effective when the relationship is carried out over a prolonged period of time.

3.2. Programmatic Elements

The apprenticeship model is ideal for addressing the needs of older youth, as it provides an opportunity to expand content knowledge, model and develop professional workplace practices, while creating a ladder of achievement and responsibility for youth to ascend..

The interdisciplinary approach, characteristic of an apprenticeship in informal science education, is especially valid to stress the features of authentic research, independently of the field of exploration: clear definition of the question/problem, assessment of the available information, identification and verification of the sources, use of careful observation and inference to produce an answer or conclusion that can be backed up by evidence.

The apprenticeship model relies strongly on adult mentorship, and it is important to fully understand the nature of this relationship for the success of the apprenticeship projects. Program developers need to help both the adult and youth participants in the apprenticeship to gain a shared understanding of what their respective roles and expectations of each other are, and what success will look like—for the project, and for the collaboration. Program developers need also to clearly identify areas of required adult expertise and to appoint different individuals to provide specific expertise. This approach both recreates the actual conditions found in any work environment and allows the youth to address multiple issues (for example, project development vs. interpersonal relations) with separate and unbiased mentors.

Acknowledgments.   YAA is a program developed by the Education and Outreach Group at the MIT Kavli Institute for Astrophysics and Space Research, in collaboration with Harvard-Smithsonian Center for Astrophysics, Catalyst Collaborative @ MIT, Timothy Smith Network, and Institute for Learning Innovation. The program is funded by the National Science Foundation (DRL 0610350).

References

Jones, K. R. & Perkins, D. F. 2005, "Youth-adult partnerships," in Lerner, R. & Fisher, C., eds., Applied Developmental Science Encyclopedia: An Encyclopedia of Research, Policies and Programs (Thousand Oaks, CA: Sage)
Halpern R. 2009, The Means to Grow Up: Reinventing Apprenticeship as a Developmental Support in Adolescence (New York, NY: Routledge)
Londhe, R., Houseman, L., & Goodman, I.F. 2009, Black Holes Experiment Gallery: Evaluation Report on Youth Collaboration (Cambridge, MA: Goodman Research Group).
Partnership for 21st Century Skills 2004, The Road to 21st Century Learning (Washington, D.C.: Partnership for 21st Century Skills)
Rooney R. 2004, Arts-Based Teaching and Learning (Washington, D.C.: VSA Arts)

Connecting People to Science
ASP Conference Series, Vol. 457
Joseph B. Jensen, James G. Manning, Michael G. Gibbs, and Doris Daou, eds.
©*2012 Astronomical Society of the Pacific*

Earth Science Mobile App Development for Non-Programmers

Daniel Oostra,[1] Sarah Crecelius,[1] Preston Lewis,[1] and Lin H. Chambers[2]

[1]*Science Systems and Applications Inc., 1 Enterprise Parkway (suite 200), Hampton, Virginia 23666, USA*

[2]*NASA Langley Research Center Mail Stop 420, Hampton, Virginia 23681, USA*

Abstract. A number of cloud based visual development tools have emerged that provide methods for developing mobile applications quickly and without previous programming experience. The MY NASA DATA (MND) team would like to begin a discussion on how we can best leverage current mobile app technologies and available Earth science datasets. The MY NASA DATA team is developing an approach based on two main ideas. The first is to teach our constituents how to create mobile applications that interact with NASA datasets; the second is to provide web services or Application Programming Interfaces (APIs) that create sources of data that educators, students and scientists can use in their own mobile app development. This framework allows data providers to foster mobile application development and interaction while not becoming a software clearing house. MY NASA DATA's research has included meetings with local data providers, educators, libraries and individuals. A high level of interest has been identified from initial discussions and interviews. This overt interest combined with the marked popularity of mobile applications in our societies has created a new channel for outreach and communications with and between the science and educational communities.

1. Outcomes

MY NASA DATA will share their experiences, resources and methods with special interest group attendees. Attendees will interact with MND staffers and learn about low or no-cost methods for developing mobile applications. Additionally, participants will examine developing mobile applications from the perspective of users that have no programming experience or knowledge. The group will have the opportunity to discuss how mobile apps can be used for education, and discuss data they would like to see available from NASA data centers in the future. Attendees will be invited to share other approaches or methods they have tried, and lessons learned from those approaches. As a result, all participants will leave with some ideas for next steps.

Connecting People to Science
ASP Conference Series, Vol. 457
Joseph B. Jensen, James G. Manning, Michael G. Gibbs, and Doris Daou, eds.
©*2012 Astronomical Society of the Pacific*

Kinesthetic Activities to Teach Challenging Topics

Darlene Smalley

*Ruth Patrick Science Education Center, University of South Carolina Aiken,
471 University Parkway, Aiken, South Carolina 29801, USA*

Abstract.   Students get engaged and increase understanding when they "become" an organelle in a cell, a process in the rock cycle, or a constellation in the zodiac. Learn by participation!

1.   Activity 1: The Living Cell

Students form a "cell" in an open area and become acquainted with the names and functions of the major organelles as they role-play how those organelles would act on raw material entering the cell. The teacher or another adult functions as the raw material and moves through the cell to different organelles.

1.1.   South Carolina Standards Addressed

Grade 5, Life Science 2–1: Recall the cell as the smallest unit of life and identify its major structures.

Grade 7, Life Science 7–2: The student will demonstrate an understanding of the structure and function of cells, cellular reproduction, and heredity.

1.2.   Materials Needed

Organelle names printed on cardstock and laminated. We have about 15 Cell Membrane cards and 5 each of the following: Nucleus, Lysosome, Endoplasmic Reticulum, Ribosome, Golgi Bodies, and Mitochondria.

1.3.   Description of Program in Which We Role-play a Living Cell

Journey into the Living Cell(Grades 5–12, Planetarium). The DuPont Planetarium takes students from the vastness of outer space to the intricacy of our inner space as students learn about cells. The program visits major cell organelles and explains their basic functions and structure. Before entering the planetarium, students become a "cell" and role-play the functions of cellular organelles.

2.   Activity 2: The Living Rock Cycle

Prior to students arriving, the teacher sets out groups of rocks and sediments in the appropriate locations and places paths between the rock types to show how they can change from one type to another. Students become part of the rock cycle, either by identifying and displaying a rock type or by holding a process card along one of the paths that lead around or across the rock cycle. The teacher then enters the rock cycle and "becomes" magma. She discusses what happens to her as she moves about the rock cycle.

2.1. South Carolina Standards Addressed

Grade 3, Earth Science 3–3: The student will demonstrate an understanding of Earth's composition and the changes that occur to the features of Earth's surface.

Grade 8, Earth Science 8–3: The student will demonstrate an understanding of materials that determine the structure of Earth and the processes that have altered this structure.

2.2. Materials Needed

1. Rocks sorted by Type: Intrusive Igneous, Extrusive Igneous, Sedimentary, and Metamorphic

2. Sediments in containers: gravel, sand, silt, and clay

3. A red mat or red fabric to represent Magma

4. Interlocking mats, carpet squares, or fabric to represent paths around and across the rock cycle

5. Paper labels for the rock types, sediments, and magma

6. Sentence strips labeled with process names:

    - Weathering and Erosion (3)
    - Heat and Pressure (2)
    - Cooling (1)
    - Melting (2)
    - Compaction and Cementation (1)

7. Rock Cycle Diagram for intro and follow-up (available upon request)

2.3. Description of Programs in Which We Use the Living Rock Cycle

Planet Earth Rocks! (Grade 3). Students observe, describe, and classify samples of igneous, sedimentary, and metamorphic rocks. Sediments and fossils are examined and described, and rock formation is discussed.

Rockin' & Rollin' (Grade 8). Students handle and classify rocks and investigate how igneous, sedimentary, and metamorphic rocks are interrelated in the rock cycle. Sediments and fossils are examined and described, and a dichotomous key is used to identify 12 rock specimens.

3. Activity 3: The Living Zodiac

The goal is for students to learn what the zodiac constellations are, why they are seasonal, and what causes seasons. Students gather in an area where zodiac constellation labels have been placed in a circle on the floor. The instructor gives pictures of the zodiac constellations to 13 students who stand in the correct locations holding their pictures so they face the center of the circle. The teacher picks a student to stand in the center of the circle as the Sun and a student to hold the Earth model. The Earth student practices orbiting the Sun inside the circle of constellations making sure to keep the North Pole pointed toward Polaris, which has been placed on the correct wall. Ask the student with the Earth model to walk slowly around the Sun again, and stop him in front of Gemini, Virgo, Sagittarius, and Pisces to discuss the solstices and equinoxes. Discuss which constellations are visible at night during each season and which constellations are hidden by the Sun's light during the day. Students who are not holding a constellation, the Sun, or the Earth may pick up a constellation label and hold it so they can see the season names. This should allow every student to participate. You could also have students take turns being the Sun, Earth, or a constellation.

3.1. South Carolina Standards Addressed

Grade 4, Astronomy 4–3: The student will demonstrate an understanding of the properties, movements, and locations of objects in the solar system.

Grade 8, Astronomy 8–4: The student will demonstrate an understanding of the characteristics, structure, and predictable motions of celestial bodies.

3.2. Materials Needed

1. Pictures of 13 zodiac constellations

2. Labels with names of zodiac constellations on one side and season of visibility on the other

3. Yellow ball to represent the Sun, or a Sun costume

4. Small Earth globe ($/sim4$") with a tilted axis

3.3. Description of Program in Which We Use this Activity

Cruising through the Constellations. Students learn the names and locations of common constellations; circumpolar, seasonal, and zodiac constellations are discussed as special groups of constellations. Students review Earth's motions and what causes the seasons as they learn why we see different constellations at different times of the year.

Contact Information

The author can be contacted at darlenes@usca.edu. For more information, visit http://rpsec.usca.edu.

Part III

Oral Contributions

Connecting People to Science
ASP Conference Series, Vol. 457
Joseph B. Jensen, James G. Manning, Michael G. Gibbs, and Doris Daou, eds.
©*2012 Astronomical Society of the Pacific*

The 2012 Transit of Venus

Paul Deans

Astronomical Society of the Pacific, 390 Ashton Avenue, San Francisco, California 94112, USA

Abstract.   On June 5–6, 2012, much of the world will experience an event that will not occur again for another 105 years—a transit of Venus. During the 18th and 19th centuries, astronomers made arduous trips to remote corners of Earth to make Venus transit observations in an attempt to calculate the Earth-Sun distance. Today a transit of Venus is simply a rare spectacle. But it is important to take care when viewing it, because observing the Sun is dangerous if proper filters for eye protection are not used.

1. First Sighting

Despite extensive searches by numerous researchers, there is no firm evidence that a transit of Venus was ever observed prior to 1639. A study of naked-eye sunspot sightings made by long-ago Chinese astronomers, who were renowned for their skywatching prowess, turned up no instances where a sunspot (that might have been the planet) was seen on a Venus-transit day. Of course, back then nobody knew that planets could transit the Sun.

In 1627, Johannes Kepler published the *Rudolphine Tables*, which consisted of a star catalog and tables of planetary motion. These tables detailed planet positions with a greater accuracy than all previous efforts. Based on the tables, Kepler realized that first Mercury, and then Venus, would cross the solar disk in late 1631.

Kepler died in 1630, but French astronomer Pierre Gassendi took it upon himself to detect both events. On November 7, 1631, Gassendi spotted Mercury on the solar face and made the first transit observation in history. He was tormented by intermittent clouds and was nearly deceived by Mercury's small size. "But when the sun shone again, I discovered further movement, and only then did I conclude that Mercury had come in on his splendid wings."

The following month, Gassendi tried to observe the transit of Venus. He missed it because Kepler's computations were not precise enough. Modern calculations reveal that the transit of December 7, 1631 actually ended about 50 minutes before sunrise at Paris, Gassendi's observing site.

According to Kepler, the next Venus transit would not occur until 1761. But in 1639 Jeremiah Horrocks, a young British amateur astronomer, reworked Kepler's Venus calculations and discovered that the planet would likely transit the Sun later that year. Because Horrocks determined this a mere month before the event, he had little time to spread the word. As a result, only Horrocks (in Hoole, a small village in Lancashire) and his friend William Crabtree (in Manchester) witnessed the transit of Venus on December 4, 1639—November 24, according to the Julian calendar, which was still in use in England at the time. In addition to being the first to see Venus silhouetted against the solar face, Horrocks used his observations to determine the diameter of Venus—it was ten times smaller than expected—and to improve the planet's orbital elements.

Figure 1.   The June 8, 2004, transit of Venus at sunrise from an overlook above the Catawba River near Connelly's Springs, North Carolina. Image courtesy David Cortner (http://www.davidcortner.com).

2. How Far the Sun

The big scientific question of the 18th century was the size of the solar system. Thanks to Kepler's third law of planetary motion, astronomers knew the relationship between a planet's distance from the Sun and its orbital period.

But though the planetary periods were known, no one knew the actual Earth-Sun separation, without which the other planetary distances could not fall into place. During a close approach of Mars to Earth in 1672, Giovanni Cassini (in Paris) and Jean Richer (in Cayenne, French Guiana) observed Mars simultaneously. Because the two astronomers were at different locations, they saw Mars in slightly different positions against the background stars—an effect called parallax. Cassini used these measurements to calculate the Earth-Mars separation, and then the Earth-Sun distance. His estimate of 140 million kilometers was low but close to the correct value.

The measurements made by Cassini and Richer were difficult to make, and astronomers were eager to find a more accurate method of determining the Earth-Sun distance. In 1677, Edmund Halley observed a transit of Mercury and realized that the solution lay in transits—especially of Venus, because it was the larger of the two inner planets. He proposed that, by recording the instant of Venus' second and third contacts from widely spaced observing sites on Earth, astronomers could calculate the distance to Venus using the principles of parallax.

Halley died 19 years before the start of the next transit pair in 1761. But his proposal inspired scientific societies in various countries to mount numerous expeditions to far-flung regions of the Earth. The goal was to secure enough cloud-free measurements, by widely separated observers, to determine the Earth-Sun distance with the highest possible precision.

Things did not work out quite as planned, even when the weather cooperated. Observers found that a turbulent Sun-Venus image (caused by Earth's unstable atmosphere), combined with a bright, fuzzy ring of light around Venus, made precise timing of Venus' contacts at the Sun's limb almost impossible.

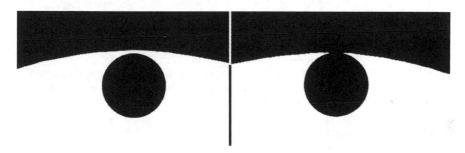

Figure 2.   An illustration showing the black drop. According to the paper "The Black-drop Effect Explained" by Pasachoff, Schneider, & Golub (2004, Proceedings of the International Astronomical Union), "... the black-drop effect was not intrinsic to Venus but was rather a combination of instrumental effects and effects to some degree in the atmospheres of Earth, Venus, and Sun..."

Even worse, timings were foiled by something called the "black drop" (Fig. 2)—a small, black extension that appeared to connect Venus' disk to the solar limb at the critical second and third contacts. The black drop caused such significant variations in the recorded contact times (even by observers sitting side-by-side) that the post-transit calculations produced a variety of Earth-Sun distances. It was not until 1824 that German astronomer Johann Franz Encke, using the new mathematical method of least squares, calculated a reasonable average value: 153.4 million kilometers.

Time does not permit a recounting of all the tales of joy and woe, success and failure, that befell the more than 150 observers at some 125 stations around the world before, during, and after the 1761 and 1769 transits. Perhaps the saddest story is that of Guillaume le Gentil, who was away from home for more than 11 years and failed to record either transit. His heartbreaking tale is recounted in a series of four stories "Le

Gentil and the Transits of Venus" by Helen Sawyer Hogg. "A (Not so) Brief History of the Transits of Venus" by Daniel Hudon is a concise summary (despite its title) of some of these expeditions, plus those of later transits.

3. The Last Tango

By the mid-1800s, more sophisticated methods of measuring the distance to the Sun threw the results of the 18th-century transit observations into question. And now it was not just the size of the solar system that mattered. In 1838, Friedrich Bessel became the first to gauge the distance to a star (61 Cygni) using the diameter of Earth's orbit as a baseline for his parallax measurement. The Earth-Sun separation was now a steppingstone to the stars, and it had to be known with the greatest precision possible.

Fortunately, another set of Venus transits was approaching (1874 and 1882), and astronomers were hoping that a new technique—photography—would solve the black-drop problem and provide accurate contact times. Expeditions were organized for the December 8, 1874 event, and astronomers from many countries, complete with vast amounts of equipment, were dispatched to the far corners of Earth. The results were unspectacular.

Visual observers encountered the same problems as their 18th-century counterparts. Photographs of the transit were generally poor, showing the same fuzziness at second and third contact that bedeviled those watching by eye and telescope. Still, multiple timings were made and the parallax calculated. The result was an improvement on previous computations, but not significantly so.

The 1882 transit provided one more chance to get it right, but enthusiasm for using Venus transits to measure the Earth-Sun distance was waning. Still, the knowledge that the next transit would not occur for another 122 years caused astronomers to try again—employing even larger telescopes and much-improved cameras. By the end of it all, the Earth-Sun distance was determined to be 149,158,000 kilometers. (Modern techniques give a distance of 149,597,870.7 km.)

On the other hand, the December 6, 1882, transit was the first of the "modern era" to be visible across Western Europe and the United States. Public interest in the event reached a fever pitch, particularly in the U.S. On transit day, telescopes proliferated on sidewalks, crowds formed, and some enterprising scope owners charged a nickel or even a dime for the privilege of a quick look at this unique sight.

4. Twice in a Lifetime

Did you catch a glimpse of the transit on June 8, 2004 (Fig. 1)? The final Venus transit of the current pair is almost upon us. Where will you be on June 5/6, 2012?

The global visibility of the 2012 transit is illustrated on the world map in Figure 3. All of North America will see some part of the transit, though for the lower 48 U.S. states, Mexico, and much of Canada, the Sun will set with Venus still on its face. Most of Europe will have the opposite experience—the Sun will rise with the transit in progress and nearing its conclusion. (The transit occurs on June 5th in North and South America, and on June 6th in the rest of the world.)

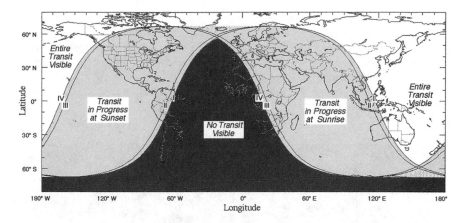

Figure 3.   Most of the world will see at least part of the 2012 transit of Venus; only parts of Africa and South America miss out. Illustration courtesy of Fred Espenak, NASA.

Fred Espenak and NASA have prepared two tables listing contact times and the corresponding altitudes of the Sun for 60 cities throughout the United States[1] and for 121 non-U.S. cities[2]. For any specific location, Steven van Roode and Franois Mignard have created a transit-calculator webpage.[3] To use it, users enter a specific address (or a general one—such as London, England) and press "locate," or drag the location marker on the map to their Venus-observing site. If any of the clocks to the right of the map are dark, it means that specific event occurs when the Sun is below the horizon. The predictions may be slightly off, so start watching early for all the contacts.

5.   Safety First

Think of the transit of Venus as an annular eclipse in miniature. This means the safety rules for observing an annular or partial eclipse of the Sun also apply to watching a Venus transit. The number one rule is: never stare directly at the Sun without using a safe solar filter (Fig. 4).

According to Ralph Chou, an Associate Professor of Optometry at the University of Waterloo (Canada), unsafe filters include all color film, black-and-white film that contains no silver, film negatives with images on them (x-rays and snapshots), smoked glass, sunglasses (single or multiple pairs), photographic neutral density filters, and polarizing filters. Looking at the Sun through a telescope, binoculars, or a camera with a telephoto lens without proper eye protection can result in "eclipse blindness," a serious injury in which the eye's retina is damaged by solar radiation.[4]

---

[1]http://eclipse.gsfc.nasa.gov/transit/venus/city12-2.html

[2]http://eclipse.gsfc.nasa.gov/transit/venus/city12-1.html

[3]http://transitofvenus.nl/wp/where-when/local-transit-times

[4]http://eclipse.gsfc.nasa.gov/SEhelp/safety2.html

Venus appears tiny against the solar disk and will be hard to see without optical aid. If you have good eyes and want to try, use the cardboard eclipse "glasses" that are commonly employed during the partial phases of a solar eclipse. You can also use #14 welder's glass, but only #14—#12 does not provide enough protection. Never place unfiltered optics to your eye when wearing eclipse glasses or using a welder's glass. The power of the magnified sunlight will burn through the eclipse glasses (or break the welder's glass) in no time.

Figure 4.   From left to right: eclipse "glasses" work for transit viewing, though you will need sharp eyes to see Venus without optical aid; binocular viewing requires solar filters properly mounted on the front lenses; a glass solar filter attached to the front of a telescope; and a homemade filter, using Baader AstroSolar Solar Filter material, on a refracting telescope (the finderscope is covered by black tape to prevent an accidental glimpse of the unfiltered Sun). All photos by the author.

Your optical gear—telescopes, binoculars, and cameras—also requires solar filters. The filters will protect your optics and ensure you do not accidentally glimpse the Sun through an unfiltered scope or camera. The solar filter must be attached to the front of your telescope, binoculars, or camera lens.

Another way to safely view the Sun (and one that is useful for a public event) is via eyepiece projection. *Sky & Telescope*'s online article[5] "Viewing the Sun Safely" explains several important precautions that must be taken to protect your telescope when using the solar projection technique.

The National Astronomical Observatory of Japan (NAOJ) has an excellent website dealing with the May 2012 annular eclipse; the "How to Observe" page[6] is very good. The NAOJ has also published a Solar Eclipse Viewing Guide in PDF format.[7] Both are excellent resources for viewing the Venus transit (which is essentially a miniature annular eclipse), and both have illustrations you can use in your own presentations.

Although care is required, it is not difficult to safely observe the Sun, and hence the transit. And whether you observe by yourself or with a group, do not miss the June 5/6, 2012 transit of Venus. You will not have another chance to see one!

---

[5]http://skyandtelescope.com/observing/objects/sun/Viewing_the_Sun_Safely.html

[6]http://naojcamp.mtk.nao.ac.jp/phenomena/20120521/obs-en.html

[7]http://www.nao.ac.jp/E/file/solar_eclipse_viewing_guide-v1.pdf

6. Transit Resources

6.1. Solar Filter Material

Safe solar filters (and material for creating home-made Sun filters) are available from the suppliers listed below. The best filter material is Baader AstroSolar Safety Film. After the transit, you can use the filters to watch sunspots on the Sun.

- Astro-Physics, Inc. (`http://www.astro-physics.com`): Solar-filter material.

- Kendrick Astro Instruments (`http://www.kendrickastro.com`): Filters for telescopes and binoculars, and solar-filter material.

- Orion Telescopes & Binoculars (`http://www.oriontelescopes.com`): Telescope filters.

- Rainbow Symphony (`http://www.rainbowsymphonystore.com`): Eclipse "glasses" and solar filters.

- Seymour Solar (`http://www.seymoursolar.com`): Filters for telescopes and binoculars and solar-filter material.

- Thousand Oaks Optical (`http://www.thousandoaksoptical.com`): Eclipse "glasses," filters for telescopes and binoculars, threaded camera filters, and solar-filter material.

6.2. Transit of Venus Information

To learn more about Venus transits (past and present), here are a few resources.

- Sheehan, W. & Westfall, J. 2004, The Transits of Venus. Prometheus Books. In this history of science, the authors chronicle the travels and explorations of scientists and adventurers who studied Venus transits in the quest for scientific understanding. A new version, *Celestial Shadows: Eclipses, Transits and Occultations*, covering transit and eclipse phenomena in general, is scheduled to be out in late 2011.

- Two transit of Venus websites contain many things transit-related, including some educational activities: `www.transitofvenus.org` and `http://transitofvenus.nl`.

- Relive the 2004 Venus transit via the TRACE spacecraft, the only satellite able to observe this transit in visible light (`http://trace.lmsal.com/transits/venus\_2004`).

- The European Southern Observatory has an extensive archive of images from the 2004 transit. Some are listed by theme; in the Chronological Order section, transit images start on page 20 (`http://www.eso.org/public/outreach/eduoff/vt-2004/photos`).

- This NASA website discusses planetary transits across the Sun and includes information about Mercury transits: `http://eclipse.gsfc.nasa.gov/transit/transit.html`).

- If you are considering holding an outreach event or an observing session to coincide with the transit, the Plan a Community Celebration webpage has many tips (`http://www.transitofvenus.org/june2012/`).

Connecting People to Science
ASP Conference Series, Vol. 457
Joseph B. Jensen, James G. Manning, Michael G. Gibbs, and Doris Daou, eds.
©*2012 Astronomical Society of the Pacific*

Engaging the Public with Astronomy in Collaboration with Outdoors/Nature Education Programs

Douglas N. Arion[1] and Sara DeLucia[2]

[1]*Carthage College, 2001 Alford Park Drive, Kenosha, Wisconsin 53140, USA*

[2]*Appalachian Mountain Club, Highland Center, Crawford Notch, New Hampshire 03575, USA*

1. Introduction

Carthage College and the Appalachian Mountain Club (AMC) have partnered to deliver hands-on astronomy programs to the public, train mentors and docents, and provide experiential learning opportunities for astronomy students. The AMC serves more than 500,000 visitors each year at its lodges, centers, and high-mountain huts; delivers a wealth of experiences in nature and the environment; and through this project will add astronomy to its education programming to provide a holistic, integrative picture of the workings of the universe to the lay public. Carthage brings professional astronomers and trained students as teachers and mentors. Carthage students and faculty are actively engaged in observational astronomy research using instruments at Kitt Peak and Steward Observatories, in performing flight experiments as part of the NASA Space Exposed Experiment Devloped (SEED) program, and in conducting outreach programs at local schools, the Carthage planetarium, and the new College observatory. This partnership will create a public education and outreach program that is unique because it will be conducted by professional astronomers, will reach large audiences, and will engage undergraduate students as educators and mentors.

2. A New Approach to Public Engagement

A substantial volume of press has addressed the scientific (il)literacy of the general population of the United States. Left unaddressed, the long-term impact of this lack of knowledge and understanding could be substantial and extremely deleterious. This is especially true in "big picture" areas of science—evolution, cosmology, and the scientific method. Consider that less than 10% of U.S. 15 year olds meet Program for International Student Assessment (PISA) performance standards in math or science; that 80% of over 10,000 students surveyed, after an astronomy course, still believe that there are phenomena that cannot be explained by science and more than 40% think the positions of planets affect human behavior; and that 40% of Americans believe that human beings were created divinely within the last 10,000 years. Conventional science education has not been able to significantly affect these issues, as evidenced by results of many studies, including the extensive research in the efficacy of astronomy education at institutions such as the Universities of Arizona and Wyoming.

In contrast, the engagement of the public with natural/outdoors activities is high, with zoos, nature centers, and outdoors programs well and enthusiastically attended. Astronomy is an integrating subject—it ties together aspects of all of these fields. This presents a golden opportunity to utilize astronomy education as a medium with already engaged audiences in the broader areas of how science is actually done, how different disciplines are understood in terms of a self-consistent scientific picture, and how current understanding is shaped by research and study in multiple disciplines to create a single, coherent story. This is now even more important as developments in any one discipline depend critically on developments in others. For example, climatology depends on both stellar evolution theory and metabolic analyses of plants and animals, and how biological evolutionary theory is dependent on solar system dynamics and the effects of impactors. Astronomy sits at the center of all of these fields where all topics merge into the total evolutionary picture of the Universe. The project we are starting integrates programming in astronomy delivered by professional astronomers and trained undergraduate student mentors into the ongoing natural and outdoors programming conducted by the Appalachian Mountain Club. Large audiences, a holistic approach to science education and outreach, and professional delivery will create significant impact on public science understanding and knowledge.

In summary, key elements of our approach include:

- intentional integration of topical astronomy programming in the natural environment;

- management and delivery of astronomy programming by *professional* astronomers;

- joint development and delivery of multi-disciplinary programming by professional astronomers and professional naturalists;

- engagement of trained physics and astronomy students as mentors and docents;

- professional development of naturalists in the discipline of astronomy;

- hands-on observing activities using naked eye, binocular, and telescopic means at both major visitor centers and backcountry/high mountain facilities; and

- cloudy night programming that utilizes new media technologies.

3. Program Elements

Public programs will be conducted to engage audiences in astronomy and the natural world through formal presentations, informal discussion, and under-the-sky observing. Three classes of programs will be conducted to address the three tiers of audiences that are served by AMC programs, as summarized in Table 1. These different groups will be addressed through different delivery approaches to achieve the greatest impact on the largest population.

Each program will include several elements, including pre-sunset talks, an overview of telescopes and observing, and actual observation through telescopes. Pre-sunset talks will give an overview of the observing experience, specifically addressing topics that link the objects to be viewed with the audience as living beings and with the natural

Table 1.   Three Tiers of Public Engagement

	Tier 1	Tier 2	Tier 3
Audience:	Visitors to AMC Facilities	Regional Population, including Schools	Special Programs: Youth Opportunities, Summer Camps
Population Size:	75,000 (Overnight) 400,000 (Day visitors)	15,000 (Regional) 1,500 (Schools)	8,000
Economic Demographic:	Middle/Upper Middle Income	Middle/Lower Income	Middle/Lower Income (Urban/Inner City)
Delivery:	Professional Astronomer	Professional Astronomer and/or AMC Staff/Carthage College Student Interns	Professional Astronomer and/or AMC Staff/Volunteers
Primary Program Timeframe:	May-September December-March	September-June	January-December

environment in which they are immersed. For example, observing a planetary nebula provides an opportunity to discuss the evolution and ultimate fate of the Sun and stellar energy production mechanisms and lifetimes/timescales, and how nebula morphology may be related to the presence or absence of planets, the effects of stellar winds, and the production and dispersal of heavier elements. Bright/dark nebulae allow discussions of dust and the presence of organic materials in space, linked to the likelihood of life elsewhere in the Universe and the seeds of life on Earth. And, of course, observing solar system objects presents opportunities to discuss volcanism, seasons, organic molecules, the habitable zone for life, and the search for extrasolar planets. Another interesting wildlife/astronomy linkage is the timing of molting/color changes in animals which is tied to the duration of sunlight; there have been instances where due to shortening of winter seasons some animals have changed coats from dark to white too soon, making them easier targets to become prey, with their camouflage working against them instead of for them. Table 2 summarizes some of the observations and the topics that can be addressed, many of which are beyond those traditionally addressed in public astronomy programs.

In the second audience tier, the AMC reaches many individuals including day visitors, attendees at special events, and school children. These audiences will be addressed primarily by AMC staff and Carthage student interns who will be provided with professional development training to prepare them to deliver astronomy content in the broad context of nature and the environment. These individuals will also have primary responsibility for program delivery in venues to support audiences that may not typically frequent AMC centers or huts, and to deliver astronomy content as part of their regularly scheduled naturalist programs. This represents a significant area for program impact by providing training and hands-on experience for AMC staff and volunteers. Working with trained student interns, astronomy content can be regularly integrated with natural-

Table 2.   Sample Objects and Topics for Discussion

Object Type	Relevant Interdisciplinary Topical Content
Planetary Nebula	Evolution/timelines of solar-mass stars, including the Sun Dispersal of metals into the ISM PN Morphology associated with companion stars and orbiting planets/planetary disks
Supernova Remnants	Production of heavy elements Connection to metal content in proto-Solar nebula/solar system objects, rocky planet material, etc. Black holes/neutron stars
Bright/Dark Nebula	Organic molecules in space Formation of stars and planet systems Carbon as the basic building block of life on Earth
Stars/Star Clusters	Stellar temperatures, lifetimes
Galaxies	The scale of the Universe The Drake Equation and potential for life elsewhere Cosmology–time/distance relation Self-consistent understanding of the evolution of the Universe (Tie-ins to Earth processes) Contrast external galaxies with Milky Way structure Cosmology–primordial hydrogen and water (same protons as from the Big Bang)
Moon	Impactors and impact on Earth evolution, including dinosaurs and the mammalian development
Sun	Solar cycle Climatology effects of solar irradiance Solar spectrum vs. chlorophyll/eye color sensitivity Sunlight duration vs. seasons/impact on plants/animals
Planets	Rocky planets, volcanology, meteorology Gas giants and planet density, solar system formation, exoplanets Life in the Universe–evolutionary processes, diversity of life
Dark Skies	Dark skies/Impact on wildlife Dark skies/Impact on human health
Satellites	GPS/General Relativity/Tectonic measurements/Cosmology

ist programming, broadening the understanding of participants of natural phenomena, and providing professional development for AMC staff and volunteers. Hopefully, this will instill interest in participants to seek out other astronomy experiences (for example, attending one of the scheduled observing sessions at AMC facilities, or to seek out such experiences in their home towns).

The third audience tier addresses the Youth Opportunity Program (YOP) at the AMC. The AMC delivers special naturalist programming through YOP to inner city and urban students who benefit tremendously from immersion in the natural world. YOP works with urban youth agencies, primarily from the Boston and New York City areas, to connect at-risk youth with the natural world through hands-on experiences in backcountry environments. Through this program, astronomy-related programming will be offered to YOP groups staying at AMC's facilities. Approximately 130 YOP groups involving nearly 1500 at-risk urban youth, ranging in age from 7 to 19 (13 on average), stay at these NH facilities annually.

Professional development and training activities will be implemented in order to deliver programming to the three tiers of audiences outlined above and to create a cadre of individuals who regularly integrate astronomy topics into naturalist programming. Given that AMC facilities are busiest during mid-summer and mid-winter, mid-spring presents the best opportunity to provide professional development activities for AMC staff and volunteers. Workshops will be held for these individuals, utilizing both in-class and under-the-sky components. It is expected that the well-established *Astronomy From the Ground Up* (AFGU) curriculum and associated text (*Skies Above, Earth Below*) will be used along with media tools to prepare them for public programs. Hands-on observing with iPads and telescopes will be conducted to train them for under-the-sky activities. When possible, AMC staff will attend AFGU workshops hosted by the Astronomical Society of the Pacific to further expand and improve on their astronomy capabilities.

Astronomy students from Carthage College will serve as interns in this program at AMC Facilities during each summer, the high season for engaging audiences. These students have all conducted professional-level astronomy research as well as public outreach activities before being engaged as docents for the proposed program. Carthage conducts several observing runs at either Kitt Peak National Observatory or Steward Observatory each year. Students are trained to operate large telescopes, to obtain and calibrate images, and analyze the images using IRAF and other image processing tools,. Carthage also has several major data-driven research programs, including studies of QSO magnesium line tracers of dim galaxies and galaxy halos using SDSS, and studies of diffuse interstellar bands in stellar spectra. Unlike typical intern programs, in this instance the students will already be trained in telescope operation, astronomical science, and public engagement, and will thus serve as mentors to AMC staff and be outstanding docents during public programs at AMC facilities. They will have the opportunity to hone their skills in communicating science to the general public, and gain from being engaged in outdoors/natural education programs that should greatly broaden their viewpoint and understanding of the integrated picture of scientific knowledge. This represents a considerable improvement in the preparation of science students.

One of the failings of many public engagement programs is that they are "point of sale" type programs only touching the participant for a short period of time with little opportunity to sustain and follow-up to create real impact. All program attendees will be registered, and contact information collected. Follow-up e-mails will be sent to each participant both to collect follow-up assessment data and to point participants to additional resources. Knowing which sessions the participants attended, the messages sent will be tailored to link them to more information on the specific objects observed/discussed and topics addressed during the program.

Evaluation of the program will be conducted by surveys of participants that will be obtained primarily on-line and supplemented with paper forms. Participants/attendees will be asked to register and sign-in for programming, providing contact data to allow for follow-up investigations, and to provide them access to additional resources and information. Students majoring in sociology at Carthage will develop the survey instruments, conduct the surveys, and collate and analyze the results. Assessment data will be utilized to determine if attendance has changed the appreciation and understanding of the validity of scientific analysis, the linkages among the disciplines presented, and their personal connection with the larger natural world. Key measures to be addressed include: (1) attitudes about importance of science in everyday life; (2) attitudes/understanding of the integrative nature of scientific understanding; (3) likelihood to continue studying/investigating topics covered during the session; (4) likelihood of recommending AMC programming to others; and (5) demographic measures of participant characteristics.

4. Summary

The new program will significantly improve public engagement in astronomy—and will do so in a way that significantly broadens the range of this engagement to personally connect participants with the greater world. The program activities are designed to make significant and permanent changes in the way that nature programs are delivered, and to set a paradigm for professional engagement in public outreach. By utilizing as a core ongoing nature and outdoor education activities, the program will intrinsically link the full range of natural phenomena into a single, coherent story.

References

Arion, D. N., Martin, E. O., Pennington, C., & Kuttruff, S. 2008, Deep Images of NGC 2371/2 in [O III], AAS
Arion, D.N. and E.O. Martin, January 2008, Narrow-Band Images of Several Planetary Nebula in [O III], AAS
Friedman, S., York, D., McCall, B., Dahlstrom, J., Sonnentrucker, P., Welty, D., Drosback, M., Hobbs, L., Rachford, B., & Snow, T. 2011, "Studies of Diffuse Interstellar Bands. V. Pairwise Correlations of Eight Strong DIBs and Neutral Hydrogen, Molecular Hydrogen, and Color Excess," ApJ, 727, 33
Program for International Student Assessment 2009 Results: What Students Know and Can Do Vol. 1, Organization for Economic Co-Operation and Development, 2009
Slater, T. F. & Slater, S. J. 2008, "Development of the Test Of Astronomy STandards (TOAST) Assessment Instrument," University of Wyoming, Bulletin of the American Astronomical Society, 40, 273

Connecting People to Science
ASP Conference Series, Vol. 457
Joseph B. Jensen, James G. Manning, Michael G. Gibbs, and Doris Daou, eds.
© *2012 Astronomical Society of the Pacific*

What Would Galileo Do? An Update on the Galileo Teacher Training Program

Brian Kruse, James G. Manning, Greg Schultz, and Andrew Fraknoi

Astronomical Society of the Pacific, 390 Ashton Ave., San Francisco, California 94066, USA

Abstract.   The Galileo Teacher Training Program (GTTP) is a heritage program of the International Year of Astronomy (IYA) and Beyond that provides professional development for teachers in grades 3–12. At the core of every GTTP workshop are activities inspired by Galileo's iconic observations, IYA and NASA resources and activities, activities related to fundamental concepts to assist teachers in meeting their curriculum goals, and the inclusion of resources adaptable for use in the classroom. GTTP actively utilizes a hands-on, inquiry based collaborative activity model of learning, modeling effective techniques for engaging student interest and promoting scientific literacy for both content and process. This article includes lessons learned from the pilot and following workshops and plans to take the concept forward in flexible and adaptive ways.

1. Background

Established in 1889, the Astronomical Society of the Pacific (ASP) strives to increase the understanding and appreciation of astronomy by engaging scientists, educators, enthusiasts and the public to advance science and scientific literacy. The International Year of Astronomy (IYA) in 2009 presented a unique opportunity to develop sustainable educational programs with the goal of serving astronomy and science education during IYA and into the future. The Galileo Educator Training Program (GTTP) was one such program targeting classroom teachers.

The vision for GTTP was to provide a program to help address well documented needs to:

- engage students more actively in science,

- foster understanding of the process of science,

- promote critical thinking, and

- encourage students to consider careers in STEM (Science, Technology, Engineering, and Mathematics) fields.

2. GTTP Workshops

The ASP worked in partnership with the New Jersey Astronomy Center for Education (NJACE) to develop a GTTP workshop model for formal educators. The workshop was developed to use Galileo's observations and IYA materials to model the process of science while helping classroom teachers meet their curriculum goals. The workshop model offered a flexible approach using and/or addressing four basic elements:

- Galileo related activities,

- fundamental concepts in astronomy,

- adaptable educational tools, and

- materials developed for IYA.

Workshops have addressed these elements through the incorporation of materials such as Dark Skies Awareness, the Galileoscope, NASA developed resources, and content developed by workshop co-hosts. Hands-on astronomy activities were drawn from these sources, as well as those compiled in the ASP publication *The Universe at Your Fingertips*.

GTTP workshop content is flexible and customizable depending on the needs of the teacher participants. The involvement of partner organizations in co-hosting workshops has also provided the opportunity for flexibility in customizing content related to specific missions. Workshops sponsored by Northrop Grumman allowed the incorporation of content related to the James Webb Space Telescope. Likewise, lunar science was emphasized in workshops supported by the NASA Lunar Science Institute.

The basic workshop model also allows flexibility in workshop duration. Depending on the situation, workshops can follow a 1-day, 1.5-day, or 2-day model. Teachers who participate in two full days of GTTP professional development earn a certificate naming them as "Galileo Ambassadors." An online "community of practice" is in development, which will allow networking between Ambassadors, as well as extended online learning opportunities.

2.1.   Workshops 2009–2011

Including the initial workshop in September 2009, six GTTP workshops have been conducted, with approximately 250 teachers participating. The following list details the six GTTP workshops:

- 2009 September 12–13 in Millbrae, California

- 2010 March 13–14 in Redondo Beach, California

- 2010 April 24 in Los Altos, California

- 2010 July 31–August 1 in Boulder, Colorado

- 2011 March 26–27 in Redondo Beach, California

- 2011 July 30–31 in Baltimore, Maryland

3. Lessons Learned

Every GTTP workshop has provided a learning opportunity for the ASP in how to refine the workshop model to better serve teachers. Some of the more important lessons learned include:

- teachers are eager to incorporate inquiry-based, hands-on astronomy activities in their classrooms;

- developing a truly national effort and network requires additional resources, particularly funding;

- developing a sustainable "community of practice" that teachers will regularly access requires additional resources, both in a digital infrastructure, and in the funding to implement and sustain the network; and

- interested potential partners are in good supply.

The key is in turning the potential partners into actual partners able to provide the resources to accomplish the program goal of reaching more teachers.

4. Galileo Educator Network

In 2011, the ASP received a NASA EPOESS grant, allowing us to take the next step in creating a national network of NASA Galileo Educators. Building off the initial successes of GTTP, the NASA Galileo Educator Network (GEN) is a distributed and leveraged teacher professional development program, led by the ASP. The purpose of GEN is to help teacher educators and teacher professional development providers engage and educate teachers, especially in grades 3–9, using effective instructional strategies, scientific practices, and NASA educational resources, all revolving around engaging and inspiring astronomy content. During each project year, a national professional development institute will be conducted, training professional development providers who subsequently develop and lead their own GEN workshops for teachers in their regions and communities. Each GEN workshop engages teachers in 15 hours of professional development, combining GTTP and NASA content and resources.

In May 2011, site leaders from seven Project ASTRO National Network sites, plus a representative from GEN partner organization the National Earth Science Teachers Association (NESTA), received two days of professional development on how to conduct a GEN workshop. These sites will conduct their teacher workshops during the 2011–12 school year. Over years two and three of the project, representatives from 36 additional sites will receive such training from the ASP and the three partner organizations: NJACE, NOAO (National Optical Astronomy Observatory) and NESTA. Participants in a GEN professional development institute will be designated NASA Galileo Educator Fellows. These Educator Fellows will then conduct their own GEN workshop in their region or community. With a goal of each trainer engaging 15 additional teachers, it is expected the GEN program will impact at least 660 teachers. Teacher graduates of a 15-hour GEN professional development workshop will be designated "NASA Galileo Educators."

The anticipated result of GEN is a "nationalized" Galileo Teacher Training Program, where a national network of trained teachers will be supported in promoting inquiry-based, hands-on student science engagement demonstrating the process of science, building critical thinking and collaborative skills, and encouraging consideration of STEM related careers.

Connecting People to Science
ASP Conference Series, Vol. 457
Joseph B. Jensen, James G. Manning, Michael G. Gibbs, and Doris Daou, eds.
© *2012 Astronomical Society of the Pacific*

Hawaii's Annual Journey Through the Universe Program

Janice Harvey,[1] Doris Daou,[2] Brian Day,[2] Timothy F. Slater,[3] and
Stephanie J. Slater[3]

[1]*Gemini Observatory, 670 N. Aʻohoku Place, Hilo, Hawaii 96720, USA*

[2]*NASA Lunar Science Institute, Ames Research Center, Building 17, Room 103,
MS: 17-1, Moffett Field, California 94035, USA*

[3]*Center for Astronomy & Physics Education Research, University of Wyoming,
Dept. #3905, 1000 E. University Ave., Laramie, Wyoming 82071, USA*

Abstract. Hawaii's annual Journey through the Universe program is a flagship Gemini public education and outreach event that engages the public, teachers, astronomers, engineers, thousands of local students and staff from all of the Mauna Kea Observatories. The program inspires, educates, and engages teachers, students, and their families as well as the community. From February 10–18, 2011, fifty-one astronomy educators from observatories on Mauna Kea and across the world visited over 6,500 students in 310 classrooms at 18 schools. Two family science events were held for over 2,500 people at the ʻImiloa Astronomy Education Center and the University of Hawaii at Hilo. The local Chamber of Commerce(s) held an appreciation celebration for the astronomers attended by over 170 members from the local government and business community.

Now going into its eighth year in Hawaii, the 2012 Journey Through the Universe program will continue working with the observatories on Mauna Kea and with the NASA Lunar Science Institute (NLSI). As a new partner in our Journey program, NLSI will join the Journey team (Janice Harvey, Gemini Observatory, Journey Team Leader) and give an overview of the successes and future developments of this remarkable program and its growth.

The future of America rests on our ability to train the next generation of scientists and engineers. Science education is key and Journey through the Universe opens the doors of scientific discovery for our students. www.gemini.edu/journey

1. Program Objectives

The Journey through the Universe program[1] as currently implemented in Hawaiʻi has three primary objectives for the Gemini Observatory:

- connect and engage learners with educators, scientists and engineers in an effective, lasting, and relevant manner;

- engage the local community at all levels; and

---

[1]www.gemini.edu/journey

- foster an environment where students can pursue science, technology, engineering, and mathematics (STEM) careers and find local support and role models for their advancement.

Figure 1.   Peter Michaud, Gemini Observatory, demonstrates carbon dioxide mirror cleaning during a Journey classroom event.

Likewise, the local Department of Education (North Hilo/Laupahoehoe/Waiakea Complex) has additionally stated their objectives for the program, which are to:

- heighten awareness of science in classrooms;

- help students meet the Hawaii Content and Performance Standards and national standards;

- provide rigor, relevance, and relationships in curriculum, instruction and assessment;

- tap into the rich resources that are available in the Hilo community;

- improve teaching staff preparation in content fields; provide professional development, in-service training sessions, networking, and articulation amongst educators, scientists, and community members who can help improve teaching; and

- educate parents and the community in space science.

2.   History

Journey through the Universe is a national initiative that was developed under the leadership of Jeff Goldstein, Director of the National Center for Earth and Space Science

Figure 2.   Students, their teachers and school administrators participate in a solar system walk during Journey through the Universe week in 2011.

Education. Dr. Goldstein's goal was to bring a nationwide network of communities that would share a common goal of bringing science, technology, engineering, and mathematics researchers into local schools. The original Journey program began in 1999 and grew into 13 communities across the country. Gemini Observatory joined the Journey network in 2004 and, in part due to the significant outreach and educational resources on Mauna Kea, provided a model community for the effective integration of Journey Through the Universe into the local classrooms. The combination of a rigorous effort on behalf of the astronomy community and the enthusiastic support of the local Department of Education, the Journey program has grown and thrived in the local community.

3.   Key Elements of the Journey Program

3.1.   Astronomers, Engineers, and STEM Educators in the Classroom

The Journey program recruits astronomy educators from the observatories on Mauna Kea, the University of Hawaii at Hilo, 'Imiloa Astronomy Education Center, NASA, and other observatories and institutions from across the country and the world. Astronomers as well as engineers are not often equipped to visit classrooms and share their personal journeys which led them to their career choice. The astronomer's workshop held before Journey Through the Universe week provides scientists and engineers assistance in planning their classroom presentations and hands-on workshops. This also provides an opportunity for dialogue between the astronomy educators and the teachers regarding the most appropriate activities for them to share with students.

Figure 3.　　Jeff Goldstein, keynote speaker at NSTA 2011 and founder of Journey Through the Universe, meets Hawaii's Journey through the Universe team.

3.2.　Master Teacher and Teacher Workshops

The national Journey Through the Universe program produced and still provides modules that include lesson plans for the teachers and master teachers. All lessons met the National Science Standards, and we aligned all modules with the Hawaii State Science Standards. Our local Department of Education invested significant monetary and personnel resources to align four of the five modules. These four modules are now rotated on a four-year-cycle and are designed to facilitate engaging, relevant inquiry-based learning.

3.3.　Ambassadors

Ambassadors accompany the visiting scientists and engineers in the classroom. They introduce the astronomy educators to the classroom and assist them with any demonstrations and hands-on activities shared with the students. The ambassador team is mostly comprised of business owners, physicians, bankers, and local astronomy staff, to identify just a few. This provides an opportunity for our local community to understand the value of STEM education in the classroom and pass that information on to others who may have an interest in working with our Journey Through the Universe program in future years.

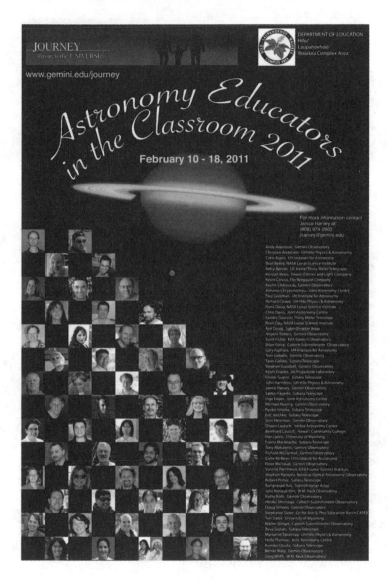

Figure 4.   The 2011 Journey Through the Universe Astronomy Educators poster.

3.4.   Family Science Events

The local community embraces the annual family science events held during Journey Week. The lecture/talks presented by the distinguished astronomers and educators are augmented by hands-on activities and are attended by thousands of learners of all ages from our local community. In past years the events have been held at our local mall, the University of Hawaii at Hilo Theater, downtown Hilo theaters and the 'Imiloa Astronomy Education Center.

Figure 5.   Journey Through the Universe 2011 teacher workshop held at 'Imiloa Astronomy Education Center.

Figure 6.   Journey Through the Universe trains ambassadors to accompany and support our science educators in the classroom.

3.5.   Chamber of Commerce Thank You Celebration

Hawaii's local Hawaii Island Chamber of Commerce and Japanese Chamber of Commerce join together annually to sponsor a thank-you celebration for our scientists and engineers, Department of Education staff and teachers, ambassadors and business partners. This event gives the Chambers a way to express their gratification to the community for promoting STEM education in the classroom and beyond. The local community is deeply invested in the Journey Through the Universe program and our governor's, mayor's, senate and house of representative's offices attend the event or send representatives and provide proclamations of support.

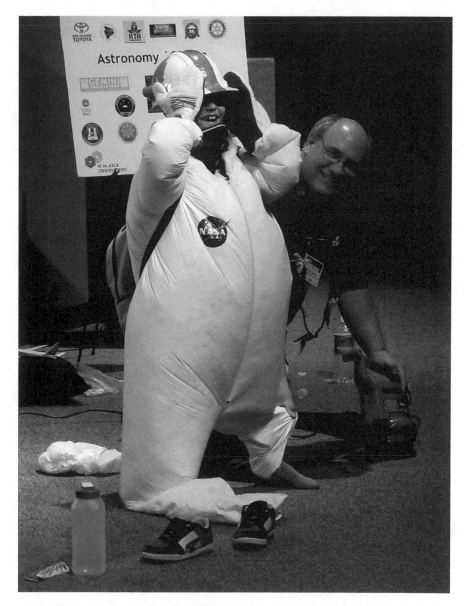

Figure 7. Kevin Caruso creates a spacesuit for a young student during a Journey Through the Universe family science event.

3.6. Community Support

Journey Through the Universe receives financial support from many local businesses including the Bank of Hawaii, Thirty Meter Telescope, Hawaii Island Chamber of Commerce, the Japanese Chamber of Commerce, Rotary Clubs, Big Island Toyota, and the Business-Education Partnership, just to name a few. In 2011, the NASA Lunar Science

Institute provided substantial monetary support and brought in their own scientists to participate in our Family Science Events and classroom presentations.

3.7. Promotions/Media Coverage

Our local media proudly offers thousands of dollars in free advertising and promotion of our Journey Through the Universe events. Our local radio stations and newspapers strongly promote the program's message of STEM education and science outreach to the community. Promotions include banners at our local airport and mall, at every participating school, and on main intersections going into the base facilities of the observatories and the University of Hawaii at Hilo. The schools place Journey information on their marquees and websites in order to promote the scheduled astronomer visits in the classroom as well as the Family Science Events held throughout the week.

4. Adapting the Journey program in other communities

The observatories on Mauna Kea offer a unique abundance of scientists for the Journey program. In most communities there are local universities or colleges that could work alongside a local amateur astronomy club to provide STEM education to their local classrooms. Astronomy is a gateway science to biology, math, engineering, chemistry and physics. Most communities have the resources to sustain a variation of the Journey program that reflects their own Department of Education's needs.

5. Future Plans

For Journey Through the Universe 2012, the partnership has expanded to include the University of Hawaii at Hilo's School of Pharmacology and their undergraduate chemical engineering program. A proposed one-week basic engineering program is being offered to the Journey Master Teachers during the 2011 Christmas break. The possibilities of adding other STEM resources and instruction are limitless.

6. Conclusion

Journey Through the Universe is Gemini Observatory's flagship outreach program. Hawaii's Journey model has proven to resonate with our local schools, businesses, and general public, and is expected to grow and impact our community for many years to come.

This model is adaptable for other communities with a variety of resources and circumstances. It is obvious that a program of this nature not only fills a need but also produces a valuable resource for educators and parents that will help guide our next generation of explorers.

Connecting People to Science
ASP Conference Series, Vol. 457
Joseph B. Jensen, James G. Manning, Michael G. Gibbs, and Doris Daou, eds.
©2012 *Astronomical Society of the Pacific*

WWT Ambassadors: WorldWide Telescope for Interactive Learning

Patricia S. Udomprasert,[1] Alyssa A. Goodman,[1] and Curtis Wong[2]

[1]*Harvard-Smithsonian Center for Astrophysics, 60 Garden St, MS 42, Cambridge, Massachusetts 02138, USA*

[2]*Microsoft Research, One Microsoft Way, Redmond, Washington 98052, USA*

Abstract. In our presentation, we demonstrated some key features of the WorldWide Telescope (WWT). Here we describe the results of a WWT Ambassadors (WWTA) Pilot Study where volunteer Ambassadors helped sixth-graders use WWT during a six-week astronomy unit. The results of the study compare learning outcomes for 80 students who participated in WWTA and 70 students at the same school and grade who only used traditional learning materials. After the six-week unit, twice as many "WWT" as "non-WWT" students understood complex three dimensional orbital relationships; tremendous gains were seen in student performance in science overall, astronomy in particular, and even in using "real" telescopes. We describe plans for expansion of the WWTA program.

1. WorldWide Telescope

The WorldWide Telescope (WWT) computer program is a "Universe Information System" that offers an unparalleled view of the world's store of online astronomical data. WWT weaves astronomical images from all wavelengths into an "interface" that resembles their natural context, the sky, while simultaneously offering deep opportunities to teach and learn the science behind the images. A 3-dimensional model of the Solar System and cosmos empowers students to visualize relationships between the Earth, Sun, Moon and beyond; to learn how their motions affect what we see in the night sky; and to understand the seasons we experience at different times of year. WWT tools are free for all non-commercial use. Figure 1 shows a screenshot of WWT, with some of the key features highlighted.

The full WWT application has been downloaded over 5 million times so far, and the "web client" and API forms of WWT have been accessed even more often.[1] WWT is evolving to become a key research tool within the online astronomy ecosystem known in the U.S. as the "Virtual Astronomy Observatory," but it also offers unprecedented opportunities for STEM outreach and education. WorldWide Telescope has been called out in the 2010 Astronomy Decadal Survey sponsored by the National Research Council (Committee for a Decadal Survey of Astronomy and Astrophysics & National Research Council 2010) as a significant contribution to the public understanding of astron-

---

[1]`www.worldwidetelescope.org/webclient/`

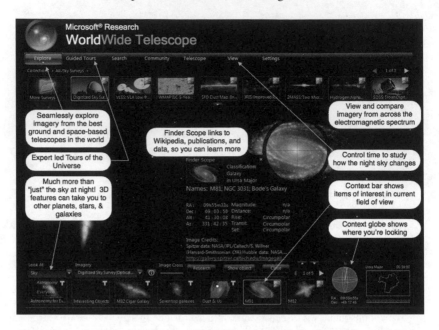

Figure 1.   Annotated screenshot of WWT, showing the user interface and high-lighting some key educational features of the program.

omy, calling it a *corporate version of previously under-funded efforts of astronomers to accomplish similar ends, [that] coordinates the world's public-domain cosmic imagery into one resource, allowing people on home PCs to explore the cosmos as if they were at the helm of the finest ground and space-based telescopes.* WWT was designed from its inception with personal inquiry, exploration, discovery, and explanation in mind, and those features have already been shown to excite STEM learners (Landsberg, Subbarao, & Dettloff 2010).

The WorldWide Telescope Ambassadors Program (WWTA) is an outreach initiative run by researchers at Harvard University and Microsoft Research. WWT Ambassadors are astrophysically-literate volunteers who are trained to be experts in using WWT as a teaching tool. Ambassadors and learners alike use WWT to create dynamic, interactive Tours of the Universe, which are shared in schools, public venues, and online. Ambassador-created Tours are being made freely available and will ultimately form a comprehensive learning resource for Astronomy and Astrophysics.

Our group has piloted the use of WWT in schools, and we present some preliminary results here. The pilot program took place in the spring of 2010 at Jonas Clarke Middle School (JCMS) in Lexington, Massachusetts. Michelle Bartley, our partner teacher, used WWT with 83 sixth grade students over the course of a 6-week long astronomy unit. Figure 2 shows some images from the classroom at JCMS during the pilot program. In the spring of 2011, we expanded the program to include all 240 sixth grade students at Clarke and another 100 students at Prospect Hill Academy, an urban charter school in Somerville, Massachusetts. All the results presented are from the completed first year pilot.

Figure 2. Students at work with WWT at Jonas Clarke Middle School in Lexington, Massachusetts, where we piloted some WWT materials in 2009–2011.

2. WWT and Moon Phases

Preliminary data from our Pilot at JCMS demonstrates the potential power of WWT's visualization environment. Our partner teacher Michelle Bartley named the Moon's phases as a topic that students typically struggle to visualize and understand, so we created an interactive WWT tour specifically to aid in the teaching of this topic. The 80 pilot students worked with the tour for only one class period, but our results show that the WWT approach holds great promise. After all the 6th grade students at JCMS completed their astronomy unit, we administered an anonymous quiz to Group A (79 students) who used the WWT Tour and to Group B (71 students) who used only traditional materials (see Fig. 3). One question was designed to test memorization skills (identify a Moon phase based on a picture), and the other question was designed to test understanding (sketch a diagram of the Earth, Sun, and Moon when the Moon is in the depicted phase). Students in both groups performed almost identically on the memorization question, suggesting that using WWT does not impact the type of learning that must be done by rote. However, there was a significant difference in student performance on the question that was designed to test their understanding. Although students in both groups struggled with Moon phases, confirming that it is a challenging concept for the sixth graders to understand, more than twice as many students in Group A than Group B were able to answer the "understanding" question correctly after working with the WWT Tour on Moon phases. Note that Group A only spent half an hour working with the tour, and perhaps even larger gains could have been made if they had had more time to devote to this topic.

Table 1 Sample size: N_A=79; N_B =71	Q1: Memorization		Q2: Understanding	
	Group A (WWT)	Group B (no WWT)	Group A (WWT)	Group B (no WWT)
Incorrect	7%	5%	38%	65%
Partially Correct	31%	33%	21%	16%
Correct	62%	61%	41%	19%

Figure 3.   Results of a "Moon Phases Quiz" administered at the JCMS pilot after their astronomy unit. Twice as many students in the WWT group (Group A) were able to correctly sketch the relative positions of the Earth, Sun, and Moon for a given Moon phase than in the non-WWT group (Group B).

3.   Student Feedback about WWT

Our preliminary pilot research shows that WWT has great potential to excite and interest students and help them learn science. After our sessions at JCMS, we administered anonymous free-response surveys. We received 72 complete surveys, and 71 (99%) were highly positive about using WWT. In questions about how WWT could help them learn science and what they would tell their best friend about WWT, the students expressed genuine excitement at being given the freedom to explore and learn in WWT's interactive environment. Figure 4 offers some representative student responses.

Student Quotes after working with WWT:

"Learning about our universe by actually seeing and exploring it makes it easier to contemplate and more fun."

"It gave me a better mental map of the universe."

"Awesome, amazing, cool, incredible (repeat 30 times)"

"You can explore the universe yourself and you don't always have to only learn from the teacher."

"This is way cooler than Call of Duty."

Figure 4.   Sample feedback from students in the JCMS Pilot, answering the questions "How is WWT helping you to learn science?" and "What would you tell your best friend about WWT?"

4. Pre and Post-test Likert Scale Anonymous Surveys

As with the Moon Quiz, we surveyed two groups—one that used WWT (Group A) and one that did not (Group B)—about their interest level and self-perceived understanding of astronomy and science before and after the astronomy unit. We used a Likert scale (1=low; 5=high) on the survey, and we present the survey analysis results in Figure 5 in terms of the effect size. Effect size measures the gain (or loss) in units of the pre-test standard deviation:

$$\text{Effect Size} = \frac{(\text{posttest average} - \text{pretest average})}{(\text{pretest standard deviation})}$$

Effect size absolute values of 0.25 or less indicate essentially no effect, 0.25 to 0.5 a small effect, 0.5 to 0.75 medium, and 0.75 or greater large effect (Cohen 1988). In Figure 5, each point plotted shows mean effect size, and the bars show plus or minus one standard error on the mean. Group A (with WWT) showed statistically significant gains on all questions asked. Group B (no WWT) showed statistically significant gains in their self-reported factual knowledge and understanding of astronomy topics, but they did not show gains in interest in astronomy or science in general. Group A self-reported a significant gain in the ability to visualize Sun-Earth-Moon relationships while Group B did not, consistent with the results of the Moon Quiz described in Section 2. One concern expressed about WWT is that the beautiful immersive environment might lead users to lose interest in using real telescopes: our data indicates that the contrary is true (note the last question in the figure).

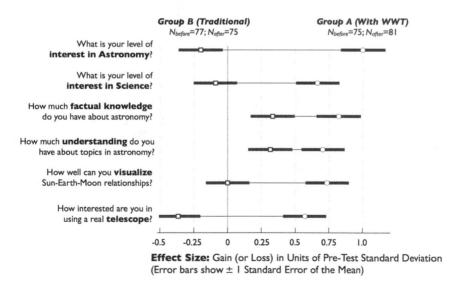

Figure 5.   A plot comparing the Effect-Size seen in two different groups at JCMS in response to various survey questions. Effect size is the gain (post-test average minus pre-test average), normalized by the pre-test standard deviation. Group A was our pilot group that used WWT; Group B was a group of similar ability students that did not use WWT.

5. Beyond the WWTA Pilot

Pending funding, we plan to expand the WWTA program to three more carefully selected, socioeconomically diverse U.S. sites, and we are developing an online community that serves as a resource for Ambassadors, teachers, and students beyond those locations. Online materials will be available through several sites (at Harvard and Microsoft) and will be integrated with existing online curriculum programs such as WGBH's Teachers' Domain and Microsoft's Partners in Learning. More information is presently available at wwtambassadors.org.

We also plan to develop a series of "visualization labs" in WWT that promote a deep understanding of key space science concepts making up the National Science Education Standards for grades 5–8 (National Research Council 1996). The topics covered will include seasons, Moon phases, eclipses, and distance scales in the Universe. Beyond specific content knowledge, the labs will be designed to familiarize students with important science skills, such as connecting observations, evidence, and explanations. These labs will be free and will require no special equipment other than a computer and an Internet connection, helping to reduce the equity gap in access to lab experiences, and they will capitalize on astronomy's special appeal among students and the public, helping to turn around those who have lost interest in science.

Acknowledgments.   The WWTA team would like to thank Microsoft Research for funding the WWTA pilot project, and we thank our partner teacher, Michelle Bartley and all her students at Jonas Clarke Middle School for their participation and contributions to this work. We are grateful to all our volunteer Ambassadors who share their expertise and passion for astronomy with the public, especially our inaugural team of Ambassadors: Dick Post, Mary Becker, and Jeremy Cushman.

References

Cohen, J. 1988, Statistical Power Analysis for the Behavioral Sciences, 2nd ed., (Philadelphia: Lawrence Erlbaum Associates)

Landsberg, R. H., Subbarao, M. U., & Dettloff, L. 2010, "WorldWide Telescope and Google Sky: New Technologies to Engage Students and the Public," Barnes, J., Smith, D. A., Gibbs, M. G., & Manning, J. G., eds., ASP Conference Series, 431, 314

National Research Council 1996, National Science Education Standards, (Washington, D.C.: National Academies Press)

Committee for a Decadal Survey of Astronomy and Astrophysics & National Research Council 2010, New Worlds, New Horizons in Astronomy and Astrophysics, (The National Academies Press)

Connecting People to Science
ASP Conference Series, Vol. 457
Joseph B. Jensen, James G. Manning, Michael G. Gibbs, and Doris Daou, eds.
© *2012 Astronomical Society of the Pacific*

The Galileo Teacher Training Program Global Efforts

Rosa Doran,[1] Carl Pennypacker,[2] and Roger Ferlet[3]

[1]*NUCLIO, Núcleo Interactivo de Astronomia, Largo dos Topázios, 48, 3 Fte, 2785-817, São Domingos de Rana, Portugal (geral@nuclio.pt).*

[2]*Hands On Universe Division: Physics/SSL, Lawrence Berkeley National Lab, Cyclotron Road, Mail Stop 50-232, Berkeley, California 94720, USA*

[3]*CNRS / UPMC, UMR7095, Institut d'astrophysique de Paris 98bis Bd. Arago, 75014, France*

Abstract.   The Galileo Teacher Training Program (GTTP) successfully named representatives in nearly 100 nations in 2009, the International Year of Astronomy (IYA2009). The challenge had just begun. The steps ahead are how to reach educators that might benefit from our program and how to help build a more fair and science literate society, a society in which good tools and resources for science education are not the privilege of a few. From 2010 on our efforts have been to strengthen the newly formed network and learn how to equally help educators and students around the globe. New partnerships with other strong programs and institutions are being formed, sponsorship schemes being outlined, new tools and resources being publicized, and on-site and video conference training conducted all over the world. Efforts to officially accredit a GTTP curriculum are on the march and a stronger certification process being outlined. New science topics are being integrated in our effort and we now seek to discuss the path ahead with experts in this field and the community of users, opening the network to all corners of our beautiful blue dot. The main aim of this article is to open the discussion regarding the urgent issue of how to reawaken student interest in science, how to solve the gender inequality in science careers, and how to reach the underprivileged students and open to them the same possibilities. Efforts are in strengthening the newly formed network and learning how to equally help educators and students around the globe.

1.   Strengthening the Galileo Teacher Training Program Network to Help Educators and Students Around the Globe

Being a global effort, GTTP partners have to deal with different challenges in different parts of the globe. In some countries the challenge is to address gender issues, namely the lack of interest of girls in science topics. In some nations the problem is to reawaken in young generations the interest for science. In developing nations problems can range from bad Internet connections to the availability of electricity. Being so different in their nature, they all share one point in common: challenges can all be addressed and solutions found if we work as a community. Making our best resources and tools available to all and discussing innovative solutions and building on already known best practices are the most effective means to overcome the so-called digital divide. We are building bridges between nations, bridges between highly technological and traditional classrooms, bridges between different cultures and races. This is the vision for the global efforts of GTTP.

Figure 1.   A mosaic of training venues across the world.

2.   Partnerships with Other Programs and Institutions

Participating in conferences around the world, it is very easy to notice the vast amount of innovative and very creative programs and challenges being offered to the public in general and the school community in particular. The institutions behind these efforts are producing amazing resources and making them freely available on the web. This is obviously amazing and impossible to even imagine one or two decades ago. However, they miss some important aspects of our modern society, the ever-increasing distance between the developed world and those striving to get themselves started in the use of digital content. For educators, the possibilities are overwhelming, and on many occasions, they can't bring themselves to choose where to invest their learning efforts. In spite the fact that the tools and resources are freely available on the web, they demand a considerable initial effort on the part of the user, and the multitude of choices often makes it difficult for educators to make a decision. GTTP is aiming to help bring these initiatives to all corners of the world to help make available the needed resources and get promoters started and engaged in some of the offered programs. GTTP believes that the best approach to address this issue is to use a cooperative approach and try to merge similar initiatives, promoted by different institutes, into the same "dressing," bringing them to our audience in the most effective way. Some examples of partnerships we are establishing by promoting in our training events and publicizing with our partners include the following programs:

IASC: The International Astronomical Search Collaboration[1] is an educational outreach program for high schools and colleges, provided at no cost to the participating schools. Students involved in this program experience the thrill of research and discovery while looking for new asteroids or confirming the path of already known ones. Students learn to analyze images and to spot possible new or existing objects.

---

[1]http://iasc.hsutx.edu/

Dark Skies Rangers: This is a program promoted by the National Optical Astronomy Observatory in which students learn about the importance of preserving the dark night skies,[2] the impact of light pollution in our daily lives, and learn how to advocate for proper lighting. After actively promoting awareness of the problem in their local communities, they become Dark Skies Rangers certified by the NOAO.

Astronomers Without Borders: GTTP is working in partnership with Astronomers Without Borders[3] to introduce an education component in programs like GAM (Global Astronomy Month), such as the GTTP MoonDays during GAM2011. A whole new set of materials that are being prepared for 2012 and beyond.

3. Sponsorship Schemes

An effective way to create, enhance, and maintain a network is by promoting regional training events followed by national and local training venues. In regional training events we try to gather promoters from different countries (in the same region or using the same language). The next step is to provide seed support for the promotion of a national/local event. In these events, promoters will spread the resources and tools introduced to them during the regional venues. In 2010 this was done with the support of the International Astronomical Union (IAU) and the Mani Bhaumik prize received by GTTP (12 nations were funded). In 2011 support was received from the Las Cumbres Observatory Global Telescope Network (8 venues were funded). Efforts now are beginning to use "crowd support" to guarantee support for many more partners.

4. New Tools for Communication

A new strategic plan for communication between partners and to publicize our proposals is being designed and implemented. We started a Facebook page for the program[4] and created two groups for the GTTP community exchange of information.[5] We also started the publication of a GTTP newsletter where partners can announce activities and events, publish results, share resources, etc. Participation in conferences and publication in journals is also part of the new strategy.

5. On Site and Video Conference Training

Training educators face-to-face is undoubtedly the best approach, especially when introducing highly innovative tools and resources. In-person training permits the introduction of modern research methodology in a classroom setting. But this is not always

---

[2]http://www.darkskiesawareness.org/DarkSkiesRangers/

[3]http://www.astronomerswithoutborders.org/

[4]Facebook page: www.facebook.com/galileoteachers

[5]www.facebook.com/groups/galileoteachers/ and
www.facebook.com/groups/gttp.iberoamerica/

Figure 2.   Training session in Venezuela using video conference.

possible, so GTTP is developing a methodology to achieve the same goals by training educators via video conference. We held training sessions at venues in Chile, Kenya, and Venezuela during 2011 with very promising results.

6.   Efforts to Officially Accredit GTTP Curricula

We are presently producing a set of GTTP curriculum suggestions ranging from offers to countries with broadband Internet in the classroom to the introduction of modern science using readily available materials. While designing these curricula we are also designing and implementing a strategy to obtain endorsement from several institutions as well as supporting promoters' efforts to obtain endorsement and accreditation at a national and local level. These are very important initiatives because they allow educators to build their personal curriculum while developing their activities with us.

7. Certification Criteria

Not all teacher training sessions addressing science can be called a GTTP training session. We make an effort to ensure the inclusion of modern science techniques and knowledge in all of them.

7.1. Session Certification

Sessions outlines have to be presented to GTTP and endorsed by at least two GTTP partners to be certified. The criteria, defined on the GTTP webpage, require that whenever possible sessions should: address elementary themes and/or concepts of astronomy; promote the use of tools and resources that require naked eye or small telescope observations; include hands-on activities with readily available materials; and use digital media, data mining, and robotic telescopes. A list of such tools and resources suggested by GTTP can also be found on the program website.[6]

The types of certification include the following:

- Participation certification – Participants of a GTTP training sessions are entitled to receive a participation certificate.

- Galileo Teacher – All participants exhibiting proof that they have applied some of the recommended resources in the classroom will receive a Galileo Teacher Certificate.

- Galileo Ambassador – Organizers of training sessions become Galileo Ambassadors.

The whole process is designed to create a continuing motivation for educators to participate in the GTTP network.

8. Integrating New Science Topics

GTTP is partnering with other science education efforts that are building very innovative resources for science education in Europe. **Discover the Cosmos** is a project funded by the European Commission that aims to demonstrate innovative ways to involve teachers and students in the use of eScience through the use of existing electronic infrastructures in order to spark young people's interest in science. The **Connecting Classrooms to the Milkyway** is another European funded project, promoted by the European Hands-on Universe consortium, that aims to develop the first European network of radiotelescopes for education, enabling schools to explore the Milkyway through the Internet and using inquiry based science education (ISBE) in schools.

In summary, GTTP is building a strong support network for educators that takes advantage of our diversity to add strength to the collaboration.

Acknowledgments. The Galileo Teacher Training Program is now supported by Global Hands-on Universe Association. The innovative resources and tools, the training network methodology were created in resemblance of the process initiated by the European Hands-on Universe (a European Commission funded project).

---

[6]Galileo Teachers website: `www.galileoteachers.org`

Poster and exhibit hall. Photo by Paul Deans.

Connecting People to Science
ASP Conference Series, Vol. 457
Joseph B. Jensen, James G. Manning, Michael G. Gibbs, and Doris Daou, eds.
© 2012 Astronomical Society of the Pacific

The ELAA 2 Citizen Science Project: The Case for Science, Equity, and Critical Thinking in Adult English Language Instruction

Melody Basham

Mary Lou Fulton Teacher's College, Arizona State University, Tempe, Arizona 85287, USA

Abstract.   This article summarizes a paper presented at the recent ASP conference *Connecting People to Science* in Baltimore 2011. This action research study currently in progress aims to explore the impact of integrating science into English language instruction (English Language Acquisition for Adults, or ELLA) serving largely Hispanic immigrants at an adult learning center based in Phoenix, Arizona.

1.   Science and Equity in English Language Learners

The National Science Education Standards (1996) states that the commitment to science is to include those populations who traditionally have not received the opportunity to pursue science. Yet, we must question if this is indeed the case for ELL (English Language Learners) students.

In the U.S. we must ask ourselves to what extent are we promoting the idea that science is only for those who achieve a certain level of English proficiency. There does appear to be an existing mindset within education concerning the ability of ELL students to learn science, while at the same time learning a second language. In Arizona and California, bilingual education was replaced with English immersion programs serving K–12 students based on the belief bilingual programs are not effective. This policy requires students to spend a significant amount of their school time learning English only before learning other content. There are many educators who believe that this policy can only widen the achievement gap placing ELL students at risk of falling behind native English speaking students in content areas such as science.

High stakes testing has further contributed to the problem as teachers struggle to teach to the test while putting less emphasis on those subjects that are not viewed as critical in meeting the testing standards. One must examine the potential ramifications of the NCLB (No Child Left Behind) policy, particularly in regards to how it is decreasing the number of hours of science being taught at the elementary level and how it is impacting what teachers teach, and in what language. Florin Mihai, a researcher in multicultural education, points out that "the state requires ELLs to be assessed in the content areas while they are learning English, not after they have reached full proficiency in English" (Mihai 2010). Yet it appears this is not the case in many K–12 districts, and it is not limited to just K–12 ELL students as it is also seen in ELAA (English Language Acquisition for Adults), which is the focus of this study.

Currently in the Arizona Adult Education State Standards we can find science and the social sciences as part of the curriculum for ABE and GED English speaking students who are at the highest levels (levels 5–6) of English acquisition. Science, however, is not an option for those at levels 1–4 who make up 75% of the students attending the new Rio Salado 7th Avenue Green Learning center based in Phoenix, Arizona, which is the setting for this study. One might argue that this is perhaps a political issue as much as it is a pedagogical issue, in that it appears there is an exclusion of certain groups from science content based primarily on perceptions of the ability and potential of the English as a Second Language (ESL) learner, whether a K–12 student or an adult learner.

This study came about partly in response to the Rio Salado College's present initiative to implement content-based English instruction to introduce the concepts of sustainability, geosciences, atmospheric, and environmental sciences to their adult learners to encourage students to pursue careers in science and green technology.

2.   Creating Pathways to Science and Higher Order Thinking

How might science promote higher order thinking in the ESL classroom that will lead the English learner to learning beyond the grammar book? What would happen if we were to introduce higher order thinking at the beginning of adult English instruction rather than the end?

English language researcher Margo Gottlieb states that the present goal of English proficiency is centered around everyday living skills; but this is not sufficient in the promoting of academic learning, or what she defines as academic language proficiency. English instruction needs to be integrated with academic content through the use of tools that that are appropriate for each student's English proficiency level (Gottlieb 2006). If content is adapted, we can create a pathway for all ELL learners to content regardless of their level or stage of language proficiency. Cognitive psychologist Lev Vygotsky was a pioneer in this line of thinking and stated that if a student's knowledge of content is organized only around everyday experience (or is non-scientific), then the student's ability to implement cognitive thinking is limited. He goes on to state that language associated with everyday experience can actually interfere with the process of achieving higher cognitive skills that are needed in science and other areas of life. Vygotsky believed that the development of scientific thought had to be taught and could be achieved through tools of mediation and the merging of thought, language, and instruction (Wells 1994).

In this study, the semiotic tool of choice is the concept map, which is being used as both an assessment tool and as a method of teaching both content and language. Vygotsky also viewed the concept map as an effective symbolic system that allowed the externalization of human thought processes and provided a dynamic process of making meaning (Aguilar 2008). Personal Meaning Mapping (PMM) is a variation of concept mapping developed by John Falk (Institute for Learning Innovation based in Maryland), and is used in informal learning environments. This method is based on current cognitive and neural science research that shows learning is a relative and constructive process (Falk 2007).

3.   The ELAA 2 Explorers: Citizen Science at the Rio Salado Green Learning Center

In a recent class with my Level 2 adult English language learners, we were engaged, or at least trying to be engaged, in the process of learning about prepositions. Somehow we ended up on the topic of our solar system. I proposed the question using the preposition *around*. "Does the Earth go *around* the Sun or does the Sun go *around* the Earth?" I was not sure what the answers would be. Out of a class of 15 students, only one Hispanic student believed the earth rotated around the sun. This student stated she knew the answer to that question because she learned it while helping her children with homework. I asked another question, "Is the sun a star?" No one was able to answer this one. "It can't be a star," came the response. Yet, they concluded it wasn't a planet either. So what exactly was it? I was met with disbelief when I explained that indeed the sun was a star. This is an example of many typical discourses that I would encounter in my classroom which led to the eventual realization that my students were indeed critical thinkers and that they did have a desire to learn more about our Earth, the universe, and their place in it.

3.1. Is There Space in the ELL Classroom for Science?

One may challenge the notion of science in the ELAA classroom as it is not yet known to what extent science will distract students from learning English, or to what degree it will impact their level of English proficiency. To get answers to these questions will most likely go beyond the current study. What I do hope to gain by the completion of the study is a better understanding of the use of tools such as the concept map in providing effective pathways for learning both English and science content, and to what degree students change their perception of science and their ability to do science.

Although science emphasizes factual knowledge and evidence-based objective research, in "citizen science" the learner is contributing to the scientific knowledge base, while at the same time forming their own meaning and drawing their own conclusions as a result of their experience. Students who do citizen science are encouraged to define what the scientific process and data they collected means to them. This, of course, makes it more of a subjective and interpretive experience and requires alternative and qualitative methods of assessment to evaluate such experiences. This action research study will be implementing a mixed method approach to assess both quantitative data as related to before and after student surveys, test scores, and qualitative data related to student perceptions gained via before and after semi-structured interviews, and before and after concept maps that will be constructed by the students.

4. Citizen Science as Cognitive Justice

Indian scholar Shiv Visvanathan presents the idea that the rights that come with citizenship also need to include "cognitive justice," which he defines as the recognition of the plurality of knowledge and the right for different forms of knowledge to co-exist (Visvanathan 1997, as cited in Leach, Scoones, & Wynne 2005). Visvanathan argues that to accomplish this, there needs to be an alternative science that enables a discourse between several different forms of knowledge that will inevitably lead to a more equitable and democratic world.

Citizen science, I believe, has the potential of serving this role by being a bridge between the scientific critical thinker, the interpretive cultural worldview of the citizen, and the English learner. Pedagogically, it can serve as a tool for promoting higher thinking. Politically, it presents a pathway to a global citizenship that has no borders.

Physicist Piet Hut proposes a "middle" ground when he discusses Husserl's "epoche," in that one needs to first let the phenomena speak for itself by suspending all judgment and all presuppositions (Hut 2001). He uses the following to illustrate his point: Galileo, when looking at how the Sun seems to revolve around the Earth, bracketed the common belief that the Earth itself is immobile. It was then easy to see that a rotating Earth and a fixed Sun would give rise to exactly the same phenomenon. By separating the phenomena from the belief structures in which these phenomena had always been embedded, he found new interpretations which opened new doors for scientific exploration (Hut 2001).

Hut further states, "the idea of stepping out of the world so to speak, in order to observe the world and your own role in it better, makes a lot of sense" (Hut 2001). Hut defines Husserl's epoche as "that of an eternal beginner," or "someone who approaches reality with a true beginner's mind...a childlike innocence that shows the world new and fresh in each moment." (Hut 2001). This, I believe, presents a powerful metaphor as we look at how participants in citizen science engage with a phenomenon from which they are not aware of the pre-existing frameworks. In this study, my Level 2 ELAA students will truly be approaching the world with a beginner's mind, and yet will be required to think as a scientist. In doing so, the cultural perspectives and voices of the students will be preserved, and at the same time serve as the foundation for which to build new perspectives and understandings.

You can join us on our journey at the ELAA 2 Explorer website.[1] You can contact the author with comments or questions at `access2discovery@gmail.com`.

Acknowledgments.   I wish to thank Dr. David Carlson for his encouragement and support in this study. Also a special thanks to Linda Putnam, Instructional Coordinator/Site Manager at the Rio Salado Green Learning Center for her passion and vision in promoting new pathways to teaching and learning for our adult English learners.

References

Aguilar Tamayo, M. F. 2008, "Novak and Vygotsky and the representation of the scientific concept," paper presented at the Third International Conference on Concept Mapping, Finland

Falk, J. H., Reinhard, E. M., Vernon, C. L., Bronnenkant, K., Deans, N. L., & Heimlich, J. E. 2007, "Why zoos & aquariums matter: Assessing the impact of a visit," Association of Zoos & Aquariums

Gottlieb, M. 2006, Assessing English Language Learners (Corwin Press)

Hut, P. 2001, "The Role of Husserl's Epoche for Science: A view from a physicist," invited paper presented at the 31st Husserl Circle conference in Bloomington, Indiana

Leach, M., Scoones, I., & Wynne, B. 2005, Science and Citizens: Globalization and the Challenge of Engagement, (Zed Books Publishing)

Mihai, F. M. 2010, Assessing English Language Learners in the Content Areas, (University of Michigan Press)

Wells, G. 1994, "Learning and Teaching Scientific Concepts: Vygotsky's ideas revisited," paper presented at the conference on Vygotsky and the Human Sciences in Moscow, September, 1994

National Science Education Standards, 1996

National Academy of the Sciences, `http://www.literacynet.org/science/scientificliteracy.html`

---

[1] `https://sites.google.com/site/elaa2citizenscienceproject/`

Connecting People to Science
ASP Conference Series, Vol. 457
Joseph B. Jensen, James G. Manning, Michael G. Gibbs, and Doris Daou, eds.
© *2012 Astronomical Society of the Pacific*

Amidst the Beauty of the Night Sky, which of the Constellations am I?

Kimberly A. Herrmann

Lowell Observatory, 1400 West Mars Hill Road, Flagstaff, Arizona 86001, USA

Abstract.

A well known constellation am I–
I *never set in the northern sky.*
Native Americans and Greeks, you see,
Wrote legends when they recognized me.
Two close stars still test for keen eyesight.
Two point to the North Star—what a light!
Look for my *galaxies*, you know where.
I am ____ _____, the _____ ____!

(©Kimberly A. Herrmann, 2003)

I have loved rhyming poetry ever since I can remember—from Dr. Seuss's *Green Eggs and Ham* and Maurice Sendak's *Chicken Soup with Rice*, to the works of Shel Silverstein and Jack Prelutsky. It was not until high school, though, that I realized that I could combine two of my loves—those for poetry and astronomy. Since then, I have written almost 100 astronomical riddles and always attempted to include as much astronomy content as possible—even in the riddles about constellations. Consequently, I have found them effective in teaching or reviewing aspects of astronomy in several venues, including college astronomy courses for non-science majors and events with elementary school students. More recently I have used 23 constellation riddles to create informative and entertaining seating slides that have been playing at Lowell Observatory and the Youngstown Planetarium. I hope to have these slides played at other venues as well and also hope to publish my riddles as a series of books someday.

1. Using Astronomy Riddles at Penn State University

As an astronomy and astrophysics graduate student at Pennsylvania State University, I used my astronomy riddles as part of two courses for non-science majors as well as posting 20 around a hallway for the casual visitor's "edutainment."

I was invited to be one of a five member team to create an on-line course that delivers the same content as a traditional introductory astronomy class in four units (Basic Astronomy and the Nighttime Sky, Our Solar System, Stars and the Milky Way Galaxy, and Extragalactic Astronomy and Cosmology) but through an *interactive science fiction story* (Herrmann et al. 2008). Each unit follows the educational adventure of a different fictional "Astro 101" student who has been "abducted" by aliens. At the end of all but one of the total 35 subsections, a "Riddler" alien poses one of my riddles to review the content of that specific section. After the correct answer is given, the Riddler proceeds to ask additional questions about the clues in the riddle. Each unit also contains four of my constellation riddles. Roughly 700 students have taken this course every semester since Spring of 2007 and it seems to be very popular with the students.

Additionally, twice I taught two sections of a one-credit astronomy lab, and I used my riddles as a fun way to teach constellations. With each lab, I assigned one riddle and, as part of the lab grade, the students were required to give: (1) the official Latin name (e.g., Leo), (2) the English explanation (e.g., the Lion), (3) the student's source (just knowing it was perfectly valid), (4) a rough sketch of the brightest stars, and (5) a list of three neighboring constellations. For extra credit, the students could answer questions about the clues. For example, the questions about the clues for the riddle at the opening of this article are:

1. What is a term for constellations that *"never set in the northern sky?"*

2. What *"Two close stars still test for keen eyesight?"*

3. What two stars *"point to the North Star?"*

4. What *"galaxies"* are contained in the boundaries of this constellation?

2. Answers for the Constellation Riddle and Clue Questions (Spoiler alert!)

In case you did not figure it out, the answer to the above riddle is Ursa Major, the Great Bear, and the answers for the clue questions are:

1. Constellations that *"never set in the northern sky"* are called *circumpolar* because, for a specific latitude range, they are close enough to a pole that they are always above the horizon. There are also southern circumpolar constellations.

2. The two close stars that *"still test for keen eyesight"* are Alcor and Mizar; however it is thought that they are not a true binary. Mizar is itself a quintuple star system and easily shows up as headlights (two stars) in a small telescope.

3. Merak and Dubhe (the two bowl stars not connected to the handle of the Big Dipper) *"point to the North Star,"* also known as Polaris.

4. Ursa Major contains many *"galaxies,"* including M81, M82, M101, M108, M109, and thousands more in the Hubble Deep Field!

3. Using Astronomy Riddles for Other Audiences

During the past three years that I have been a postdoctoral fellow at Lowell Observatory, I have been using my astronomy riddles somewhat differently. I still use some (primarily the 13 solar system ones) with classes, but now it is normally with elementary school students as part of The Lowell Observatory Navajo-Hopi Astronomy Outreach Program (see the article about this program, including its own riddle, p. 257 in these proceedings). I also know that several K–12 teachers who have attended week-long summer astronomy professional development workshops at Penn State have requested (and been given) copies of a selection of my constellation riddles to take home and use with their classes (the riddles posted around the hallway have attracted the attention of some attendees). At least one of these teachers contacted me to let me know that she used them to inspire her students to write their own riddles, but unfortunately I have had no contact with any of the other teachers.

I have given several popular talks about astronomy by just going through selections of my riddles. Also, for the past four years I have not given a research presentation (even at international conferences attended by some of the world's experts in their fields) without starting with a content-summarizing riddle.

4. Slideshows of 23 Constellation Riddles

Most recently, I have created automatically playing slideshows, complete with instrumental music and animations of the riddle text, images, and clue explanations. For the past two years, these riddles have routinely played as widescreen "seating" slides before feature presentations at Lowell Observatory, and more recently, at the Youngstown Planetarium. At Lowell, there is one set of ten riddles for each month which play in the evenings and four sets (one for each season) of eight riddles which play during the day.

The slides for each riddle took ~8 hours to create, all on my own time. First I collected images from the Internet to illustrate the riddles and tried to use pictures from Wikimedia Commons and screenshots from Stellarium as much as possible since those are freely useable images. Whenever possible I credited the source of an image. I tried to use a different color and font color for the text of each riddle but used white Comic Sans for all the clue explanations. Next I animated the individual components, consistently checking that a reader should have enough time to look at each section before the next part appeared.

Figures 1 and 2 show an example set of slides for the Ursa Major riddle. However, there is significantly more animation than is illustrated here. In each riddle, no more than one riddle line appears at any one time and sometimes only parts of a line show up at a time. For example, in the second slide, "Native Americans" dissolves in first, followed by the corresponding image, then "and Greeks," and the corresponding image, then "you see," then the fourth line, and lastly the explanation text. Similarly, on the first slide each of the four images of the Big Dipper shows up one at a time in a counter-clockwise motion followed by the "because I'm circumpolar!" explanation. Furthermore, the images on slides 3–5 come up a little at a time. The text of the answer always drops down from the top one letter at a time after the constellation image is revealed. See Table 1 for a list of the 23 constellations that currently have slideshows, as well as the months when they play at Lowell Observatory and some astronomy concepts highlighted per riddle.

Each riddle is set to instrumental music, which significantly adds to the effect. Lowell Observatory subscribes to Broadcast Music, Inc., so I only used music that is registered through that association. I always tried to choose a piece of music that was relevant to the constellation, such as music from Disney's animated *Hercules* for the Strongman and tracks from *Dragonheart* for Draco. Often I defaulted to music by Enya when I could not think of an obvious music choice. Oddly enough, this was the case for Ursa Major and I ended up using "The Memory of Trees." Since then I have learned of a legend about three birds (instead of hunters) following the bear and succeeding in killing it only just before the start of autumn, causing the bear's blood to fall down from the sky and redden the leaves of trees as well as the robin's breast.

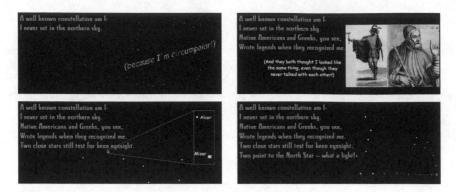

Figure 1.   The first four of eight sample slides for the Ursa Major riddle. See the text for more details about the additional animations on each slide. All the images came from Wikimedia Commons or were created using PowerPoint tools.

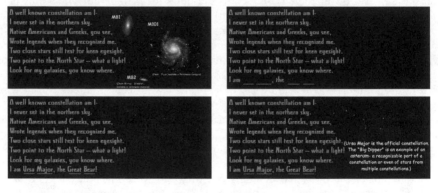

Figure 2.   The last four of eight sample slides for the Ursa Major riddle. All the images came from Wikimedia Commons or Stellarium.

5.   Concluding Remarks

It is interesting to note that the first riddles I wrote were actually ten about birds—in French! In my junior year of high school, my French teacher assigned each member of my class to write (and illustrate) a children's book. While I did not have any ideas for a short story at the time, I thought it would be fun to write some riddles. I first wrote the riddles (in AABB rhyming format with 9 syllables per line) in French and then translated them as closely as possible into English, while still maintaining the poetic format. The following summer I started writing constellation riddles. Since I had already illustrated the bird riddles, I recently constructed fewer than 20 small books of these riddles which I printed myself and then had spiral bound. However, it is a dream of mine to have my bird and astronomy riddles published someday.

Acknowledgments.   I would like to the thank the ASP for giving me a chance to share some of my astronomy riddles with conference attendees.

Table 1.   Table of constellations presented in slideshow format at Lowell Observatory

Constellation	Month(s): Main Astronomy Concepts
Andromeda	Sept–Feb: Local Group galaxies, Mythology
Aquila	July–Nov: The Milky Way, Summer Triangle, Coat Hanger
Auriga	Dec–May: Changing night sky, Open Clusters
Boötes	April–Aug: Arcturus: finding it & stellar properties
Canis Major	Jan–April: Sirius (several properties), M41
Cassiopeia	Sept–March: Mythology, Cass A, EM Spectrum
Cepheus	Sept–Dec: Cepheid variables, Circumpolarity
Corona Borealis	May–June: Mythology, Eclipsing binaries
Cygnus	June–Nov: Deneb, Albireo, North American & Veil Nebulae
Draco	May–Aug: Thuban, Precession, Rastaban
Gemini	Dec–May: M35, NGC 2392, Pollux (exoplanets), Castor (binaries), Ecliptic
Hercules	May–Aug: Zenith, Mythology, M13 & M92, Ras Algethi, Asterism
Leo	Feb–July: Ancient Astronomy, Alternate cultures, Regulus, Asterism
Lyra	June–Nov: Vega, Binary Stars, Ring Nebula
Orion	Dec–April: Orion Nebula, Mayan astronomy, Betelgeuse, Rigel
Pegasus	Sept–Jan: Mythology, M15, 51 Pegasi (exoplanets)
Perseus	March: Perseids, Double Cluster, Mythology, Algol, Eclipsing Binaries
Sagittarius	Aug–Sept: Changing sky, Milky Way, Asterism, Nebulae, Clusters, Ecliptic
Scorpius	July–Aug: Mythology, Antares, Star Clusters
Taurus	Oct–April: Pleiades, Crab Nebula, Aldebaran, Hyades
Ursa Major	Jan–Dec: Cross Cultural Mythology, Alcor & Mizar, Pointers, Galaxies
Ursa Minor	Jan–Dec: Circumpolarity, Latitude, Precession, Polaris as 49th brightest star
Virgo	April–June: Ecliptic, Cross Cultural Mythology, M104, Virgo Cluster, Spica

References

Herrmann, K. A., Hunter, D. A., Bosh, A. S., Johnson, M., & Schindler, K. 2012, "The Lowell Observatory Navajo-Hopi Astronomy Outreach Program,", Connecting People to Science, J. B. Jensen, J. G. Manning, M. G. Gibbs, & D. Daou, eds., ASP Conference Series, 457, 257

Herrmann, K. A., Palma, C., Charlton, J. C., & Narayanan, A. 2008, "Astro 001: Interactive, Online, and with a Sci-Fi Storyline," EPO and a Changing World: Creating Linkages and Expanding Partnerships, C. Garmany, M. G. Gibbs, & J. W. Moody, eds., ASP Conference Series, 389, 379

National Earth Science Teachers Association exhibit. Photo by Paul Deans.

Connecting People to Science
ASP Conference Series, Vol. 457
Joseph B. Jensen, James G. Manning, Michael G. Gibbs, and Doris Daou, eds.
© 2012 Astronomical Society of the Pacific

Improving the Pipeline of Women in STEM Fields: Addressing Challenges in Instruction, Engagement, and Evaluation of an Aerospace Workshop Series for Girl Scouts

Carolyn D. Sealfon[1] and Julia D. Plummer[2]

[1]*Department of Physics, West Chester University, West Chester, Pennsylvania 19383, USA*

[2]*Arcadia University, 450 S. Easton Road, Glenside, Pennsylvania 19038, USA*

Abstract.   The Women in Aerospace and Technology Project (WATP) is a collaborative effort between the Girl Scouts of Eastern Pennsylvania, the American Helicopter Museum, Boeing Rotorcraft, Sikorsky Global Helicopters, Drexel University, West Chester University, and Arcadia University. The program aims to increase the representation of women in STEM (Science, Technology, Engineering, and Math) fields; the evaluation team identified a secondary goal to assess growth in participants' understanding of scientific inquiry. Girls, grades 4–12, were invited to join Girl Scout troops formed at the American Helicopter Museum to participate in a series of eight workshops on the physics and engineering of flight. Five college women majoring in physics and engineering were recruited as mentors for the girls. Lessons were written by local aerospace industry partners (including Boeing and Sikorsky); the mentors then taught the lessons and activities during the workshops.

To evaluate the impact of this project, we collected data to answer two research questions: 1) In what ways does the program impact participants' attitudes towards science and interest in pursuing science as a career? 2) In what ways does the program impact participants' understanding of the nature of scientific inquiry? In this article we summarize results from two sources of data: before and after survey of attitudes about science and end-of-workshop informal questionnaires. Across the seven months of data collection, two challenges became apparent. First, our assessment goals, focusing on scientific interest and inquiry, seemed misaligned with the workshop curricula, which emphasized engineering and design. Secondly, there was little connection among activities within workshops and across the program.

1. WATP Program Description

The American Helicopter Museum in West Chester, Pennsylvania started the Women in Aerospace and Technology Project (WATP) to help address the underrepresentation of women in science and technology fields. Two new Girl Scout troops were created with a science and technology focus. These troops met at the museum every other Saturday afternoon for two hours. Local engineers from Boeing Rotorcraft and Sikorsky Global Helicopters created lesson plans and activities for the Girl Scouts and took them on field trips to visit their factories. The lesson plans were implemented and taught by women science and engineering majors at Drexel University and West Chester University. These college students, in turn, were mentored by women physics or engineering faculty and had the opportunity to job-shadow engineers at Boeing and Sikorsky. The long-term vision for the program is to offer an appealing and continuous pathway for women to enter aerospace or technology, starting in 4th grade, through college, to eventual employment in industry.

2. Research Questions and Methodology

To assess the effectiveness of this program in its inaugural year, we focused on two research questions: 1) In what ways does the program impact participants' attitudes towards science and interest in pursuing science as a career? 2) In what ways does the program impact participants' understanding of the nature of scientific inquiry?

To evaluate the program's impact on participants' attitudes towards science, we used the Developing Attitudes Towards Science Measures (DASM) instrument developed by Kind, Jones, & Barmby (2007) as a before and after survey. This survey consists of 46 Likert questions (see Appendix), subdivided into seven factors: science in school, self-concept in science, hands-on activities, science out of school, future participation in science, importance of science, and school in general. Students ranked each statement on a scale from 1 to 5 where 1 corresponds to "strongly disagree" and 5 corresponds to "strongly agree." The factors dealing with in-school attitudes would not have been directly influenced by the WATP, and thus changes in these factors serve as a baseline for comparison to changes in relevant factors, such as "future participation in science." Pre and post-surveys were completed by 20 of the participants.

We were also interested in how the program impacted participants' understanding of the nature of scientific inquiry. We gathered data, including the Views of Scientific Inquiry instrument (Schwartz, Lederman, & Lederman 2008), field notes from the program, and end-of-workshop questionnaires to help address this question. In this paper, we focus on the written responses to two open-ended end-of-workshop questions. After each of the five workshops at the American Helicopter Museum, girls were asked: 1) What new questions do you have about [workshop topic]? 2) What new investigation or experiment would you like to try? Between 17 and 27 participants responded to these questions each week.

Table 1.   Categories for student questions (end-of-workshop question #1)

Category	Descriptors
Factual	Recall of information, close-ended questions. Example: *Who flew the fastest helicopter?*
Relationship	Question that attempts to find connections between two or more concepts. Example: *How fast does a helicopter's propellers have to spin to get enough thrust to move the vehicle into the air?*
Investigation	Question could lead to an investigation; may be a comparison, cause & effect, predictive, or descriptive. Example: *If you pushed the nose of a plane down, would you speed up?*
Design	Question refers to how something is made, designed, or tested for functionality. Example: *Can you make an airplane out of sturdy foam?*

The first question was used to examine the type of questions participants proposed. Prior research has examined the type of questions posed by students, the impact of teaching questioning skills on students' abilities, how students' questions relate to other factors, and the interplay between teachers' response and students' perception of students' questions (Chin & Osborne 2008). Students' questions "indicate that they have been thinking about the ideas presented and have been trying to link them with other things they know" (Chin & Osborne 2008). Thus, they can be used to interpret student thinking, and interest and can be used to evaluate higher-order thinking (Dori & Herscovitz 1999). Prior research has also found that research questions posed by students can be used to assess their understanding of questions that are ready for an inquiry-based investigation (Chin & Kayalvizhi 2002). A coding scheme was developed, based on prior literature (Chin & Brown 2002) and patterns in the data, to categorize the types of questions posed by the students (Table 1).

The second question was used to identify participants' ability to suggest their own investigations, based on the experiences from that day's workshop. Designing investigations is a key scientific practice in the new Framework for K–12 Science Education (National Research Council 2011) and is central to proficiency in scientific inquiry. A second coding scheme was developed to categorize students' responses to this question through the patterns that emerged in student responses (Table 2).

Table 2.   Categories for student investigations (end-of-workshop question #2)

Category	Descriptors
Investigation	Student suggests a question or problem statement that would lead to an investigation. Example: *If you put helium in the balloon powered rotor, will it go higher or lower than something else?*
Complex Design	Student specifies the purpose of their design. Example: *I would like to try and build a robot that will do things that a human can't do.*
Simple Design	Student suggests something they would like to design or build but not the purpose behind that suggestion. Example: *I would like to make a robotic helicopter.*
Interest	Student suggests something they want to learn without describing a way they could investigate. Example: *Can a helicopter fly backwards?*
Method of learning	Student suggests a way to think they could find out about something they are interested in that is not a scientific investigation or experiment. Example: *We could go inside a helicopter and learn about the controls.*

3. Results

3.1. Attitudes Assessment

The DASM survey indicates no significant change from pre-test to post-test (see Figures 1 and 2). However, the consistency from pre-test to post-test indicates some general trends in participants' attitudes. For example, students overwhelmingly enjoy hands-on activities, and express slightly negative views towards becoming a science teacher.

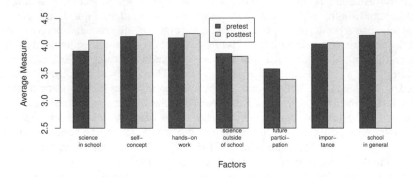

Figure 1.   Average pre-test and post-test measures for each of the seven attitude factors from the DASM on a scale from 1 to 5, where 5 represents the most positive attitude towards science. Dark gray bars represent pretest averages and light gray bars represent post-test averages (see Appendix for the questions in each factor).

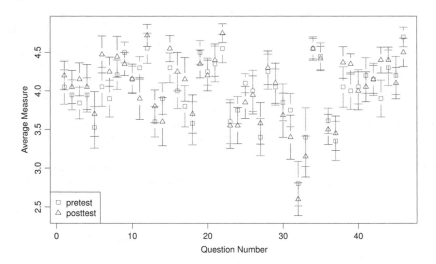

Figure 2.   Average pre-test and post-test measures with standard deviations for each of the 46 questions on the DASM. Squares denote average pre-test scores and triangles denote average post-test scores on a scale from 1 to 5, where 5 represents attitudes most positive towards science (see Appendix for individual questions by number).

3.2.   Nature of Science

Table 3 shows the range of students' questions after each workshop. While there was some fluctuation, potentially due to a combination of the individual workshop topic or overall time in the project, some consistent patterns emerged looking across the weeks. The most frequent type of question posed by students was factual, while a smaller fraction of questions were more complex relational, investigative, or design-based questions. However, about a third of the responses were "no question," suggesting that students either were not connected or interested enough in the topic to pose a question, or were not practiced in posing their own questions. There is some indication that the percentage of students without questions decreased by the end of the program, though it is unclear if this is due to the intervention increasing their interest/ability to pose questions, or if this is due to attrition of students who were not interested in continuing in the program. Based on these findings, we conclude that students need training on how to ask questions. Further, we may also consider the lack of questions as an indication of the level of engagement and interest in the workshops.

Table 3.   What new questions do you have about [workshop topic]?

Category	Week 1	Week 2	Week 3	Week 4	Week 5	Total
Factual	27%	27%	32%	40%	33%	30%
Relationship	0	18%	5%	30%	19%	13%
Investigation	3%	5%	0	10%	0	3%
Design	17%	5%	5%	0	0	6%
Other	13%	5%	11%	10%	29%	13%
None	40%	41%	47%	20%	19%	35%
Total # of responses	30	22	19	17	21	109

Table 4 shows the range of responses generated by participants about the type of new investigation they were interested in attempting. The findings suggest that most participants did not have a clear understanding of how to formulate a potential investigation, and that the workshops were not successful in improving this level of knowledge. The types of questions/statements posed by the students are somewhat consistent with prior research that found that elementary students, who worked together to generate their own questions for investigations, either pursued a variation on the preceding classroom activity or invested a new question based on their own personal interest and ideas (Keys 1998). Participants in this study built off the themes of the workshops by proposing design-based problems (instead of investigations), as well as suggesting concepts in which they were interested that were not necessarily ideas they could investigate or problem-solve.

Our initial design of this assessment question did not consider that students' experience was often focused on design problems rather than investigations. In future work, the assessment should better reflect the design of the workshops. Further, these results suggest that the participants also need more explicit instruction on how to propose their own investigations and design problems.

4.   Discussion and Future Directions

As both authors are scientists, our assessments focused on the nature of scientific inquiry. However, the lessons and activities designed by industry emphasized engineering and design. Thus our assessments were mismatched to the program.

Table 4.   What new investigation or experiment would you like to try?

Category	Week 1	Week2	Week 3	Week 4	Week 5	Total
Investigation	27%	10%	35%	28%	5%	21%
Complex design	11%	9%	10%	19%	14%	13%
Simple design	26%	5%	15%	10%	5%	23%
Interest	26%	32%	10%	14%	19%	21%
Method	0	0	10%	19%	0	5%
Other	11%	14%	10%	5%	5%	9%
None	0	32%	10%	5%	0	9%
Total # of responses	30	22	20	21	21	109

We believe the program would benefit from better connections among activities within workshops and across the program. Part of the power of science and engineering is the interrelationships among ideas, yet these connections often need to be explicitly addressed for students to recognize them. Many of the lessons and activities were presented as isolated components of the program.

Lastly, the initial workshops involved more lecture and demonstrations with fewer hands-on activities. The lessons could have been improved by building more directly on current research on best-practices in science and engineering education. Further, student mentors would have benefited from additional support in developing their own pedagogical content knowledge. Workshops could also be designed to better train participants in posing questions and proposing investigations or design problems.

Acknowledgments.   We thank all of the partners in the WATP project, including Michael Spletzer, who started and managed the program, the American Helicopter Museum, the volunteer engineers from Sikorsky and Boeing, the student mentors, and the participating Girl Scouts.

References

Chin, C. & Brown, D. E. 2002, "Student-generated questions: A meaningful aspect of learning in science," International Journal of Science Education, 24, 5, 521

Chin, C. & Kayalvixhi, G. 2002, "Posing problems for open investigations: What questions do pupils ask?" Research in Science & Technological Education, 20, 2, 269

Chin, C. & Osborne, J. 2008, "Students' questions: a potential resource for teaching and learning science," Studies in Science Education, 44, 1, 1

Dori, Y. J. & Herscovitz, O. 1999, "Question-posing capability as an alternative evaluation method: Analysis of an environmental case study,", Journal of Research in Science Teaching, 36, 4, 411

Keys, C. W. 1998, "A study of grade six students generating questions and plans for open-ended science investigations," Research in Science Education, 28, 3, 301

Kind, P., Jones, K., & Barmby, P. 2007, "Developing attitudes towards science measures," International Journal of Science Education, 29, 7, 871

Schwartz, R. S., Lederman, N., & Lederman, J. 2008, "An instrument to assess views of scientific inquiry: The VOSI questionaire," International Conference of the National Association for Research in Science Teaching

National Research Council 2011, "A framework for K–12 science education: Practices, crosscutting concepts, and core ideas"

Appendix: DASM Attitude Survey Questions

Survey adapted from (Kind, P., Jones, K., & Barmby, P. 2007)

Factor 1: Science in school

1) We learn interesting things in science lessons.
2) I look forward to my science lessons.
3) Science lessons are exciting.
4) I would like to do more science at school.
5) I like science better than most other subjects at school.
6) Science is boring.

Factor 2: Self-concept in science

7) I find science difficult.
8) I am just not good at science.
9) I get good grades in science.
10) I learn Science quickly.
11) Science is one of my best subjects.
12) I feel helpless when doing science.
13) In my science class, I understand everything.

Factor 3: Hands-on activities

14) We often do hands-on activities in my science lessons.
15) Doing hands-on activities in science is exciting.
16) I like doing science hands-on activities because you don't know what will happen.
17) Doing hands-on activities in science is good because I can work with my friends.
18) I like doing hands-on activities in science because I can decide what to do myself.
19) I would like more hands-on activities in my science lessons.
20) We learn science better when we do hands-on activities.
21) I look forward to doing science hands-on activities.
22) Doing hands-on activities in science is boring.

Factor 4: Science outside of school

23) I would like to join a science club.
24) I like watching science programs on TV.
25) I like to visit science museums.
26) I would like to do more science activities outside school.
27) I like reading science magazines and books.
28) It is exciting to learn about new things happening in science.

Factor 5: Future participation in science

29) I would like to study more science in the future.
30) I would like to study science at university.
31) I would like to have a job working with science.
32) I would like to become a science teacher.
33) I would like to become a scientist.

Factor 6: Importance of science

34) Science and technology is important for society.
35) Science and technology makes our lives easier and more comfortable.
36) The benefits of science are greater than the harmful effects.
37) Science and technology are helping the poor.
38) There are many exciting things happening in science and technology.

Factor 7: School in general

39) I really like school.
40) I would recommend my school to other students.
41) I find school boring.
42) I feel that I belong at my school.
43) Most of the time I wish I wasn't in school at all.
44) I get on well with most of my teachers.
45) I am normally happy when I am in school.
46) I work as hard as I can in school.

Welcome Reception at the Maryland Science Center. Photo by Paul Deans.

Connecting People to Science
ASP Conference Series, Vol. 457
Joseph B. Jensen, James G. Manning, Michael G. Gibbs, and Doris Daou, eds.
©*2012 Astronomical Society of the Pacific*

The Unknown Moon: Eliminating Misconceptions and Strengthening Lunar Science Literacy in the Classroom

M. Alexandra Matiella Novak

Johns Hopkins University Applied Physics Laboratory, 11100 Johns Hopkins Road, Laurel, Maryland 20723, USA

Abstract. Our Moon is an ideal tool for teaching about space science and Earth's place in our solar system. The Moon remains the most studied object in our solar system and the only other body, besides Earth, from which humans have collected field samples. Despite our long history of studying the Moon, there are still many questions that remain unanswered. Most students believe that we know everything there is to know about the Moon, but in actuality it still remains very much "unknown." For example, recent radar observations of the lunar poles suggest the presence of water ice, but the quantity remains unknown. Additionally, remote-sensing analysis of the Moon's regolith suggests the presence of chemicals that can be used as resources when manned-missions return to the Moon, but how we would access those resources remains unknown. These studies and many more need to be shared with students in a way that regenerates excitement for future exploration of our Moon and our solar system. Common lunar misconceptions and ideas for establishing Lunar Science Literacy Concepts (LSLC) will be discussed in this paper.

1. Introduction

Our Moon, our closest neighbor in this vast universe, is one of the most studied bodies in the solar system, and the only other body, besides Earth, from which humans have collected field specimens. Despite the fact that we've studied the Moon so closely, for so long throughout human history, it still remains very much "unknown." Many questions about the Moon still persist within the lunar scientific community such as, is there water on the Moon? And, what types of technologies are required for a permanent manned-outpost on the Moon? The answers to these questions will help prepare the next generation of astronauts and solar system explorers, as the Moon is a naturally good site for testing new exploration technologies and serving as a waypoint between Earth and other planets.

In order to effectively teach lunar science in the classroom, we must first address misconceptions that have found their way into our students' psyches. According to the National Research Council (NRC 1997), successful ways to correct misconceptions are to: 1) anticipate the most common misconceptions; 2) encourage students to discuss their conceptual framework using evidence and tests; 3) address misconceptions with demonstrations and lab work; 4) use assessment and reassessment to validate student concepts. All of these techniques can be easily integrated with lunar science and discussions on the Earth-Moon relationship in the classroom.

Another tool for correcting misconceptions is to establish a baseline lunar literacy framework that outlines the absolute basic concepts students must understand about the Moon and its relationship to Earth. This paper offers a preliminary list of Lunar Science Literacy Concepts (LSLC) developed in collaboration with current lunar scientists. The LSLC can be used as an introduction to lunar science, or as a discussion point for investigating more complex science topics and current research questions.

As the human race continues to investigate the Moon and make new discoveries (the most common misconception being that we already know everything there is to know about the Moon), we begin to see that the Moon is even more closely related to Earth than previously thought. This makes the important task of teaching lunar science even more feasible as we can use the Moon's relationships and similarities to Earth as tools to make the Moon more familiar and more a part of our daily lives. Returning manned missions to the Moon has already been identified as an important step in future space exploration (Spudis 2005), and preparing a new generation of lunar and solar system explorers is in America's best interest. Current research and non-manned exploration of the Moon is aggressive and more lunar exploration missions are planned for the future. The information we are gathering from these missions is in preparation for returning Americans to the Moon and keeping our nation ahead of the rest of the world in technology innovation and future space exploration.

2. Teaching Lunar Science in the Classroom

The idea of teaching lunar science in the classroom is not a new one. The National Science Education Standards (NRC 1996) Earth and Space Science Standards can be interpreted to allow for teaching lunar science at every level K–12 (Table 1).

Table 1.   Earth and Space Science Standards (NRC 1996)

Levels K–4	Levels 5–8	Levels 9–12
Properties of Earth materials	Structure of the Earth system	Energy in the Earth system
Objects in the sky	Earth's history	Geochemical cycles
Changes in Earth and sky	Earth in the solar system	Origin and evolution of the Earth system
		Origin and evolution of the universe

These science standards are already well addressed in the classroom, and incorporating lunar science into curriculum developed for these standards is straightforward. However, common misconceptions associated with lunar science can undermine the overall measurement of how well these standards are taught. For example, if a student does not understand that the Moon orbits the Earth, then that has consequences for how well the student then understands objects in the sky (K–4), Earth in the solar system (5–8) and the origin and evolution of the Earth system and the universe (9–12).

A better understanding of lunar science and the Earth-Moon relationship can only serve to strengthen students' understanding of all Earth and Space Science standards. For this reason, the Moon is an ideal and appropriate tool for addressing and assessing how well students are learning basic science concepts. However, in order to assess how well the concepts are being addressed, a list of critical concepts must first be produced.

2.1. The Moon as a Tool for Teaching

There are several reasons the Moon is an ideal tool for teaching in the classroom. First, it is an object in space that is easily found and identified. When teaching about objects in space, the inability of the student to easily identify that particular object may hinder the learning process and make the object studied seem more abstract and not directly related to the student's environment. The Moon, however, is an object a student can easily identify, and this aids in building a personal experience with the Moon and learning about lunar science. Even with a pair of moderately strong binoculars, an observer can identify features on the surface of the Moon.

Second, the most obvious fit for teaching lunar science is within disciplines related to science. The scientific nature of observing and studying the Moon makes it an appropriate topic to encourage interest in the science, technology, engineering and mathematics (STEM) disciplines. Yet discussions about the Moon and its relationship to Earth can also be addressed in non-science subjects such as art, history, government and social studies. Very often, institutions managing solar system exploration missions have an artist who uses scientific data to create accurate representations of what these far away objects must look like up close. Additionally, Americans have a unique historical and cultural connection to the Moon, being that we are the only nation to have landed humans on the Moon, and that significant accomplishment is a source of national pride.

2.2. Identifying and Correcting Misconceptions

In order to correct misconceptions about the Moon, they should first be identified through a pre-assessment activity. Trauth-Nare & Buck (2011) suggest using a Tchart to elicit students' prior knowledge and interest in science topics. Figure 1 shows an example of what a Tchart would look like for general lunar science.

Topic: The Moon	
What I know about the Moon and its relationship to Earth	Questions I have about the Moon

Figure 1.   Example T-chart for assessing prior knowledge of the Moon

For a more thorough pre-assessment, the T-chart can be broken down into categories that force the students to think about particular concepts. For example, the students are asked to identify what they know about the geologic formation of the Moon, or the Moon's movement in relation to Earth. Pre-assessment is critical in identifying and correcting misconceptions, and it is just as important as assessing the students' knowledge after instruction (Trauth-Nare & Buck 2011).

This type of activity also identifies some common misconceptions that students share. Identifying and anticipating common misconceptions is useful because the instructor can have classroom presentations and activities ready for use immediately after the pre-assessment, and little new preparation is needed from one class to the next. Table 2 identifies some common misconceptions students have about the Moon. Although the list is incomplete, it does offer an idea of misconceptions that are most likely to be encountered in the classroom.

Table 2.   Common Misconceptions in Lunar Science

1. The Moon has a dark side.
2. The Moon makes its own light.
3. The phases of the Moon.
4. The Moon affects seasons on Earth.
5. The Moon is "dead," or not geologically active.
6. We already know everything we need to know about the Moon.

Misconceptions can be corrected, but they must be corrected by their owners (Gooding & Metz 2011). In the classroom, the owners of these misconceptions are the students, and guiding them towards correcting their misconceptions requires the instructor's help. Techniques for correcting misconceptions suggested by NRC (1997) were discussed in Section 1. In addition to those, we recommend the following strategies to correcting misconceptions specific to lunar science.

First, the physical relationship between the Earth and Moon is three dimensional, and should be taught this way as well. It is very difficult to teach multi-dimensional concepts, especially to younger children, in a two dimensional format (like on a chalk board or a piece of paper). Allowing students to explore the three dimensional relationship between the Earth and the Moon also provides opportunities for hands-on activities and demonstrations.

Second, connecting lunar science to Earth science will make the Moon more familiar to students, and therefore recalling lunar science concepts will be easier. There are many similarities among the geologic processes that have shaped the Moon and those that have taken place, or are taking place, on Earth. Many comparisons can be made between lunar features, such as craters and volcanoes, and those same features on Earth. These comparisons, also called "comparative planetology," can also strengthen the idea that we formulate hypotheses of how planets evolved by picking features on other planets and making comparisons to how those same features formed on Earth.

Third, new discoveries about the Moon are frequently released, and these should be shared to highlight that the Moon is still very much a mystery and we have much to learn about it. Even though humans have explored the Moon, the area explored was relatively small—the astronauts from Apollo 11 explored an area of the Moon barely the size of a baseball field. Spacecraft are sending us new and exciting information about areas on the Moon never studied before and we should be motivated to further explore these areas with manned missions.

2.3.   Lunar Science Literacy Concepts

Another tool for teaching lunar science in the classroom is to engage the students with concepts that are considered critical for understanding the Moon, and Earth's relationship to the Moon. Table 3 lists some of these concepts, developed alongside scientists that are involved in current lunar science research. The concepts are divided into categories to illustrate the flexibility in which these concepts can be incorporated into the classroom.

Table 3.   Lunar Science Literacy Concepts (LSLC)

Categories	Concepts
Physical Characteristics (STEM classes)	-The Moon is covered by craters and volcanic material -The Moon has no atmosphere, but there are atomic particles near the surface -The Moon's gravity is 1/6 that of the Earth's -The Moon reflects light, it is not a source -The Moon has permanently shadowed regions -The Moon's temperature ranges from +120 degrees C in the day to -150 degrees C at night -The Moon is not as dry as we once thought
Relationship to Earth (Earth science classes)	-The Moon orbits around the Earth -The "near" side of the Moon is always facing Earth, but the Moon does rotate around its own axis -The Moon is about 250,000 miles away from Earth -The Moon causes ocean tides on Earth -Similar geologic processes that have taken place on Earth have also taken place on the Moon
Place in History (History, Social Studies classes)	-American Neil Armstrong was the first man on the Moon -Twelve men have walked on the Moon -1972 was the last time man walked on the Moon

This list of concepts is incomplete and will surely be updated as new technologies are developed and discoveries are made. Already, recent lunar missions have made new discoveries and have sent back images of places on the Moon that have never been seen before. For example, the Mini-RF instrument onboard the Lunar Reconnaissance Orbiter uses radar technology to "see" into permanently shadowed craters. Not only are the floors of these craters now visible to us, but more investigations of these areas have suggested that water ice may exist on these crater floors. Establishing the presence of water on the Moon would be a major discovery since water on the Moon could be used as a natural resource by future human explorers.

3. Conclusion

Teaching lunar science in the classroom is simple and can be appropriately combined with existing science education standards. Students are already well familiar with some lunar science concepts, but may still have some misconceptions. Identifying these misconceptions is the first step towards correcting them, and this can be done easily through pre-assessment of the students' prior knowledge. Furthermore, introducing the students to a set of lunar science concepts, for example the LSLC suggested in this paper, can further solidify their knowledge of the Moon and positively impact their level of science literacy.

Acknowledgments.   The author would like to thank the NASA Lunar Science Institute and the Mini-RF instrument science team for their support.

References

Gooding, J., & Metz, B., 2011, From Misconceptions to Conceptual Change, The Science Teacher, 34
National Research Council 1996, National Science Education Standards (National Academies Press)
National Research Council 1997, Science Teaching Reconsidered (National Academies Press)
Spudis, P. 2005, Solar System Science and Exploration, Johns Hopkins APL Technical Digest, 26, 4, 315
Trauth-Nare, A. & Buck, G. 2011, Assessment for Learning, The Science Teacher, 34

Connecting People to Science
ASP Conference Series, Vol. 457
Joseph B. Jensen, James G. Manning, Michael G. Gibbs, and Doris Daou, eds.
© *2012 Astronomical Society of the Pacific*

Introducing Astrophysics and Cosmology as Part of Multidisciplinary Approaches to Liberal Arts Courses Addressing "The Big Questions" of Human Experience

Joseph C. Wesney

Sacred Heart University, 5151 Park Avenue, Fairfield, Connecticut 06825, USA, wesneyj@sacredheart.edu

Wesney Educational Consulting, 6 Orchard Street, Cos Cob, Connecticut 06807, USA, joe@wesney.com

Abstract. There is an opportunity to bring college students to the exploration of the grandeur and wonder of the universe through the design and crafting of courses for the university and liberal arts curricula that would develop multidisciplinary perspectives within the frames of reference of astrophysics and cosmology. There is broad interest within colleges and universities to provide courses that examine "The Big Questions" of human experience from a variety of perspectives. The study of the history of discoveries and insights that we have gained through the development of astrophysics and cosmology provides course options for students to use to explore these questions. Such hybrid courses enable students to approach the questions of origins, human existence, appreciation of the natural world, appreciation of the universe at large, and the significance of our evolving comprehension of the universe from a variety of disciplinary perspectives, including those that border on the astrophysical and cosmological domains. There are within such courses opportunities to examine historical, philosophical, theological, and cultural perspectives as they intersect with our scientific understanding of where and who we are. The first of these courses at Sacred Heart University has been developed and presented for the past two years as part of the new Core Curriculum. The development of that course, entitled *The Journey in the Physical Universe*, will be discussed, and insights will be shared.

Starting in 2007, faculty members from the College of Arts and Sciences at Sacred Heart University (SHU) in Fairfield, Connecticut initiated a program for entering students, known as the Common Core curriculum. The description of this program as presented on the SHU website[1] is:

"Common Core: The Human Journey immerses students in a coherent and integrated understanding of the arts and sciences. It develops critical/analytical skills and abilities as well as engages students in the Catholic intellectual tradition as rigorous intellectual inquiry.

The Human Journey focuses on four fundamental and enduring questions of human meaning and value: 1. What does it mean to be human? 2. What does it mean to live a life of meaning and purpose? 3. What does it mean

---

[1]http://www.sacredheart.edu/pages/18771_common_core_the_human_journey.cfm

to understand and appreciate the natural world? 4. What does it mean to forge a more just society for the common good?"

SHU students must participate in certain courses, some of which they may select from a menu of courses offered as part of the Common Core (CC). In their freshmen year students must take foundation courses designed to build skills of academic discipline, improve clarity of thought and sharpen skills in research and communication. These include courses such as: *HICC 101: The Human Journey: Historical Paths to Civilization* and *ENCC 102: Literary Expressions of the Human Journey*. In their sophomore and junior years the students may select from a variety of courses in the *CC 103A: Social and Behavioral Sciences—The Human Community: The Individual and Society* and/or the *CC 103B: Natural Sciences—The Human Community and Science Discovery*. During their senior year the students engage in their Capstone Common Core Course, *RS/PHCC 104: The Human Search for Truth, Justice, and the Common Good*. Over the entire experience students will have taken seven or more courses in which they will examine the "four fundamental and enduring questions" of the Common Core in the context offering a rich variety of subject areas to sample and broaden their academic and intellectual horizons. Simultaneously, they will be able to hone their skills as scholars, by being engaged in academic practices related to what is known as the Catholic intellectual tradition.

"The Catholic intellectual tradition is a collection of characteristically Catholic concepts, habits, and values that have been developed in a variety of disciplines and through an intercultural conversation lasting over 2000 years. Among the Catholic intellectual tradition's central claims are that (1) humans exist in relation to a Triune God; (2) God's presence is mediated through the particular; (3) morality is objective and knowable; (4) human knowledge can be connected into a coherent whole; (5) faith and reason work together to provide understanding of the world; and (6) humans have inviolable dignity and are responsible toward the common good." (Adapted from a summary description, entitled "THE HUMAN JOURNEY: Describing the Catholic Intellectual Tradition" by Brian Stiltner, Associate Professor of Religious Studies, SHU. Additional detail regarding the CIT can be found at the SHU website.[2])

Within this academic setting the author of this paper embarked in the Fall Semester of 2008 on the development of a physics-related course to complete the set of courses under the *CC 103B: Natural Sciences—The Human Community and Science Discovery* group. Since courses in biology and chemistry had been developed previously, and were up and running, physics was the last to be added. The research, design, and development of the curriculum for the course, entitled *PYCC 103: The Journey in the Physical Universe*, took place during the Fall Semester 2008, and a seven-student pilot section was launched in the Spring Semester 2009. Formative and summative assessments of the course were conducted, and appropriate changes were made for full implementation in the 2009–10 academic year, with one regular section offered each semester. Formative and summative evaluations of curriculum and instructional design have been carried out since that time, with changes made as needed, and the course is now beginning its third year of operation, with two sections available to students each semester.

---

[2]http://www.sacredheart.edu/pages/234_the_catholic_intellectual_tradition.cfm

The design of the curriculum for *The Journey in the Physical Universe* is focused on five overall course goals which are related to the Catholic intellectual tradition and each of the four fundamental and enduring Common Core questions. These are then woven into the fabric of the course as threads of continuity as each of the topics relating to the physical universe are examined from not only the scientific perspectives, but also from the historical, philosophical, and theological perspectives as well. In this manner the students are able to consider the multiple dimensions of our understandings, and how those understandings have changed over the history of human inquiry about the universe within which we live and are a part of.

The formal statements of the five course goals are presented in the Syllabus as follows:

"Through the *PYCC 103: The Journey in the Physical Universe* course, students will be able to demonstrate their use of aspects of the Catholic Intellectual Tradition (CIT) to gain:

1. an understanding of the ways in which we have historically altered our view of the physical universe through the applications of the methods of scientific reasoning, and the ways that these views have informed our understanding of God and His Creation;

2. the ability to examine how the ways in which our understanding of the interaction of physics and faith provide us with the keys to help us realize what it means to be human;

3. insight into the way in which an understanding of the physical universe, in the context of faith, can help people enhance their ability to live deeper lives of meaning and purpose;

4. an increased appreciation and understanding of the natural world as seen from the frame of reference of the sciences, particularly as seen in the context of physics and astronomy; and

5. a realization that the ways in which we choose to utilize our knowledge of the physical universe, and to apply the technology derived from that knowledge to meeting human needs, is a reflection of wise stewardship involving the interaction of faith, knowledge, and reason to serve the common good."

In order to engage the students in substantive reading related to these goals, the following books were selected for use as course texts:

Francis S. Collins, *The Language of God: A Scientist Presents Evidence for Belief*, New York: Free Press, 2006.

Ian G. Barbour, *When Science Meets Religion*, New York: HarperCollins Publishers, Inc., 2000.

Hugh Ross, *Why the Universe is the Way It Is*, Grand Rapids, MI: Baker Books, 2008.

John Polkinghorne, *Quantum Physics and Theology: An Unexpected Kinship*, New Haven: Yale University Press, 2007.

Krista Tippitt, *Einstein's God: Conversations About Science and the Human Spirit*, New York: The Penguin Group, 2010.

The division of the coursework for the semester is distributed over 26 75-minute class sessions. Each session begins to address a primary question, based on one of the five course goals, and then opens several secondary questions for the students to think about as we engage in the various in-class discussions, conduct various demonstrations, consider various models, and see various video segments related to the topics being considered. Out-of-class assignments involve the students in reading, discussion, viewing and listening to on-line audio or video resources, naked-eye sky observations and conducting research on topics of interest. Each activity generates some type of written or calculated response which the students submit electronically via email. Each in-class activity and out-of-class activity calls for a period of reflection, and a written response summarizing what of greatest significance has been learned, and what personal reaction the student has to that. These multi-sensory assignments to "read, view, observe, reflect, react and write" activities have been considered to be extremely valuable components of the students' learning experiences, as expressed in their summative course assessment.

The course content is subdivided over the semester as follows, with each, major goal-related segment covering four or five days, and each class meeting labeled as a letter of the alphabet:

I. What are the dimension of physics and the dimension of faith, and how do they relate to our knowledge of the universe?

A.   Epistemology: How do we know what we know? What is faith, and what is science?

B.   Where are you (u)? What is the large-scale size and structure of the universe (U)? What does the size and scale of the U have to do with our sense of human significance? What does the size and scale of the U have to do with our characterization of God?

C.   How has our understanding of size and scale in the universe changed over time? Modeling the Earth, the Earth-Moon System, and the Solar System.

D.   What is the scale of the universe in time? How has our view of this time-scale changed over time? What is the universe like on the small scales of space and time?

II. How does our knowledge of the physical universe, in the context of our faith, relate to our humanity?

E.   What does it mean to be "human?"

F.   How do we, on the human-scale, fit into the universe of space and time?

G.   How does faith inform our understanding of our humanity and our place in the U?

H.   Where does the human journey begin on the: *space* dimension? *Time* dimension? *Spiritual* dimension?

III. How does our awareness of and knowledge of the physical and the spiritual universe enable the realization of meaning and purpose in our lives?

I.   What is reality?

J.   Why are we here?

K.   What is truth? (VERITAS?)

L.   How does the human "desire to know" relate to: our search for knowledge, our spiritual quest, and our understanding of the universe?

M.   The Midterm Exam. Once the first three segments of the course have been completed, the students write their Mid-Term Exam responses. They are able to utilize their notes, written reflections, and written reactions to respond and to examine concepts that have been presented through in-class experiences or through their out-of-class work. Through this exam, students indicate that they are able to pull together concepts of space, time, and our humanity to realize how our understanding of the universe has developed over the history of human exploration. Once again the significance of paradigm shifts seems to have greater meaning in the retrospective opportunity offered by this exam.

During the second half of the course, the details of some of the major concepts of physics are used to consider developing understanding of the universe on all scales of consideration, from the scales of everyday experience, to the cosmological scales, to the quantum mechanical scales. The changes in our human understanding of who and where we are from ancient cultures, to pre-Copernican conceptions of natural philosophy, to the birth and development of classical physics, to the revolutionary discoveries of atomic, nuclear, and quantum mechanical physics, to the discoveries of modern astrophysics and cosmology, to the major questions that lie before us today all enable students to see how conceptions of "reality" have grown and developed. The importance of the development of scientific models of the design and functioning of the universe is then related to the developments of technology which extends our intellectual reach, and enables our application of that technology to the improvement of the "common good." As can be seen in what follows, the work in section IV of the course focuses on the development of the physics of the natural world as it allows our greater appreciation of the universe on all levels. It is then followed in section V of the course by the consideration of applications.

IV. How does our knowledge of the principles of physics enable us to appreciate the natural world?

N.   Scientific models and reality. Early conceptions of the motions of objects on Earth, the motions of the Earth itself, and the motions of objects in the heavens.

O.   The "mechanical universe" of physics: Copernican, Galilean, and Newtonian Models.

P.   Fundamental forces ("interactions") in the universe, and the nature of matter/energy.

Q.   The behavior of particles and the behavior of waves—sound, earthquakes, light.

R.   The nature of light—wave/particle duality, and the "misbehavior" of particles—quantum physics.

S.   Atomic and nuclear physics, quantum physics, and theology.

T.   The origin and nature of the universe. The origin of the Solar System.

U.   Why does the universe function on the basis of fundamental mathematical laws?

V.   Is the universe a product of chance or a matter of purpose? Can we know?

V. How does our knowledge of and application of the principles of physics in the context of our faith, enable us to improve our society for the common good?

W.   How do science, faith and our place in the world relate? Do science and faith conflict? What about the case of Galileo? What distinguishes science, faith, and religion?

X.   How can we use scientific knowledge and technology for the "improvement" of society? What about the case of nuclear energy? Nuclear weapons? What guides our decision making?

Y.   What role does faith have in dealing with our personal future and the future of society? What about the cases of natural and man-made disasters? How are we best able to cope with reality? (H. G. Wells' story "The Star")

Z.   What is our source of hope? Will science and technology save us from disasters? How should we define wise stewardship? Are you simply a product of blind chance or a person of purpose?

The PYCC 103 course concludes with a summative Final Exam designed to enable students to draw together the concepts and ideas related to the course, and to synthesize responses reflecting deeper thought and understanding than they had on entry to the course, of the universe and themselves within it.

Students' reactions to the course as expressed in their responses to course evaluations reflect a broad range of opinion and perspective. Most seem to be very satisfied with the experiences that they have had, and the amount that they have learned. Several have commented on the value that they have derived from having to think deeply about the universe as we know it today, and about consideration of their own "worldview" with respect to their spiritual relationships. Some have said that they have never worked so hard, but found it to be so intellectually satisfying. A year or two ago, one student commented, "This is the first course that I have taken, where I feel that I have gotten my money's worth."

Connecting People to Science
ASP Conference Series, Vol. 457
Joseph B. Jensen, James G. Manning, Michael G. Gibbs, and Doris Daou, eds.
© 2012 Astronomical Society of the Pacific

Spectra, Doppler Shifts, and Exoplanets: A Novel Approach via Interactive Animated Spreadsheets

Scott A. Sinex

Department of Physical Sciences and Engineering, Prince George's Community College, Largo, Maryland 20774, USA

Abstract.   Students investigate spectral line generation, discover the red- and blue-shifts of the spectral lines of moving objects, and analyze the periodic signal for exoplanet discovery. All of this is accomplished using pre-built animated spreadsheets.

1.   Introduction

Atomic spectroscopy is a powerful tool for determining the elemental composition of materials. In astronomy, atomic spectroscopy is done remotely using the light emitted by stars and combining this with the light passing through other materials in space to derive composition and even temperature (Robinson 2007). When atomic spectroscopy is united with other concepts such as the Doppler Effect, astronomers have an even more powerful tool to derive information. In this article, an approach used with astronomy students in a general education course is presented using interactive animated spread-sheets. The students will discover how spectral lines are generated, how the Doppler shift affects lines from moving stars, and how the sinusoidal signal of an exoplanet can be analyzed. The approach brings some astronomical tools and concepts to students using an off-the-shelf piece of software, Microsoft Excel, without belaboring the mathematics. All the spreadsheets are constructed using computational formulae with much of it hidden under the graphs—no programming is required (Sinex 2007, 2011). A number of the form tools, which function on both Mac and PC platforms, are used to add interactivity. The three spreadsheets discussed are available at the Astro Excelets website.[1]

2.   Discovering Atomic Spectra

Using gas discharge tubes and diffraction glasses, students are introduced to atomic emission line spectra. They observe the spectra of hydrogen, helium, neon, mercury, and nitrogen, and then identify two unknowns using the spectral drawings they made. Students quickly see that each element has a unique spectral fingerprint, which is demonstrated further using a Java applet.[2] This applet contains the absorption and

---

[1]http://academic.pgcc.edu/~ssinex/excelets/astro_excelets.htm

[2]http://jersey.uoregon.edu/vlab/elements/Elements.html

192   *Sinex*

emission spectra of the elements of the periodic table. Now how do you explain the generation of the spectra lines?

In the Generating Atomic Line Spectra spreadsheet, students review the relationship between energy and wavelength and explore the visible spectrum of light including the UV and IR regions at the extremes. Then students go to the H spectrum tab to examine how absorption lines and emission lines are generated by an electron changing energy levels in a hydrogen atom. Students move a single electron to see the generation of a single line. The wavelength of each line is given. After seeing how each line is generated and what electron transition is involved, all the visible lines can be shown (Fig. 1). For the absorption spectrum, the continuous spectrum of visible light is in the background. Comment boxes (little red triangles in upper right corner of cell) are used to supply information, hints, and explanations to aid students. Students discover how the electronic transitions generate the various spectral lines and the difference between absorption and emission lines. Since the energy levels are different for each element, an element has a distinctive spectrum that allows its identification. Students examine this on the element spectra tab.

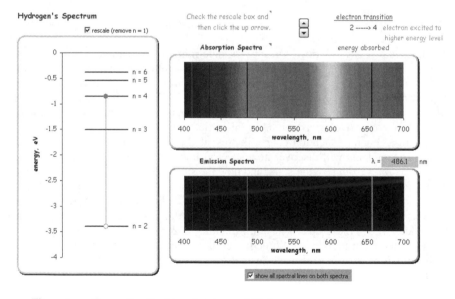

Figure 1.   Generating the Line Spectrum of Hydrogen

3.   Spectroscopy in Motion – The Doppler Shift

What happens to the spectral lines if the observed object is moving? To an observer on Earth, an object may be moving away from the Earth (increasing distance between Earth and object), toward the Earth (decreasing distance), or parallel to the Earth's motion (no change in distance). In the Spectroscopy in Motion: A Method to Measure Velocity spreadsheet, students review element spectra and mixtures to identify and then they are introduced to the effect of the Doppler shift. Figure 2 illustrates how a student would discover the red- and blue-shifts of spectral lines by changing the velocity and direction of the motion.

Students discover how astronomers can tell if the star is moving away (red-shift) or toward (blue-shift) the Earth. With a click of a check box, students can examine the calculations and see how the radial velocity is determined. On the relativity tab (not shown) students can investigate the relativistic velocity as well.

Hubble (1929) essentially discovered that most objects in the universe are moving away from each other. This indicates the expansion of the universe.

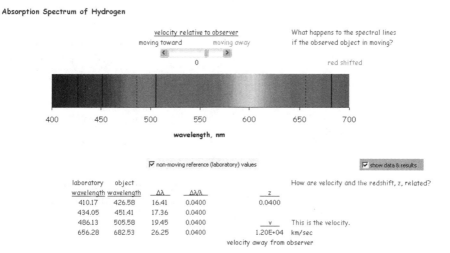

Figure 2.   Discovering the Doppler Shift

Determining Hubble's Law is a major use of redshift data for a variety of astronomical objects. Hubble's Law shows that the radial velocity is a linear function of the distance the object is from Earth, or, the farther an object is from the Earth, the faster it recedes. The radial velocity is determined by the redshift of spectral lines. Astronomers can then use this to compute the distance. To get accurate distances, we need an accurate value for the Hubble constant. The Hubble tab has students examine a number of data sets for deriving the Hubble constant and allows then to quickly use the constant to estimate the age of the universe (Figure 3).

4.   Spectroscopy in Motion – Exoplanet Detection

In the Spectroscopy in Motion II: A Method for Finding Exoplanets spreadsheet, students examine what the Doppler signal can reveal about possible exoplanets around a star. This spreadsheet starts with the concept of center of mass since the occurrence of any exoplanet(s) will cause the center of mass to shift from the star's center. This causes the star to wobble which will induce a periodic signal in the radial velocity (for a near circular orbit viewed edge on, the signal is sinusoidal). Students discover this on the signal detection tab and can explore how the period, p, and amplitude, K, influence behavior (Figure 4). Students further fit the data for 51 Pegasi b with a sine wave while minimizing the sum of the squared error to discover the exoplanet's mass in multiples

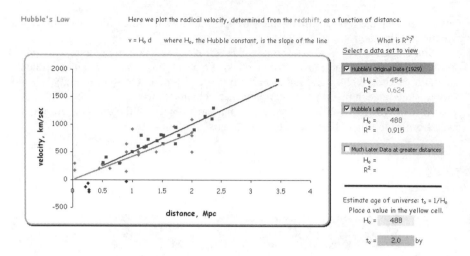

Figure 3.   Modeling Hubble's Law Data

of Jupiter and its orbital distance in AU, which they compare to the actual values. All the mathematics is camouflaged in the spreadsheet and the instructor can decide if it needs to resurface for discussion or not.

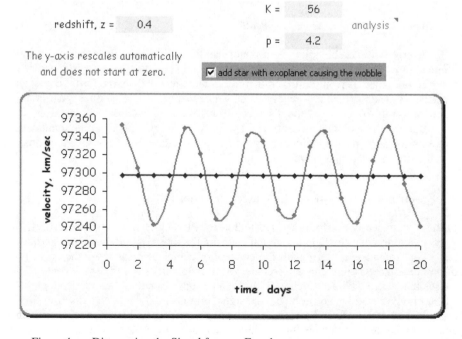

Figure 4.   Discovering the Signal from an Exoplanet

Once the simple system is examined, instructors can decide if more complicated orbital systems need to be explored. A variety of websites provide the orbital information. We go one step further by looking at how our solar system would behave if an alien astronomer were to examine it from afar.

What would the Doppler shift data look like for our solar system as observed by an alien astronomer? How many planets could be detected in our multi-planet system? On the our solar system tab (Figure 5), students examine the individual signal for the four Jovian planets. When they set the scale to maximum for comparing to Jupiter's signal, students see how minor the other planets are (LoPresto & McKay 2004, 2005). Only Saturn's signal has any significant effect on the total signal, so are alien astronomer might assume we only have two large planets orbiting our Sun.

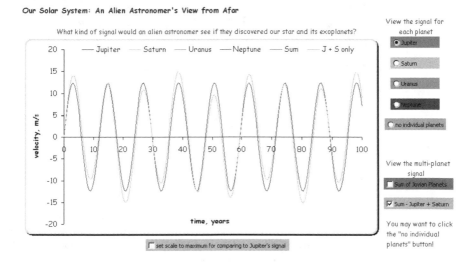

Figure 5.   Viewing Our Solar System from Afar

5.   Student Assessment

The activity and spreadsheet for the Spectroscopy in Motion: A Method to Measure Velocity were used in an introductory astronomy laboratory in the fall of 2010, and the results for 20 of 21 students are shown in Figure 6. A mean score of 40 out of 50 was obtained.

6.   Constructing Interactive Excel Spreadsheets or Excelets

The wherewithal for producing Excelets can be found at the Developer's Guide to Excelets (Sinex 2011), which includes a tutorial, illustrated instructions, and many more examples. Take the interactive features tour to see what you can do in Excel. The forms toolbar provides a variety of features (spinners, scroll bars, checkboxes, etc.) that are easy to use, and when combined with logical functions, lookup tables, conditional formatting, and a number of simple tricks provides a wealth of interactivity and dynamic

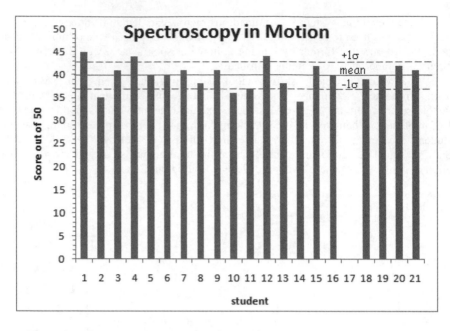

Figure 6.   Student Performance (n = 20 students)

display. The use of the forms toolbar allows Excelets to function on both PC and Mac platforms. All of this is done using computations (formulas and available functions) in the cells. The use of comment boxes adds explanation, hints, and answers for students as well. Always look under the graphs, as many of the tricks, such as turning lines on graphs on and off or tracer points are explained there.

7.   Some Final Thoughts

The interactive animated spreadsheets produce an engaging pedagogy for students in the classroom. Through the predict-test-analyze and explain method of questioning, students can discover a vast array of concepts. The mathematics can be incorporated in as the instructor deems necessary. Using these interactive spreadsheets as part of assessment is the next step (LoPresto 2010).

The spreadsheets and accompanying activities discussed, plus others that explore topics in astronomy, are available for free download.[3]

Acknowledgments.   The author wishes to thank Barbara Gage of Prince George's Community College for providing comments on this article.

---

[3]http://academic.pgcc.edu/~ssinex/excelets/astro_excelets.htm

References

Hubble, E. 1929, "A Relationship between Distance and Radial Velocity among Extra-Galactic Nebulae," Proc Natl Acad Sci USA, 15, 3, 168

LoPresto, M. C. 2010, "Using Visual Assessments and Tutorials to Teach Solar System Concepts in Introductory Astronomy," Astronomy Education Review, 9

LoPresto, M. C. & McKay, R. 2004, "Detecting Our Own Solar System from Afar," Physics Teacher, 42, 4, 208

LoPresto, M.C. & McKay, R. 2005, "An Introductory Physics Exercise using Real Extrasolar Planet Data," Physics Education, 40, 1, 46

Robinson, K. 2007, Spectroscopy: The Key to the Stars (Springer-Verlag London Limited), 160

Sinex, S.A. 2007, "Excelets: Excel's Excellent Adventure"

Sinex, S.A. 2011, Developer's Guide to Excelets, `http://academic.pgcc.edu/~ssinex/excelets` (accessed 03 June 2011).

Part IV

Poster Contributions

Connecting People to Science
ASP Conference Series, Vol. 457
Joseph B. Jensen, James G. Manning, Michael G. Gibbs, and Doris Daou, eds.
©2012 Astronomical Society of the Pacific

American Geophysical Union Education and Public Outreach Programs: Empowering Future Earth and Space Scientists

Bethany Holm Adamec and Pranoti Asher

American Geophysical Union, 2000 Florida Avenue NW, Washington, D.C. 20009, USA

Abstract.   The American Geophysical Union (AGU) is an international non-profit scientific association with more than 60,000 members, working on a broad spectrum of scientific topics that span all of the Earth and space sciences. AGU's educational programs capitalize on the intrinsic allure of the Earth and space sciences, and their fundamental relevance to daily life. Through education- and career-focused events at annual AGU meetings, national conferences on science education reform, professional development workshops for teachers, special programs for pre-college and post-secondary students, awards for science educators, and printed and electronic resources, AGU offers an array of opportunities that expose students, teachers, and life-long learners to the freshest, most accurate scientific knowledge and the excitement of discovery.

1.   About AGU

AGU is the world's largest organization of Earth and space scientists, with over 60,000 members around the globe. Since its founding, AGU has been dedicated to furthering its science through the individual efforts of members and in cooperation with other national and international scientific organizations. These goals are met through publishing scientific journals and other technical publications, sponsoring scientific meetings of various sizes throughout the year and a variety of other educational and scientific activities.

2.   Education at AGU

AGU's educational programs capitalize on the intrinsic allure of the Earth and space sciences, and their fundamental relevance to daily life. Through education- and career-focused events at annual AGU meetings, national conferences on science education reform, professional development workshops for teachers, special programs for pre-college and post-secondary students, awards for science educators, and printed and electronic resources, AGU offers an array of opportunities that expose students, teachers, and life-long learners to the freshest, most accurate scientific knowledge and the excitement of discovery.

The education-specific goals illustrated in Figure 1, along with specific strategies, will help AGU meet its overall organizational goal related to workforce or "talent pool" development. Particular emphasis is being placed on building partnerships and collaborations that will increase the effectiveness of AGU's outreach efforts related to education.

Figure 1.   AGU Education and Public Outreach Goals.

3.   Education at the AGU Fall Meeting

Among the education workshops and other events that will be held at the 2011 Fall AGU Meeting[1] are the Bright Students Training as Research Scientists (BrightSTaRS) poster session and luncheon with AGU leadership, where middle and high school students affiliated with summer science programs present their research.

Additionally, in partnership with the National Earth Science Teacher's Association, AGU runs teacher workshops on Earth and space science at meetings. These Geophysical Information for Teachers (GIFT) workshops allow secondary science teachers to hear about the latest geoscience research from the scientists making the discoveries, explore new classroom resources for their students, and visit exhibits and technical sessions of the AGU meeting for free.

Another important event at Fall Meeting is the Heads & Chairs workshop, where heads and chairs of Earth and space science departments meet to discuss current issues

---

[1]http://sites.agu.org/fallmeeting/

facing higher education. This year, the morning program will focus on recruiting under-graduate majors, fundraising and development, and small group discussions on topics suggested by participants. The afternoon program is titled "Advancing Women and Un-derrepresented Minorities in the Academic Geosciences;" participants will discuss the current status of these groups and barriers to hiring and promotion.

Finally, AGU conducts a range of family science events including an annual public lecture concerning a current "hot topic" in Earth and/or space science, and Exploration Station, an afternoon-long hands-on science program administered by member scientists.

4. Education Beyond Fall Meeting

In addition to Fall Meeting programs, AGU Education provides opportunities for AGU members to participate in outreach activities and programs and supports national STEM education issues by collaborating with other organizations on public policy initiatives. The department also collaborates on grants and proposals concerning science, technology, engineering, and math education with AGU members, other geoscience organizations, and outreach/media entities.[2]

Acknowledgments.   We wish to thank the Astronomical Society of the Pacific for inviting us to partner with them on this important education and outreach-focused conference.

---

[2]For more information, see http://www.agu.org/education/

Participants in Year of the Solar Sys tem workshop. Photo by Paul Deans.

Connecting People to Science
ASP Conference Series, Vol. 457
Joseph B. Jensen, James G. Manning, Michael G. Gibbs, and Doris Daou, eds.
© *2012 Astronomical Society of the Pacific*

"From Earth to the Solar System:" Public Science Exhibitions for NASA's Year of the Solar System

Kimberly Kowal Arcand,[1] Megan Watzke,[1] Julie Fletcher,[2] and Daniella Scalice[2]

[1]*Chandra X-ray Center, Smithsonian Astrophysical Observatory, 60 Garden Street, Cambridge, Massachusetts 02138, USA*

[2]*NASA Astrobiology Inst., NASA Ames Research Center, Moffett Field, California 94035, USA*

Abstract. Launched in May 2011, "From Earth to the Solar System" (FETTSS) is a public science program that brings planetary science, astronomy, and astrobiology images to audiences in non-traditional science outreach locations. FETTSS seeks to sustain and build upon the success of the award-winning International Year of Astronomy 2009 project "From Earth to the Universe." FETTSS utilizes a similar grass-roots-type of approach to emphasize the point that science-learning experiences can be anywhere. Exhibiting a curated collection of print-ready images of the Solar System, FETTSS aims to spark socially-based engagement and enhance exploration of astronomical content through free-choice learning outside the walls of (but also in partnership with) science centers or planetariums. The research component of FETTSS investigates casual versus intentional audiences, the possibility for participants to reshape their identity or non-identity with science through public events, and additional audience demographics.

1. Introduction

The "From Earth to the Solar System" (FETTSS) project exhibits images of our solar system in public venues and other free, community-based locations. FETTSS is a derivative effort of "From Earth to the Universe" (FETTU), which was a major project of the International Year of Astronomy 2009. FETTU has put astronomy content into more than 1,000 locations since 2009, including parks, metro stations, airports, hospitals, and libraries. This has attracted, at least partially, a non-self-selected audience for science outreach, and shown, by the limited evaluation of a subset of the FETTU locations, that inspiration and small learning gains can occur in such environments (Arcand & Watzke 2011). In 2011, FETTSS continues the implementation of the FETTU model, with a focus on solar system images and subject matter tied to the study of astrobiology.

1.1. Categorizing Public Science

Public art can be defined in the humanities as artwork that has been created to be placed in an accessible, often outdoor, public location. Often, the works involve site specificity, community involvement, and collaboration.

Projects such as FETTU and FETTSS—and others described below—may be able to play a similar role in communicating science with the public. The authors use the term "public science" to describe this model (Arcand & Watzke 2011).

Science outreach projects that have been conducted outdoors or in another type of public or accessible space (from public park to metro stop) often attempt to reach new audiences. By hosting these events in alternative places for communicating science (Norsted 2010), they might particularly attract non-intentional or casual visitors (Crettaz von Roten 2011). Public science events may also be considered as neutral territory for participants who do not have to actively seek out locations for engagement with science content (Riise 2008).

Looking through the lens of this public science definition, the authors would include the following as some examples:

- Science City was an outdoor exhibition that utilized the street, fences, buildings and other public structures in New York City (Cole & Cutting 1996);

- Science on the Buses decorated city buses with large informational science posters inside or outside, taking science concepts outside museum and planetarium walls;

- science festivals such as the USA Science and Engineering Festival, the World Science Festival, and others present science in a public sphere through large outdoor programs with exhibits, discovery stations, talks and activities; and

- Music and Astronomy Under the Stars is a program at public parks that accompanies music performances such as at the Newport Folk Festival (Lubowich 2010).

2.   NASA's Year of the Solar System and FETTSS

FETTSS was conceived as a response for NASA's Year of the Solar System (YSS) and adds to the now-defined category of public science. The content of the FETTSS project weaves together themes in multi-wavelength astrophysics, astrobiology, and planetary science, and includes images from amateur astronomers, field scientists, NASA missions, and international astronomy organizations.

The FETTSS organizational structure follows the FETTU-style, grass-roots approach (Russo & Christensen 2010) that allows local organizers to print their own version of the exhibition for their venues. The FETTSS project supplies high-resolution electronic files to be displayed in any way that makes sense for a given venue. This allows organizers of each FETTSS exhibit to consider issues of site specificity and to involve their local communities. The free images and related materials have been approved for non-commercial outreach use. Identifying resources for the printing, installation, and other logistics are the responsibility of the local hosts.

In addition, the Chandra X-ray Center/Smithsonian Astrophysical Observatory and NASA Astrobiology Institute are coordinating a traveling FETTSS exhibit. The exhibit (which consists of 15 outdoor/all-weather double-sided image stands, bilingual in English and Spanish) is being loaned free of charge to venues that commit to hosting the project while providing some basic programming around the exhibit content—such as organizing a star party, scavenger hunt, or question and answer session.

Figure 1. Left: FETTSS was launched at a high-traffic shopping mall in Corpus Christi, Texas, USA (May 9–31, 2011), and featured events being programmed around Space Week activities as coordinated with the Corpus Christi Museum of Science and History and the National Center for Earth and Space Science Education. Photo credit: Rick Stryker. Right: FETTSS went on to be hosted outside the National Air and Space Museum in Washington D.C., with intern-led tours as pictured here (June–July 2011), and was also included as part of the Astronomy Night on the National Mall activities that drew about 1,500 participants. Credit: Smithsonian National Air and Space Museum.

3. Evaluation

The FETTSS research questions will follow up on previous public science project results as well as results from the ongoing Aesthetics & Astronomy research project. These questions include: Who are we attracting with science displays in these "everyday situations?" Are there more incidental visitors than intentional visitors with public science events? Are we attracting a less-science-initiated audience than other science venues might attract? Do any participants follow up with their local science center, library, or other resources? Is there any reshaping of the participant's identity (or non-identity) with science through public science events such as these? This evaluation may help shed light on whether public science events can be effective ways of reaching new audiences.

4. Conclusion

FETTSS provides an opportunity to disseminate scientific information to the public in unique ways. FETTSS has the potential to draw in diverse populations in terms of age, gender, level of prior knowledge, and cultural and socio-economic identity. There are implications then for increasing capacity for STEM outreach and public science literacy. For example, increased exposure rates of general populations to astronomy outreach in the FETTU project led to inspiration and small learning gains, and millions of people were reached relatively inexpensively. FETTSS is poised as a useful test case on the viability of this specific public science model.

Acknowledgments. This material is based upon work supported by the National Aeronautics and Space Administration under proposal NNX11AH31G issued through the Science Mission Directorate.

References

Arcand, K. K. and Watzke, M. 2011, "Creating Public Science With the From Earth to the Universe Project," Science Communication 33, 3 (in press)

Cole, P. R. & Cutting, J. M. 1996, "The Inside Story of Science City—An Outdoor Public Science Exhibition," Curator: The Museum Journal, 39, 4, 245

Crettaz von Roten, F. 2011, "In search of a new public for scientific exhibitions or festivals: the track of casual visitors," Journal of Science Communication, 10, 1, 1

International Astronomical Union, Final Report, ISBN: 978-3-923524-65-5

Lubowich, D. 2010, "Music and Astronomy Under the Stars 2009," Barnes, J., Smith, D. A., Gibbs, M. G., and Manning, J. G., eds., Science Education and Outreach: Forging a Path to the Future, ASP Conference Series, 431, 47

Norsted, B. A. 2010, "Take Me Out to the Ball Game: Science Outreach to Non-traditional Audiences," Barnes, J., Smith, D. A., Gibbs, M. G., and Manning, J. G., eds., Science Education and Outreach: Forging a Path to the Future, ASP Conference Series, 431, 170

Riise, J. 2008, "Bringing Science to the Public," D. Cheng, M. Claessens, T. Gascoigne, J. Metcalfe, B. Schiele, S. Shi, eds., Communicating Science in Social Contexts (Brussels, Springer), 301

Russo, P. & Christensen, L. L., eds. 2010, International Year of Astronomy 2009

Connecting People to Science
ASP Conference Series, Vol. 457
Joseph B. Jensen, James G. Manning, Michael G. Gibbs, and Doris Daou, eds.
© *2012 Astronomical Society of the Pacific*

Developing Teenage Youth's Science Identity Through an Astronomy Apprenticeship: Summative Evaluation Results

Ross Barros-Smith,[1] Irene Porro,[1] and Emmalou Norland[2]

[1]*MIT Kavli Institute for Astrophysics and Space Research, 77 Massachusetts Avenue, Cambridge, Maryland 02139, USA*

[2]*Cederloch Research LLC, 330 S. Fair Street #1004, Arlington, Virginia 22202, USA*

Abstract. We report on the results from the summative evaluation of the Youth Astronomy Apprenticeship (YAA) covering three years of implementation of the program. YAA is a year-long, out-of-school time initiative that connects urban teenage youth with astronomy as an effective way to promote scientific literacy and overall positive youth development. The program employs the strategies of a traditional apprenticeship model, common in crafts and trade guilds as well as in higher education. During the apprenticeship, youth develop knowledge and skills to create informal science education projects; through these projects they demonstrate their understanding of astronomy and use their communication skills to connect to general audiences. For some youth, participation extends across multiple years and their responsibilities for program implementation become multifaceted. Through exposing youth to astronomy investigations and providing opportunities to connect with audiences outside their program and communities, YAA expands scientific literacy to include assuming a science identity. We subscribe to the concept of science identity that describes personal ownership and integration of science into an individual's sense of self through processes of comprehension and personal meaning making. In the YAA context, science identity extends to and includes assuming an actual science advocacy role. Our methods for measuring the development of a science identity included assessments of a youth's perceived and actual understanding of science (cognitive construct), leadership in science (behavior construct), and commitment to science (affective construct).

1. Program Overview

In the YAA model, equal effort is put into pursuing science learning for academic enrichment and in stressing the link between employable skills and the skills developed in science and other professional fields—such as performing arts and museum exhibit development. YAA progressively develops a youth's science knowledge and 21st century employable skills through several stages: Afterschool Program, Summer Apprenticeship Program and YAA Outreach, and YAA Internship.

2.   Summative Evaluation

2.1.   Methodology

The summative evaluation was conducted from October 2008 through September 2009. For analysis of quantitative data of youth outcomes, participants were organized in three groups defined by program tenure and responsibility level in the program: youth new to the program, returning apprentices, and youth assistants and interns. Youth outcomes were measured at three distinctive points in the program year: March, near the beginning of the afterschool program; June, at the beginning of the summer apprenticeship program; and August, at the end of the summer apprenticeship program. Data were analyzed within and across these tenure groupings to determine: 1) changes in program outcomes across the program year for youth in each level of tenureship, and 2) differences in program outcomes across the three levels of youth tenureship at each of the three points in time when measures were taken. After data were collected, items were subjected to multiple data reduction techniques and three major constructs emerged as the top level of outcomes: *Leadership in Science*—A behavioral construct combining aspects of positive youth development with aspects of being an advocate for science; *Understanding Science*—A cognitive construct of scientific habits of mind and knowledge of science and scientific principles; and *Commitment to Science*—An affective construct combining aspects of being an advocate for science and anticipating a future in science.

2.2.   Results

Youth characterized as being new to the program had completed participation in an implementation of the afterschool program and a single iteration of the summer apprenticeship by the conclusion of the evaluation. As anticipated, newcomers had the lowest scores of all the participants on all the measures.

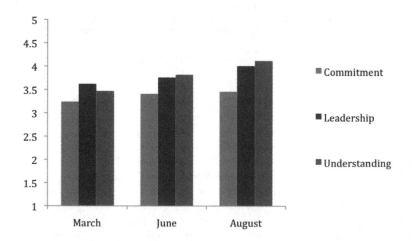

Figure 1.   Primary outcomes for youth new to the program, reported on a five-point scale where 5 is the most positive outcome.

Youth with previous program experience had completed at least one additional summer apprenticeship. Initially, the group's measured commitment to science was already stronger than the commitment of youth new to the program at the end of the year.

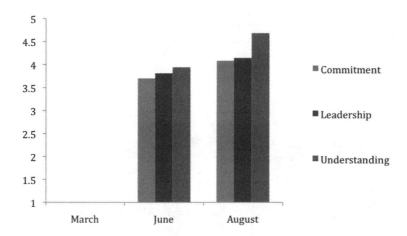

Figure 2.   Primary outcome for youth with previous program experience reported on a five-point scale where 5 is the most positive outcome.

For youth who served as youth assistants and youth interns (previous participants with current and possibly past leadership responsibilities), scores on leadership and understanding increased throughout the year.

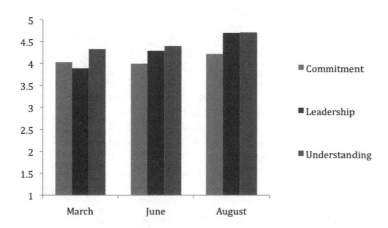

Figure 3.   Primary outcomes for youth returning to the program in leadership roles reported on a five-point scale where 5 is the most positive outcome.

2.3. Advocacy of Science and Development of a Science Identity

If a science identity is defined as a synthesis of personal ownership and integration of science into an individual's sense of self through processes of comprehension and personal meaning making, it is reflected in the youth's own perception of being an advocate for science. All program participants experienced a rise in their perceived sense of a role as advocates of science; however, youth assistants and interns, with advanced program responsibility, reported the greatest gains across the program and after completion of the summer program.

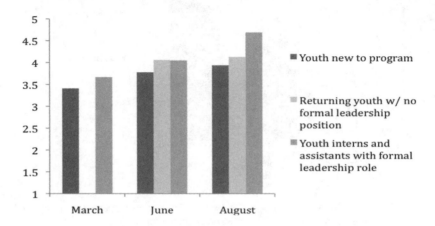

Figure 4.   Measure of science advocacy for all program participants on five-point scale where 5 is the most positive outcome.

3. Conclusions

Working with high school youth presents unique opportunities to informal science educators that are not available with other age groups. High school youth integrate an understanding of content knowledge and a new science identity with the process of personal identity development that carries on into adulthood. Focusing on high school youth also allows for applying the development of real-world professional skills as a platform for creating an understanding of content knowledge in a manner that emphasizes both these skills and knowledge. Data from the Summative Evaluation of the YAA program demonstrate that confidence in advocacy for science is connected to time spent in the program. This underscores the need for a program environment that sustains youth participation over time and continuously builds upon their past positive outcomes. The apprenticeship model is ideal for addressing the needs of older youth, as it provides an opportunity to expand content knowledge and model and develop professional workplace practices while creating a ladder of achievement and responsibility for youth to ascend.

Acknowledgments.   The Youth Astronomy Apprenticeship is sponsored by the Education and Outreach Group at the MIT Kavli Institute for Astrophysics and Space Research with support from the Kavli Foundation and the Harvard-Smithsonian Center for Astrophysics. The program is funded by the National Science Foundation (DRL 0610350).

Connecting People to Science
ASP Conference Series, Vol. 457
Joseph B. Jensen, James G. Manning, Michael G. Gibbs, and Doris Daou, eds.
©*2012 Astronomical Society of the Pacific*

Creative Writing and Learning in a Conceptual Astrophysics Course

Rhoda Berenson

New York University Liberal Studies Program, 726 Broadway, New York, New York 10003, USA

Abstract. Creative writing assignments in a conceptual astrophysics course for liberal arts students can reduce student anxiety. This study demonstrates that such assignments also can aid learning as demonstrated by significantly improved performance on exams.

1. Introduction

I often offer an optional assignment to the liberal arts students in my History of the Universe course that requires them to write an imaginative story incorporating specific scientific vocabulary. My goal has been to lessen student anxiety by providing an opportunity to submit a creative, at-home contribution to exams. I have recently discovered that writing these stories also can have a positive effect on a student's performance on the in-class exam. This would not come as a surprise to any of the proponents of the "Writing to Learn" movement, but it did come as a surprise to those students who told me how much the assignment helped them understand the concepts.

2. The Study

This study compares the results for two in-class exams for two groups of students. The first in-class exam consisted of a mix of multiple-choice questions, true/false with explanation, numerical calculations, and short essays. There was no optional assignment for the first exam. The in-class portion of the second exam was similar in format to the first, but students were given the opportunity to submit a take-home assignment that would count as 25% of the exam with the in-class portion counting for 75%. This resulted in two groups—one that chose to do the writing assignment and one that chose not to do it. Results were combined for seven classes over four semesters.

3. The Assignment

Tell the life story (biography or autobiography) of a sun-like star, starting with its formation within a nebula to its final days. Your story must be *creative*, but it should also demonstrate your scientific understanding of the many stages in the star's life. It should be understandable to someone who never studied the evolution of stars or the laws that govern that evolution. It must include 18 of the terms in the table below plus any ad-

213

ditional scientific terms that are relevant to the story. All scientific terms must be used correctly and *defined in context*. It must also include two numerical evaluations.

Scientific terms: accretion; apparent brightness; binary stars; carbon; conservation of energy; degeneracy pressure; fusion; gravitational equilibrium; gravitational potential energy; helium; H-R diagram; magnetic field; nebula; nucleus; nova; photon; quark; strong force; supernova; temperature; thermal energy; thermal pressure; weak force; white dwarf.

Grades: creativity — 5 points; correct use of scientific ideas and terms (including definitions) — 18 points; correct numerical evaluations — 2 points.

Although this assignment is similar to those suggested in some astronomy textbooks, such as the one used for this course, the requirement for specific vocabulary guides the student in writing a more detailed and comprehensive story.

4.   The Results

On the second exam, 101 students chose to write the story and 116 chose not to write. Those who chose to write had a mean grade on the first exam of 81.2 with a standard deviation of 10.9, while those who chose not to write had a mean grade of 84.5 with a standard deviation of 12.5. The results of the pooled-variance t-test indicate that the two groups had performed significantly differently on the first exam ($t = +2.01$; p-value $= 0.045$). Although both groups included students with grades ranging from very high to very low on the first exam, on average the grades were significantly lower for those who opted to write the essay on the second exam. This might be expected from students who are looking to improve their grades.

On the in-class portion of the second exam, the 101 students who chose to write had a mean grade of 85.2 with standard deviation 10.3 while the 116 students who chose not to write had a mean grade of 84.9 with standard deviation 11.8. Using the pooled-variance t-test, the null hypothesis of no difference in performance cannot be rejected at a 0.05 level of significance ($t = -0.23$; p-value $= 0.815$). It is clear that those who opted to do the writing improved their scores to match those of the initially stronger group who chose not to write the essay.

In fact, using the grades on the first exam as a baseline, when examining changes in grades between the first and second exam, the writing group significantly outperformed those who did not opt to write, raising their grades, on average, by 3.98 points as compared to 0.39 points for the non-writers.

5.   Conclusion

A creative writing assignment can improve students' understanding of concepts of astrophysics as evidenced in their performance on exams. As a bonus to the instructor, the stories are a lot of fun to read.

References

Bennett, J., Donahue, M., Sohneider, N., & Voit, M. 2010, The Cosmic Perspective (Addison-Wesley), 6th ed.

Connecting People to Science
ASP Conference Series, Vol. 457
Joseph B. Jensen, James G. Manning, Michael G. Gibbs, and Doris Daou, eds.
© *2012 Astronomical Society of the Pacific*

"Teachers Touch the Sky:" A Workshop in Astronomy for Teachers in Grades 3–9

Bonnie J. Buratti

Jet Propulsion Laboratory, California Institute of Technology, 4800 Oak Grove Dr., 183–401, Pasadena, California 91109, USA

Abstract. Eight times during the past two decades, JPL technical staff, assisted by master teachers, conducted a one-week workshop for teachers in grades 3–9. In these workshops, the teachers are walked through hands-on activities that are all based on current projects in astronomy and space science at JPL. The activities are inquiry-based and emphasize the scientific method and fundamental math and science skills. Each year the workshop focuses on a NASA theme: in 2011 it was the *Dawn* Mission to the asteroid 4 Vesta, as orbit insertion occurred right before the workshop. Several activities are based on the Lawrence Livermore Lab's Great Exploration in Math and Science (GEMS) guides. Teachers tour JPL's facilities such as the Space Flight Operations Center, the Spacecraft Assembly Facility, and the Mars Yard. The integration of the lessons into the teachers' own curricula is discussed, and a field trip to JPL's Table Mountain Observatory is included. Teachers learn of the resources NASA makes available to them, and they have the opportunity to talk to "real" scientists about their work. Teachers receive an honorarium for participation plus classroom materials.

1. Background and Scope

Using the natural fascination the public holds towards its work, NASA encourages and funds its scientists to do education and public outreach (EPO) to both children and adults. Space science is especially interesting to students, less threatening to teachers than some other sciences, and interdisciplinary in nature. These features make it the ideal vehicle for teaching basic scientific concepts to children in a concrete and captivating manner. For nearly two decades, scientific staff at the Jet Propulsion Laboratory have conducted a one-week hands-on inquiry based workshop for about 20 teachers in grades 3–9. The staff has been assisted by master teachers, helping especially in the area of integration of the activities into curricula that meet the California and National Science Standards. The grade level was chosen because this age is a critical one, when students begin to lose interest in scientific and technical fields. The philosophy of the workshop is to use inquiry-based activities that lead teachers and students to follow the scientific method as they become mini-investigators. Basic mathematical skills and scientific principles, such as graphing, making contour plots, and understanding dependent and independent parameters, are emphasized. The main focus is the primacy of experimentation in science.

Each year the workshop is based on underlying themes, one of which is a current NASA mission (this year it was the *Dawn* encounter at 4 Vesta, and in previous years it was *Cassini*, *Deep Space 1*, or Mars missions). The other themes are scientific, such as "cratering processes," "small bodies," "life in the universe," "erosion on Earth and the planets," etc. At least one field trip, usually to JPL's Table Mountain Observatory (TMO), where guest lectures and tours occur, is part of the week. Several guest lectures delivered by JPL's top scientists and tied to the themes of the workshop occur throughout the week. Finally, Steve Edberg does an evening star party with his collection of telescopes. Follow-on activities with the teachers occur throughout the year, and a Saturday conference to bring the teachers together and compare notes on implementation of activities in the classroom—and to listen to another lecture and do another hands-on activity—occurs in late fall. The teachers represent about 1500 students annually. The workshop was developed and has been led by the author.[1]

2. Activities

Table 1 is a summary of the activities that occurred during the workshop this year (8–12 August, 2011); it is typical of workshops held in the past. The hands-on activities are both ones we have developed ourselves ("Building a Quadrant" to sight angular size, originally written by Martha Hanner and Tarltonette Binion Mason, an eight grade Pasadena teacher; a cratering activity; a mass extinctions activity originally developed by Dave Morrison), and GEMS activities.

GEMS (Great Explorations in Math and Science) is collection of hands-on lesson plans developed by the Lawrence Hall of Science at the University of California at Berkeley.[2] Over the years, we have used many of the guides, including "Earth Moon and Stars," "Moons of Jupiter," "River Cutters" (with a unit on water erosion on Mars that we've added), and "More than Magnifiers." In 2011 we walked the teachers through "Oobleck; What do Scientists do," a real-life, real-time replication of the scientific method with a convention and revision of scientific models. "Color Analyzers," an introduction to the electromagnetic spectrum, was also done later in the week. GEMS guides are aligned with California and national standards, have been extensively reviewed by teachers and tested in classrooms, and are easily integrated into the science curriculum for elementary and middle schools.

We also rely heavily on activities developed through the ASP's Project Star, including the simple telescope that is available at little cost through the program (see Fig. 1). Each teacher builds a scope, and then goes outside to do a daytime astronomy program (observing sunspots and watching them rotate; counting craters on the Moon if it's up; plotting the shadow of a gnomen throughout the day, etc.). In years past, other popular activities have included the Venus/Titan topography box, an exercise to teach Radar science and basic math skills such as topographic mapping, and a rendition of the Erastothenes experiment measuring the size of the Earth. The latter activity is especially effective if done in conjunction with a school district located at another latitude.

As the point of the workshop is to encourage teachers to do hands-on activities in classrooms, the teachers are each given a copy of all the lesson plans and supplies.

---

[1]http://www.jpl.nasa.gov/education/index.cfm?page=273

[2]http://lhsgems.org/index.html

Table 1.   A summary of activities for "Teachers Touch the Sky," 2011

Monday 8/8/11	Tuesday 8/9/11	Wednesday 8/10/11	Thursday 8/11/12	Friday 8/12/11
8:30–9:00 Coffee, juice, pastries, sign-in, and socializing	8:30–9:00 Coffee, juice, pastries, sign-in, and socializing	8:30–9:00 Coffee, juice, pastries, sign-in	8:30–9:00 Coffee, juice, pastries	8:30–9:00 Coffee, juice, pastries, sign-in
• 9:00 Tour of JPL • 10:00 Introduction and purpose of workshop • Scientific misconceptions (seasons and phases of the moon) • 11:00 The cosmic distance scale (with hands on activities for all grades)	• 9:00–10:00 Finish Oobleck • 10:00–10:30 David Seidel: Introduction to JPL's Educators Resource Center • 10:45 Guest speaker: Dr. Paul Weissman: *"Small Bodies in the Solar System"* • 11:45 Get lunch (if you didn't pack one)	• 9:00 Kitchen comets • 9:30–10:30 Introduction to telescopes and construct a small telescope (Steve Edberg) • Daytime astronomy • Introduction to spectroscopy; Herschel's demonstration	• 9:00 Safety video for JPL's Table Mountain Observatory (TMO) • 9:30 Leave for field trip to TMO (Bring a bag lunch and a drink). Tours/talks by Heath Rhoades and Mick Hicks	• 9:00 NASA Explorer Schools: Mast. Teach. Lori Cirucci (Bethlehem, PA USD) Sharing: teachers demo their own best activities • 11:00 *Where the Hot Stuff is: Volcanoes on the Earth and Solar System* Dr. Rosaly Lopes
Lunch 12:00–1:00	12:00–1:00 Brownbag w/ scientists	Lunch 12:00–1:00	Lunch 12:00–1:00	Lunch 12:00–1:00
• 1:00 Red shift demo: Mast. Teach. Mike Lichtman (GUSD) • 1:30 "Sighting angular size": Make a quadrant • 3:30 Oobleck: What do scientists do? GEMS guide	• 1:00–2:00 *Dawn: the Mission to Ceres and Vesta:* Dr. Marc Rayman • 2:15–3:15 Cratering experiment • 3:15–4:00 Mass extinctions on the Moon and Earth	• 1:00–2:00 GEMS color analyzers • 2:15–3:15 *Cassini: the Mission to Saturn:* (Trina Ray) • 3:15 Curriculum integration (Mike Lichtman)	• 4:00 PM (approx.) Return from field trip.	Evaluation for NASA Adjourn by 2:00 PM

3.   Evaluation and Leveraging

If a scientist goes into a classroom for an hour and gives an age-appropriate talk, she in-fluences a single class on a single day. If a teacher is trained in a workshop, he changes

Figure 1.   Teachers participating in a daytime astronomy unit after building the ASP's Project Astro telescopes. Teachers view sunspots, craters on the Moon (if the Moon is up), and (rarely) Venus.

his teaching style every day and every year to influence and inspire thousands of students. Over the years, "Teachers Touch the Sky" has reached approximately 15,000 students, over 80% of whom are Latino or African-American. Teacher training workshops are inherently leveraging activities. Other forms of leveraging include partnerships we have developed with the JPL Educator's Resource Center, which advertises our workshop and provides supplies and resources. School districts also provide us with teachers, and in the future we plan on doing follow-on activities at schools such as walking through GEMS guides during inservice days.

As part of the workshop, each teacher fills out an extensive four-page evaluation, and future workshops are improved based on the evaluations. The survey questions ask the teachers their comfort level and confidence in teaching science before and after the workshop. On average the scores increased from 3.7 to 4.7 (on a scale of 1–5, with 5 being the highest). In the overall evaluation of the workshop, the teachers gave an average score of 4.8, with 4.9 being earned for "relationship with teachers," "communication with teachers," "motivating teachers to implement activities in classrooms," "integration of space science into existing curriculum," and "field trips and tours." In past years, we needed improvement on curriculum integration, and had master teachers Michael Lichtman and Georgia Romo do an extensive integration program for us. In the past, the field trip to Table Mountain Observatory was also weak, so we planned more activities with focused tours and mini-lectures during our day up there.

"Teachers Touch the Sky" has been immensely successful, but it is limited and modest, being funded by a small NASA EPO supplement to the author's main research programs. In the future we plan on making it into a more robust program, offering local school districts the opportunity to have master teachers who graduated from our program train and mentor other teachers in bringing experimental, inquiry-based science into the school curriculum. In-service training days and more "conferences" back at JPL for graduates of the program would also offer a lasting relationship for the 132 teachers who have graduated from the program.

Acknowledgments. This work was carried out at the Jet Propulsion Laboratory, operated by the California Institute of Technology under contract to the National Aeronautics and Space Administration.

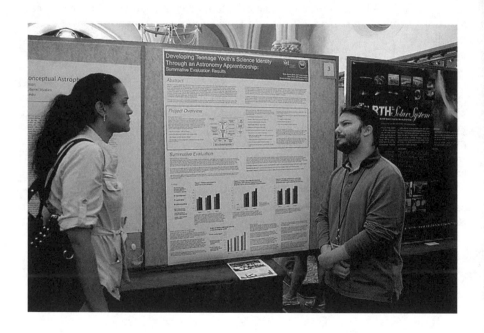

Participants discuss posters. Photo by Joseph Miller.

Connecting People to Science
ASP Conference Series, Vol. 457
Joseph B. Jensen, James G. Manning, Michael G. Gibbs, and Doris Daou, eds.
© *2012 Astronomical Society of the Pacific*

Workshops in Science Education and Resources (Project WISER): A Model for Building Content and Pedagogical Skills in Space Science for Elementary and Middle School Teachers

Sanlyn R. Buxner,[1] David A. Crown,[1] Larry A. Lebofsky,[1,2]
Steven K. Croft,[1,3] Thea L. Cañizo,[1] Elisabetta Pierazzo,[1] Steve Kortenkamp,[1]
Alice Baldridge,[1] and the Project WISER Team[1]

[1] *Planetary Science Institute, 1700 E Fort Lowell Rd, Suite 106, Tucson, Arizona 85719, USA*

[2] *Steward Observatory, University of Arizona, 933 N. Cherry Ave., Tucson, Arizona 85721, USA*

[3] *Pima Community College, 4905 E. Broadway Blvd., Tucson, Arizona 85709, USA*

Abstract. The Planetary Science Institute (PSI), in partnership with the Tucson Regional Science Center, is offering a series of professional development workshops targeting elementary and middle school science teachers in Southern Arizona. Facilitated by a team of earth and space scientists and educators, these workshops provide teachers with in-depth content knowledge of fundamental concepts in astronomy, geology, and planetary science. Each workshop is designed around core content big ideas (Wiggins & McTighe 1998), including all materials and assessments. During workshops, teachers participate in hands-on exercises using images and maps, and they conduct their own experiments.

1. Introduction

Over the past three years, the Planetary Science Institute (PSI) has developed a series of professional development workshops targeting elementary and middle school teachers to help increase their content knowledge in astronomy, geology, and planetary science. The two key goals of the workshops are to increase teachers' knowledge of current science and to give them more confidence in teaching these topics in their own classrooms. Each workshop has been developed around core big ideas (Wiggins & McTighe 1998), including presentations, activities, and assessments. Teachers participate with scientists and educators during three class sessions and complete homework activities designed to help them see the connection between the workshop topics and the Arizona Science Standard for the grade level they teach.

2. Current Professional Development Workshops

PSI has developed five workshops for teachers that build teachers' content knowledge in astronomy, geology, and planetary science. Lectures, activities and assessments are built around core big ideas.

2.1. Moon-Earth System

The Moon-Earth System workshop incorporates lectures, discussions, hands-on activities, and homework to help teachers understand historical and current exploration of the Moon, and to learn about future efforts to study the Moon. Topics and activities include: a) observing the Moon using ground-based photos and simulations to help teachers understand the Moon as viewed from their backyards; b) photogeology (using a variety of image data sets from spacecraft missions); c) impact experiments; d) examining planetary surfaces to determine relative and absolute ages; e) a discussion of how studying the geology of the Earth gives insights into studying the Moon; and f) learning about current and possible future exploration of the Moon.

2.2. Exploring the Terrestrial Planets

The Exploring the Terrestrial Planets workshop incorporates lectures, discussions, hands-on activities, and homework to help teachers understand past, current, and future exploration of terrestrial planets in the solar system. Topics and activities include: a) comparing physical properties of terrestrial planets, b) introducing the Moon-Earth system and global datasets of the terrestrial planets, c) a discussion of plate tectonics and geologic processes that create and modify planetary surfaces, d) using pennies to model half-life, and e) a photogeologic mapping exercise.

2.3. Impact Cratering

The Impact Cratering workshop incorporates lectures, discussions, hands-on activities, and homework to help teachers understand the history and significance of impact craters on Earth and other planets in the solar system. Topics and activities include: a) an introduction to impact craters on the Earth; b) exploring virtual field trips of impact craters on the PSI website; c) learning about impact rocks using PSI Impact Rock Kits; d) conducting impact experiments; e) discussing dating planetary surfaces with craters; and f) calculating the chances of impact for a planet.

2.4. Asteroid-Meteorite Connection

The Asteroid-Meteorite Connection workshop incorporates lectures, discussions, hands-on activities, and homework to help teachers understand the nature of small bodies in the solar system and how we can use them to understand the history of the solar system. Topics and activities include: a) investigating unexplored worlds; b) a tour of the solar system and a discussion about its formation; c) what lies beyond the solar system; d) investigating asteroid and meteorite spectra; e) learning about the composition of asteroids; f) measuring the densities of common compounds; g) exploring the descriptions and geohistories of asteroids Ceres, Vesta, and Hebe; and h) using PSI Meteorite Kits to examine samples from space. These activities culminate in the activity "Stories Rocks Can Tell," which integrates all the topics presented in the workshop.

2.5. Volcanoes of the Solar System

The Volcanoes of the Solar System workshop incorporates lectures, discussions, hands-on activities, and homework to help teachers understand the roles that volcanoes have played in shaping the surfaces of planetary bodies and the importance of past and current volcanic activity in the solar system. Topics and activities include: a) an introduction to volcanoes and volcanic processes; b) conducting an activity to identify igneous rocks; c) online research of volcanoes in Arizona; d) hands-on exploration of volcanic rocks using the PSI Volcanic Rock Kit; e) learning about volcanoes on other planetary bodies in the solar system; f) observing volcanic summit craters; g) exploring volcanic landforms around the Earth; and h) learning about volcanoes on Mars.

3. Workshops in Development

In addition to the five workshops currently developed (as of Fall 2011), we are developing two additional workshops related to planetary science. The Deserts of the solar system workshop will introduce teachers to deserts on Earth and Mars, using examples from the local Sonoran Desert environment. The Astrobiology and the Search for Extrasolar Planetary Systems workshop will introduce teachers to the concepts of habitability, looking for life on other planets, and looking for habitable planets outside the solar system.

3.1. Rock Kit Training Sessions

We are also developing a series of short training sessions so that educators can learn to properly use rock and meteorite kits that teachers have been introduced to in our workshops. After completing training, teachers and other community educators will be able to check out the kits for use in their classrooms, science fairs, star parties, and educational and social events.

Each kit contains supplemental materials including scientific background, supporting presentations, and additional ideas for using the kits in the classroom. These rock kits will also be available for check-out by scientists and others involved in education and outreach activities. Currently, we are collaborating with other institutions offering professional development workshops to extend the reach of this effort in classrooms and museums.

4. Evaluation Outcomes and Lessons Learned

Seventy-two teachers from 39 schools have participated in PSI workshops. Workshop participants represent schools with minority student populations ranging from 46% to 95%. One measure of success of our program is that over 50% of teachers have attended two, three, four, or five of our workshops. Teachers consistently cite hands-on activities, modeling of scientific process, and interaction with scientists as the three top benefits of the workshops. Additionally, they report an increase in the knowledge of science content, increased understanding of how science is actually conducted, and a greater confidence in their ability to teach earth and space science after their participation in the workshops.

Teachers report a great benefit of taking the workshops in any sequence they choose. They cite scheduling conflicts as the greatest barrier to being able to participate in a set sequence of workshops. This creates the need for each workshop to be a stand-alone experience and results in a wide-range of prior knowledge of participants.

Teachers consistently report wanting activities to take directly into their classrooms. Because the workshops are designed to provide teachers with science content through authentic research, teachers are encouraged to adapt/modify workshop activities to the level of instruction that is developmentally appropriate for the age of their students and aligned to the grade level requirements of the Arizona Science Standard. Additionally, the development of the rock kit trainings sessions is a direct result of teachers' requests to use the materials in their classrooms.

5. Workshop Dissemination

Through the Planetary Science Institute network of scientists, workshops will be offered in Wisconsin and Texas during 2012. Input from the results of these out-of-state workshops will be used to further refine the program.

Acknowledgments.   Our current workshops are supported by the National Aeronautics and Space Administration under Grant NNX10AE56G: *Workshops in Science Education and Resources (Wiser): Planetary Perspectives*, issued through the Science Mission Directorate's Education and Public Outreach for Earth and Space Science (EPOESS) Program.

References

Wiggins, G. & McTighe, J. 1998, Understanding By Design, (Alexandria, Virginia; Association for Supervision and Curriculum Development)

Connecting People to Science
ASP Conference Series, Vol. 457
Joseph B. Jensen, James G. Manning, Michael G. Gibbs, and Doris Daou, eds.
© *2012 Astronomical Society of the Pacific*

Thinking and Acting Like Scientists: Inquiry in the Undergraduate Astronomy Classroom

Bethany E. Cobb

The George Washington University, 2121 I Street, NW, Washington D.C. 20052, USA

Abstract. Students can benefit from a more authentic scientific experience in introductory astronomy laboratories. Rather than simply following step-by-step instructions to replicate well-known results, students in inquiry labs are forced to think critically and fully engage in the scientific process. Developing inquiry labs and activities can, however, be a challenging task. I present here some resources available for undergraduate-level educators who are interested in bringing inquiry into their classrooms. Even minor changes to current traditional labs can provide students with the opportunity to answer scientific questions more independently. I also introduce the idea of substituting scientific "poster" sessions for traditional "lab reports" to provide students with immediate feedback as well as exposure to their peers' work and thinking. Allowing students to think and act more like scientists can increase their interest and engagement in science and enhance their basic understanding of the scientific process.

1. Introduction

Engaging in inquiry in the classroom is an extraordinary way to increase student interest in science and to improve student understanding of how authentic scientific research is performed. However, effective and practical inquiry activities can be challenging (and time consuming!) to develop and implement. Fortunately, there are many excellent resources available for undergraduate-level educators who are interested in bringing inquiry into their classrooms. Some of these resources provide fully developed inquiry activities that can be put to use immediately in your classroom, or can be modified easily to fit your particular classroom needs. Other resources provide instructions for developing inquiry activities that require significant pre-class instructor preparation, such as building models for in-class use. These resources can also provide you with just pure inspiration—so that you can develop new inquiry activities suitable for your particular classroom situation. I describe here my use of inquiry activities in my undergraduate astronomy classroom. These activities were adapted from or inspired by many different sources. Working in collaborative groups, students in my classroom perform inquiry activities using real or simulated astronomical data and/or models, which allows them to think like scientists. The students then get to act like scientists by presenting results and conclusions to their peers in scientific poster sessions. Students also have the opportunity to think and act like scientists by simulating the process of developing scientific proposals and critiquing these proposals as members of the review committee. Using inquiry in the classroom is a great tool for improving student engagement and learning, and these resources will make it much simpler for you to bring inquiry into your own classroom.

2. Rethinking the Traditional Lab

In "traditional" labs, students typically follow step-by-step instructions to collect data and then analyze that data using methods prescribed in the lab manual. The results to be achieved in this process are often clearly laid out in the background or intro-duction section of the lab manual. Basically, students are given the answer, given the question, and then required to replicate the answer. This is, of course, not how "real" science is done and, therefore, provides students with no insights into the true scientific process. In the inquiry-based lab, students are required to take a more active role in determining what questions to ask and how to collect and analyze the data to answer those questions. This can be challenging for the students, as it requires more active thinking and creativity than simply following the steps in a traditional lab. However, these labs also allow students the thrill of making independent discoveries and engage them more thoroughly in the scientific process. These labs can also be challenging for the instructor. These labs can be harder to supervise and students lacking experience in this process may require significant coaching and assistance. Students may also fail to obtain a satisfactory answer to the question they posed—but that, too, is part of real science! Embracing this challenge and adding even a few inquiry-based labs to your curriculum can significantly enrich the students' experience of modern science. Even if a lab cannot be fully inquiry-based, slight modifications can allow students to solve problems more independently. Imagine a lab covering the inverse-square law of light. Instead of just telling the students that $F \propto 1/r^2$ and then asking them to verify this by measuring flux from a light bulb at set distances, why not give the students the experi-mental setup and ask them to use the setup to determine how distance effects light? Let them independently develop the methods of data taking and analysis and thus require them to think and act more like scientists.

3. Posters vs. Lab Reports

Traditional labs generally end with students turning in lab sheets (e.g. questions straight-forwardly asked in the lab manual and then answered in a few sentences). In some cases, students may turn in more formal "lab reports." These serve as the standard eval-uation tools to confirm that students have completed the assignment and are building knowledge. Again, however, they require little independence of thought or creativ-ity from the students. A method of evaluating student learning that requires more in-dependence from the students is to require them to present their methods, data, and conclusions in "poster" form (digital posters, paper posters, and posters on white-boards/chalkboards can all work). Students can then present their posters to other groups in the class (or the class as a whole) allowing for immediate feedback from both peers and instructors. Having to explain and defend their work both in written and oral form helps students internalize the material. Furthermore, immediate questioning can help students determine which parts of the material remain unclear to them. Students will also be exposed to novel thinking when they hear how other groups approached and tackled the problem at hand.

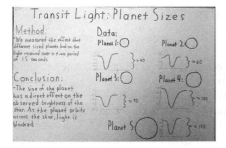
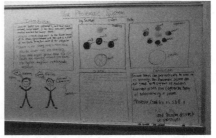

Figure 1.   Posters created by astronomy students to report on the results of a transiting planets inquiry lab (left) and an examination the importance of Galileo's observations of the phases of Venus in strengthening the Copernican Revolution (right).

4.   Inquiry Lab Examples, Resources & References

4.1.   Transiting Planets

Students performing the Transiting Planets lab at GWU are pictured in Figure 2. This inquiry lab was developed by N. J. McConnell et al. (2010), and the lab uses a relatively simple model (built from parts available at most craft/hardware stores) to simulate a transiting planet. Students use a light meter in the place of a telescope to answer student-developed questions such as, "can the size of the planet be determined?" or "can you tell if a planet has a moon?".

Figure 2.   Students performing a transiting planets inquiry lab.

4.2.   "Engaging in Astronomical Inquiry"

The Engaging in Astronomical Inquiry lab manual by S. J. Slater, T. F. Slater and D. J. Lyons (2010) contains about a dozen inquiry-based labs based on digital data and therefore the only equipment required to complete these labs is an Internet-connected computer. These labs explore some cutting-edge astronomy topics including the Hubble Ultra Deep Field, Extrasolar Planets, and Galaxy Zoo. The labs employ a "backwards faded scaffolding approach" that breaks down the steps of scientific inquiry to introduce them to students one step at a time, e.g., students first develop the ability to draw conclusions from data before being asked to devise novel scientific questions.

4.3.   Role Playing: Proposing Scientists & Review Committees

In the role plaing activity developed by B. E. Cobb,[1] students are given a scientific goal (e.g., "detect a new extrasolar planet") and a list of possible telescopes/instruments to build and are then separated into teams at competing "universities." A team member from each "university" makes a proposal to a review committee of classmates. The committee then decides how to allocate the funding "budget." The activity cycles through several rounds so that each student plays the role of proposing scientist and committee member at least once. Data "obtained" from the funded telescopes or instruments is presented to the class at the end of each round so for each new proposal cycle students must identify what new or additional data is required to make progress toward completing the scientific goal.

4.4.   The Copernican Revolution & The Phases of Venus

In the Phases of Venus lab developed by B. E. Cobb, students take on the role of pre-Galilean scientists. Using a very simple Sun/Venus model (light bulb and ping-pong ball), students ask questions about how the phases of Venus would appear from Earth in the Ptolemaic vs. Copernican systems. They must develop a hypothesis, collect model data, and then compare the model data to Galileo's actual observations of the phases of Venus in order to draw a conclusion about whether the heliocentric system or geocentric system best matches the observational data. (A whiteboard "poster" from this activity is pictured in the "Posters vs. Lab Reports" section above.)

Acknowledgments.   Bethany E. Cobb is supported by an NSF Astronomy and Astrophysics Postdoctoral Fellowship award (AST-0802333).

References

McConnell, N. J., Medling, A. M., Strubbe, L. E., Moth, P., Montgomery, R. M., Raschke, L. M., Hunter, L., Goza, B. K. 2010, "A College-Level Inquiry-Based Laboratory Activity on Transiting Planets," ASP Conference Series, 436, 97, arXiv:1009.3940
Slater, S. J., Slater, T. F., Lyons, D. J. 2010, "Engaging in Astronomical Inquiry," (Freeman). ISBN: 1429258608

---

[1]For more details on this activity inspired by C. D. Bailyn or the Phases of Venus activity, email bcobb@gwu.edu

Connecting People to Science
ASP Conference Series, Vol. 457
Joseph B. Jensen, James G. Manning, Michael G. Gibbs, and Doris Daou, eds.
© *2012 Astronomical Society of the Pacific*

Hubble's Early Release Observations Student Pilot Project: Implementing Formal and Informal Collaborations

Bonnie Eisenhamer, Holly Ryer, and Dan McCallister

Space Telescope Science Institute, 3700 San Martin Drive, Baltimore, Maryland 21218, USA

Abstract. The Hubble Space Telescope's Early Release Observations (EROs) were revealed to the public on September 9, 2009, and K–12 students and educators in five states across the country were able to join the celebration. To date, students and educators in Maryland, Ohio, New York, California, and Florida have participated in the Hubble Space Telescope's ERO Pilot Project. This is an interdisciplinary project created by the Space Telecope Science Institute's (STScI) Office of Public Outreach in which students use skills from subject areas such as language arts, science, art, and technology to research the four ERO objects and create compositions. In recognition of their participation, the students' compositions are displayed at host institutions in each state (a museum, science center, school, planetarium or library) during a special public event for participating students, their families, and teachers. As part of its evaluation program, STScI's Office of Public Outreach has been conducting an evaluation of the project to determine the viability and potential of conducting large-scale, formal/informal collaborative projects in the future and to share lessons learned. Lessons learned will be applied to a new interdisciplinary project, the James Webb Space Telescope Student Innovation Project.

1. Project Description

The Hubble Space Telescope's Student Early Release Observation (ERO) Pilot Project is an interdisciplinary project in which students explore and research Hubble's four ERO objects. They then select one of the four EROs and create a composition that communicates their research findings. Just like scientists, students describe what else they hope to learn about the ERO object they selected, why it is important for Hubble to investigate it scientifically, and why their object provides artistic value. Compositions can take various forms depending on the grade and skill level of the students. In recognition of their participation, the students' work is displayed at an area host institution (a museum, science center, school, planetarium, or library) during a special public event for participating students, their families, and teachers. In addition, select student projects are chosen for display at the Space Telescope Science Institute (STScI). The Hubble Space Telescope's Student ERO Pilot Project has been implemented in Ohio, California, New York, Maryland, and Florida. The purpose of the project is to promote enthusiasm for space exploration and introduce students to the latest Hubble images following Hubble Servicing Mission 4.

2. Project Activities

First, we worked with the consultant educators during the summer of 2009 to determine a basic framework for the project and activity guidelines. A key goal of the project was to create an activity that would be accessible to all grade-levels and types of learners. From past experience, we knew that the project should be interdisciplinary and open-ended, while still engaging students in a research opportunity related to the EROs and Hubble science. In this way, educators of all grade-levels and content areas within the pilot states would have the opportunity to participate, and would have the flexibility to implement the project in their classrooms in ways that suited their curricula and instructional programs.

Next, states were identified for the pilot effort and a tentative schedule was developed for implementing the project in these states. Specific criteria were utilized for identifying the targeted states such as: the state has already embedded Hubble education materials into curricula/initiatives as determined by our ongoing impact study, STScI staff is knowledgeable of the education community within the state as a result of past projects or collaborative efforts, there are contacts within the state who are motivated and committed to assisting with recruitment efforts and project implementation, there is a potential to reach underserved/underrepresented populations. Next, a plan was developed for identifying and recruiting specific schools and institutions within each state to participate.

In the meantime, a team at the STScI Office of Public Outreach began developing project support materials such as activity guidelines, project FAQ's (answers to Frequently Asked Questions), and criteria for the student compositions. A special, password-protected portal was developed specifically for project participants and was added to the Amazing Space Website. The portal contains the project resources for participating educators and links to research materials for students. In addition, a data collection tool was developed for capturing preliminary participant and event data in each of the four states.

Project recruitment began during the fall of 2009. Office of Public Outreach staff worked with contacts in the states of Maryland, Ohio, and Florida to recruit host institutions for the ERO public events, determine a date and time for the public events, and recruit schools/educators for project participation. The consultant educators focused their recruitment efforts in New York and California, and served as liaisons in these two states. Because of their experience and knowledge of the project participants in these two states, the education consultants continued to support the project by assisting with project evaluation activities, such as collecting attendance data for project events and conducting follow-up interviews with project participants.

3. Project Participation

On November 19, 2009, a Hubble ERO Pilot Project Event was hosted by the Maryland Science Center. During the event, students from participating schools in the Baltimore area were able to see their work displayed and explain their projects to parents and STScI staff. The event also featured the unveiling of a multiwavelength image of the galactic center of the Milky Way, a lecture about Hubble Servicing Mission 4 and Hubble's EROs, and a special 13-minute preview of the IMAX movie "Hubble 3D." This event was attended by over 100 parents, students, and educators.

On February 22, 2010, a STEM (Science, Technology, Engineering, and Mathematics) learning day was held in recognition of student participation in the Hubble Student ERO Pilot Project at Hamilton STEM Academy in Columbus, Ohio. Two assemblies were held during the day. During the evening, a nighttime event at the school brought together several hundred students, parents, and educators from four STEM Academies in the Columbus area. The nighttime event, entitled "An Amazing Space Night," showcased about 100 student ERO projects created by STEM Academy students in grades three through six. In addition, the event featured hands-on activity stations for building rockets and hover crafts, a Star Lab planetarium, and a STEM Club robotics activity.

From February through June 2010, similar events were held at three different venues in New York. Twenty-two California schools participated in the project from November 2009 through June 2010, with three major ERO events being hosted at the Adelanto County School District Office, the San Bernardino County Museum, and the Merced County Courthouse Museum.

In the spring of 2011, Stewart Middle Magnet School and Blake Magnet High School students and staff in Tampa, Florida worked together to create Hubble-centric projects and musical and dance performances. On May 4th and 5th, the schools hosted a two-day event during which student work was showcased. Participation in the Hubble ERO Pilot Project has introduced students to the ERO images while reinforcing connections between the schools' students and staff.

4.   Preliminary Findings and Lessons Learned

While implementing the Hubble ERO Student Pilot Project, we have learned that interdisciplinary activities work for the K–12 formal education community. Activities that incorporate Hubble discoveries, while addressing skills in language arts, communication, fine arts, or technology provide flexibility for exposing students to science content while addressing core curricular requirements.

Flexibility in student requirements allows a variety of grade levels and types of learners to be involved. Student compositions have ranged from pictures with captions, posters, physical models, PowerPoint presentations, reports and essays, dance recitals, musical performances, and videos. So far, students in second through twelfth grades have been involved in the project. In addition, the flexibility of the project criteria and requirements allowed for multiple institutions, schools, and grade-levels to collaborate in creating innovative programs that they otherwise could not have produced on their own.

We have also learned that the types of venues selected to host public events are contingent upon the needs of the communities/areas served by the project. For example, in upper-state New York, schools and district offices displayed student work and served as host institutions for public events due to limited access to a nearby, large museum or science center. While a key project goal was to implement a formal/informal collaborative project that would help establish or broaden connections between museums and schools, we found that identifying a museum or science center was not always feasible for all of the targeted areas of this pilot effort. As a result, host institutions for the ERO public events have varied and have included schools, school district offices, and boards of education in addition to museums and science centers.

Finally, we learned that projects implemented in different states benefit from a local coordinator or liaison to assist with project logistics, communication, and the collection of student work. Staying in contact with multiple schools and informal venues in several states can present a challenge for STScI staff members that are coordinating efforts from a distance. A local liaison not only has more background and experience with the target audience in a particular state, but also has more availability to address project questions and issues as they arise. In addition, a local liaison can assist with recruitment, promotional, and data-collection efforts.

Connecting People to Science
ASP Conference Series, Vol. 457
Joseph B. Jensen, James G. Manning, Michael G. Gibbs, and Doris Daou, eds.
© *2012 Astronomical Society of the Pacific*

Comet Inquiry in Action: Developing Conceptual Understanding of Comets through Stardust and Deep Impact Mission EPO Activities

Lori Feaga,[1] Elizabeth Warner,[1] John Ristvey,[2] Whitney Cobb,[2] and Aimee Meyer[3]

[1]*University of Maryland, Department of Astronomy, Bldg. 224 Farm Dr., College Park, Maryland 20742, USA*

[2]*Mid-continent Research for Education and Learning (McREL), 4601 DTC Boulevard, Suite 500, Denver, Colorado 80237, USA*

[3]*Jet Propulsion Laboratory, 4800 Oak Grove Drive, Pasadena, California 91109, USA*

Abstract.   NASA Discovery Program missions to comets—Deep Impact and Stardust, and their extended missions—are the rich source that their respective Education and Public Outreach teams mine to convey investigative concepts to K–12 students. Specially designed curricular activities strive to be engaging and represent science authentically. Even more, they unpack complex science content so students' conceptual understanding can develop. Multimedia elements—interactives, interviews, and games—enhance an educator's toolbox of materials used to reach diverse audiences and deepen understanding.

1. Overview

Best practice in education identifies the process of inquiry as core to developing student understanding. "With an appropriate curriculum and adequate instruction...students can develop the skills of investigation and the understanding that scientific inquiry is guided by knowledge, observations, ideas, and questions" (NRC 1996). What better model is there to study than the process planetary scientists and engineers engage in to design and fly a mission?

NASA Discovery Program missions to comets—Deep Impact and Stardust, and their extended missions—are the rich source that their respective Education and Public Outreach teams mine to convey investigative concepts to K–12 students. Specially designed curricular activities strive to be engaging and represent science authentically. Even more, they unpack complex science content so students' conceptual understanding can develop. Multimedia elements—interactives, interviews, and games—enhance an educator's toolbox of materials used to reach diverse audiences and deepen understanding.

How? By telling the mission story. Background information and context within the solar system help students get grounded and inspire questions: why go to small bodies in the first place? When the fundamental questions are apparent, the next step is clear: how do we get there? Once there: how is data gathered and how do scientists interpret it? Finally, we return to the starting place: what questions are answered—and what new ones arise?

Reaching students of all ages and backgrounds, in diverse settings ranging from urban to rural, elementary school to college, and in classrooms, after-school programs, and summer camps, Deep Impact and Stardust education materials tell the story of what it takes to design a mission and execute it. In doing that, it inspires a new generation to engage in the wonders of the solar system and ponder their involvement in future explorations and exciting discoveries.

2. Outcomes

- Students and educators learn about NASA's Discovery Missions Deep Impact and Stardust, and their extended missions, EPOXI and Stardust-NExT, and the preliminary findings of these missions.

- Participants understand how education and public outreach (EPO) materials from these missions can be used by teachers to develop the students' conceptual understanding.

- Participants understand how the EPO materials and mission design are parallel through the process of inquiry.

- EPO materials for these missions were successfully used in multiple venues with varying audiences in conjunction with recent mission events.

Acknowledgments.   This work was supported by NASA's Discovery Program; Stardust NExT contract NM0711001 to JPL, Deep Impact and DIXI contracts NASW00004 and NNM07AA99C to the University of Maryland, JPL contract 1331581 to McREL, and NASA Grant No. 09-EPOESS09-0044 issued through the SMD for *Small Bodies, BIG Concepts*.

References

NRC 1996, National Science Education Standards, (National Academy Press)

Connecting People to Science
ASP Conference Series, Vol. 457
Joseph B. Jensen, James G. Manning, Michael G. Gibbs, and Doris Daou, eds.
© *2012 Astronomical Society of the Pacific*

Visualizing Planetary Magnetic Fields (and Why You Should Care)

Matt Fillingim,[1] Dave Brain,[1,3] Laura Peticolas,[2] Darlene Yan,[2] and
Kyle Fricke[2]

[1]*Space Sciences Laboratory, University of California, 7 Gauss Way, Berkeley,
California 94720, USA*

[2]*Center for Science Education, Space Sciences Laboratory (CSE@SSL),
University of California, 7 Gauss Way, Berkeley, California 94720, USA*

[3]*Now at the Laboratory for Atmospheric and Space Physics and Department of
Astrophysical and Planetary Sciences, University of Colorado, Boulder,
Colorado 80309, USA*

Abstract. Since they are invisible to our eyes, planetary magnetic fields are difficult
to visualize. However, we can learn much about a planet, its interior, and its history
from studying its magnetic field. A challenge, then, is how can we effectively communicate the structure of planetary magnetic fields to the public, i.e., how can we help
the public visualize planetary magnetic fields. An additional challenge is how can we
effectively communicate the importance of studying planetary magnetic fields to the
public, or why should they care. We address these challenges by developing a series
of presentations about magnetic fields and their importance given on visually engaging spherical displays. We are also creating scientifically accurate three-dimensional
models of planetary magnetic fields.

1. Introduction

The magnetic fields of the large terrestrial planets, Venus, Earth, and Mars, are all vastly
different from each other. These differences can tell us about the interior structure,
interior history, and even give us clues to the atmospheric history of these planets.
Unfortunately, unless space can be permeated with tiny iron filings, magnetic fields are
invisible. As the saying goes, "out of sight, out of mind."

This leaves us with two questions. How can we best communicate the structure
of these planetary magnetic fields to the public? How can we best communicate the
importance of studying planetary magnetic fields?

We address these questions in two different ways: 1) by developing (and evaluating) a series of presentations given on visually engaging spherical displays in conjunction with hands-on activities, and 2) by creating scientifically accurate three-dimensional
models of planetary magnetic fields.

This work is part of a larger effort entitled "Seeing the Invisible: Educating the
Public on Planetary Magnetic Fields and How they Affect Atmospheres" funded through
a NASA ROSES Supplemental Education Grant. Our collaborators include science and
education professionals from the Space Sciences Laboratory (SSL), the Center for Sci-

ence Education at the Space Sciences Laboratory (CSE@SSL), and the Lawrence Hall of Science (LHS) at the University of California, Berkeley.

We will describe our presentations, 3D models, and future plans below.

2. Presentations

Our presentations are a combination of planetary images on engaging spherical displays, coupled with visual demonstrations. Our first presentation, entitled "Goldilocks and the Three Planets," is targeted to an elementary school age audience. The focus of this presentation is mainly on differences in the atmospheres of Venus, Earth, and Mars, and why Earth can support life. We employ an analogy with the familiar story of Goldilocks and the Three Bears: Venus is too hot, Mars is too cold, Earth is just right for liquid water to exist. It is the presence and stability of liquid water that made Earth habitable; liquid water helped remove the carbon dioxide from Earth's early atmosphere. To illustrate this point, we employ a demonstration. We take an ordinary rock, something like limestone works best, and place it in a container of vinegar. Bubbles begin to appear in the vinegar. These are bubbles of carbon dioxide. This shows that carbon dioxide is present inside rocks. Chemical reactions that take place in water take carbon dioxide out of the air and put it in rocks.

We tested and evaluated this presentation during the spring of 2010 on the six-foot diameter "Science on a Sphere"[TM] at the Lawrence Hall of Science in Berkeley, California. As a result of this testing and evaluation, we decided to remove the original demonstration illustrating the phases of matter with the rock, vinegar, carbon dioxide demonstration mentioned above. The earlier demonstration did not effectively communicate the importance of liquid water in changing the atmosphere early in Earth's history.

Our second presentation is targeted to a middle school age audience. It is entitled "Lost on Mars (and Venus)." This presentation focuses on differences in the magnetic fields of Venus, Earth and, Mars. To do this, we developed "global compass maps" to be shown on the Science on a Sphere. These maps show the direction a compass would point on the surface of the planet. In this format, it is very easy to see the differences between the magnetic fields of Earth (all compass arrows point north), Mars (compass arrows point in many different directions, particularly near crustal magnetic anomalies), and Venus (no arrows since there is no measurable magnetic field).

Differences in the "global compass maps" are due to differences in how the magnetic fields are formed. First we discuss the idea of a planetary dynamo. A planetary dynamo can create a planetary magnetic field if there is 1) a conducting (metal) layer in the planet that is 2) rotating and 3) convecting (churning). Earth's magnetic field arises from a planetary dynamo operating deep inside the Earth in its liquid iron outer core. It is made of iron, so it is a conductor. It is rotating because Earth is rotating, and it is convecting because it is a liquid being heated from the bottom.

The magnetic field of Mars, on the other hand, comes from surface rocks that trap or remember the magnetic field from long ago. This tells us that Mars does not currently have a planetary dynamo operating in its interior, but that it used to. Mars is smaller than Earth, so it's interior has cooled off faster. It no longer satisfies the third condition necessary for a planetary dynamo. Mars once had a planetary dynamo that the surface rocks remember, but it is no longer operating. We've learned something about the history of Mars!

Venus lacks any measurable magnetic field, so it does not have an operating planetary dynamo. However, Venus is about the same size and mass as Earth, so we expect the internal conditions to be similar. Clearly they are not. Additionally, the surface rocks don't appear to remember a past planetary dynamo. This is attributed to the surface being very young (as evidenced by its many volcanic features) and hot (the hottest planetary surface in the Solar System). The interior of Venus must be different than the interior of Earth, but we are not entirely sure how. This is a topic of current planetary science research.

As part of the presentation, we have small balls (several inches in diameter) with magnets placed at the poles (Earth), magnets scattered around the surface (Mars), or no magnets (Venus). These balls are designed to be used with small pieces of metal, small washers, and bent staples to trace out the magnetic field configuration.

We tested and evaluated this presentation during the spring of 2011 on the LHS's Science on a Sphere. The evaluation results showed that the target audience did gain understanding in the differences between the magnetic fields of Venus, Earth, and Mars as well as the origin of these differences.

Our final presentation will be targeted to a high school age audience. This presentation is still a work in progress. However, we plan to focus on the effects of the differences in the magnetic fields of Venus, Earth, and Mars. In the absence of a global, planetary dynamo driven magnetic field (such as is the case at Venus and Mars), the solar wind can slowly strip away the upper atmosphere. A strong, global planetary magnetic field (such as at Earth) can protect the atmosphere from the solar wind. The presence (or absence) of a planetary magnetic field is important for the long term atmospheric climate evolution of a planet.

In addition to finalizing this third presentation, our future plans include adapting this series of presentations from the six-foot Science on a Sphere format to a portable, table top spherical display system for traveling presentations. Through this project, in conjunction with other projects at the Center for Science Education, we have purchased a "Magic Planet"[TM] portable digital video globe from Global Imagination. This video globe will allow us to take these (and other) presentations into classrooms to reach students who do not traditionally go to the Lawrence Hall of Science.

3.   3-D Models

Recently, we finished construction of a scientifically accurate, three-dimensional scale model of the magnetic field of Mars interacting with the interplanetary magnetic field (IMF). We use rigid wires to represent magnetic field lines. Three types of magnetic field lines are present: lines (wires) which intersect the surface of Mars at two points ("closed" magnetic field lines emanating from the crust), lines which intersect the surface of Mars at one point while the other reaches the edge of the box (magnetic field lines that are "open" to the IMF), and lines which intersect the box at both ends (IMF lines). This model can be used in conjunction with the presentations as an additional visualization tool.

The pictures below show the views of the front and back of the model. The front of the model shows surface features of Mars (volcanoes and canyons); the back of the model shows a cut-away view of the interior of Mars. The crustal magnetic field lines are shown to be originating near the surface rather than deep in the interior. For scale, the length of the case is approximately 2 feet.

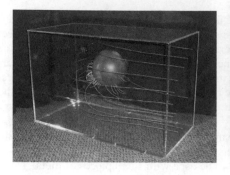

Figure 1.   Photographs of the front (left) and back (right) of our 3-D wire model
of the magnetic field of Mars

Finally, our future plans also include constructing similar wire models of the magnetic fields of Venus and Earth. These display models will be part of the traveling
presentations and will show scientifically accurate representations of each planet's magnetic field. These models will be important in communicating the shape and scale of
the magnetic field in relation to the planet.

Acknowledgments.   We would like to thank Sue Guevara and Gretchen Walker
at the Lawrence Hall of Science for their assistance with the Science on a Sphere.
In addition, we would like to thank Maia Werner-Avidon and Dawn Robles from the
Center for Research, Evaluation & Assessment at the Lawrence Hall of Science for
her work evaluating our presentation. This work is supported in part by NASA grant
NNX09AD43G.

Connecting People to Science
ASP Conference Series, Vol. 457
Joseph B. Jensen, James G. Manning, Michael G. Gibbs, and Doris Daou, eds.
© *2012 Astronomical Society of the Pacific*

The Universe at Your Fingertips 2.0 DVD-ROM: A Collection of Hands-on Activities, Resource Guides, Informational Articles, and Videos for Teaching Astronomy

Andrew Fraknoi,[1] Brian Kruse,[1] Suzanne Gurton,[1] Anna Hurst Schmitt,[1] Leslie Proudfit,[1] and Dennis Schatz[2]

[1]*Astronomical Society of the Pacific, 390 Ashton Ave., San Francisco, California 94112, USA*

[2]*Pacific Science Center, 200 Second Ave., N. Seattle, Washington 98109, USA*

Abstract. A new edition of the ASP's key educational publication *The Universe at Your Fingertips* has been issued in DVD-ROM format, containing 133 classroom-tested, hands-on activities (organized by subject), 43 articles with background information about topics in astronomy, 9 articles on teaching and learning space science in the 21st century, 17 guides to the best published and web resources on key topics, 12 short instructional videos, and a host of images.

1. History of Project

Starting in the early 1990's, the Astronomical Society of the Pacific (ASP) produced a resource notebook of astronomy activities and guides for teachers. First, this loose-leaf notebook accompanied the Society's national workshops on teaching astronomy in grades 3–12, but ultimately an expanded version became part of Project ASTRO, the Society's flagship program that trains and pairs volunteer amateur and professional astronomers with 4th–9th grade teachers in year-long classroom partnerships.

Starting in 1995, *The Universe at Your Fingertips* became the Society's best-selling item—ultimately tens of thousands of copies were distributed to schools, colleges, planetaria, museums, and informal science groups. A second volume, *More Universe at Your Fingertips*, followed a few years later, and the two volumes ultimately added up to almost 1,200 pages—so massive, it became affectionately known as the "ASP doorstop."

As the ASP developed new activities for its various programs and as the information and resource guides became dated, it was high time to put out a new edition. But with all the activities we had in mind, it would have been over 2,000 pages—expensive to print, hard on trees, and impossible to carry without serious weight training. But modern technology came to the rescue, and the whole package now fits on a single DVD-ROM, with everything ready to print as needed at the user's convenience.

The 133 activities come from Project ASTRO, Astronomy from the Ground Up, Family ASTRO, Night Sky Network, *AstroAdventures*, NASA missions, and many other astronomy education programs around the country. We are very grateful to the many dozens of projects and individuals who gave us permission to use these materials and to the many hundreds of educators who helped us test them in classroom situations.

2. Contents of the DVD-ROM

- 133 classroom-tested, hands-on activities (organized by subject)

- 43 articles setting out background information about topics in astronomy

- 9 articles on teaching and learning space science in the 21st century

- 17 guides to the best published and web resources on key topics

- 12 short instructional videos for some of the key activities and ideas

- some of the best modern astronomical images and a guide on how to find many more

3. Some of the Activities on the Disk

- Observing & Modeling the Moon's Phases

- The Reasons for the Seasons

- Sky Time: Kinesthetic Astronomy

- Pocket Sun Clock

- The Earth's Shape and Gravity

- Revolutionary Venus

- Mars Opposition Dance

- In the Footsteps of Galileo: Jupiter's Moons

- Should Pluto be a Planet: Student Symposium

- Making and Mapping a Volcano

- Impact Cratering

- Do Fish Believe in Water? Do Students Believe in Air?

- The Toilet Paper Solar System Scale Model

- Bike Years Versus Light Years

- The Pocket Solar System

- Making a Comet in the Classroom

- Vegetable Light Curves and Asteroids

- Star Finding with a Star Finder

- Three-D Constellations

- The Earth's Revolution and the Zodiac

- Convection and Miso Soup

- Maunder Mystery: Missing Sunspots

- So Many Stars, How Do You Count Them?

- Transit Tracks: Searching for Planets Around Other Stars

- Your Galactic Address

- How Many Objects in the Hubble Deep Field?

- Modeling the Expanding Universe

- Invent an Alien

- Red Hot, Blue Hot: Mapping the Invisible Universe

- Digital Images

- Light Pollution & Sky Brightness

- Did We Really Land on the Moon?

- Schoolyard Medicine Wheel

- The Toad in the Moon

- The Top 10 Tourist Sights in the Solar System

- Astronomy in the Marketplace

- Finding Your Way to Mars, Pennsylvania

4.   Organization of the Disk

4.1.   Arrangements through Menus

We have used menus on the DVD-ROM to help arrange the materials in a variety of ways:

Organized by Topic in Astronomy
Organized by Primary Setting or Skill Emphasized
Organized Into Sequences

4.2.   Sequences of Activities

We have put together ten sequences of activities which answer the sorts of standards-based questions that reflect the curriculum in many classrooms across the U.S.:

- How does the appearance of the Moon change with time?

- How does the Sun move through the sky each day and each year?

- What are the reasons for the seasons?

- How did we learn the size and shape of planet Earth?

- How can we get a sense of the scale (of sizes and distances) in our solar system?

- What do we find when we explore the other planets in our solar system?

- How can we find our way around the night sky?

- How can we understand the vastness of the universe (numbers of objects and distances)?

- How do astronomers know the motion and age of the universe?

- How do we search for signs of intelligent life elsewhere in the cosmos?

4.3. Multimedia Types

The disk also includes higher-resolution images that go with a number of the articles and activities, a set of sound files for the activity on the search for extra-terrestrial intelligence, and 12 short videos that explain some of the projects and model how to do some of the activities.

5. Who Can Make Use of the Disk:

- Teachers in grades 3–12

- College instructors of Astro 101 courses (many have adapted the activities into labs or small-group collaborative projects)

- Those who do teacher training workshops

- Family Science leaders

- Planetarium and museum educators

- Amateur astronomers who do outreach

- Nature center educators and park rangers

- Astronomers who volunteer in a classroom in a local school

6. How to Obtain a Copy

The DVD-ROM is made available through the ASP's non-profit AstroShop on-line.[1] Disks cost $29.95 individually and special bulk rates are available for larger orders. A special password protected website has been set up for registered users, with updates, corrections, and information on new developments in astronomy and education. Instructions for accessing the site and getting the password are on the disk.

The author can be contacted at afraknoi@astrosociety.org.

---

[1]www.astrosociety.org/uayf

Connecting People to Science
ASP Conference Series, Vol. 457
Joseph B. Jensen, James G. Manning, Michael G. Gibbs, and Doris Daou, eds.
© *2012 Astronomical Society of the Pacific*

Developing STEM Leaders Through Space Science Education and Public Outreach

Michael G. Gibbs and Dianne Veenstra

Capitol College, 11301 Springfield Road, Laurel, Maryland 20708, USA

Abstract.   Capitol College, located in Laurel, Maryland, established the Center for Space Science Education and Public Outreach with the mission to assist in educating future leaders in the science, technology, engineering and math (STEM). This presentation shares emerging best practices through innovative methods to create awareness regarding STEM outreach programs and activities related workforce development and career pathways.

1.   The 2011 Capitol College Developing STEM Leaders Program

Through the Capitol College Center for Space Science Education and Public Outreach, the Developing STEM Leaders Program was a free, semester-long series of workshops composed of high-impact programs to advance Capitol College students in science, technology, engineering, and mathematics (STEM) education and careers. The workshops were designed to be beneficial to undergraduate students studying in any of the college's degree programs.

2.   Workshops

The program consisted of a series of interactive workshops:

Workshop 1: Developing the Future Leader. The speaker was Kristin Waters, Assistant Registrar at Capitol College. The session provided students with life lessons, tips on finding passion and motivation, managing passion in a professional and personal environment, and remaining passionate for the future. Students were able to reflect on their goals and aspirations, and received tools to assist them with future goals.

Workshop 2: The NASA TRMM Mission and Velcro Sat Team. Julio Marius, NASA Mission Manager and Mission Director at the Goddard Space Flight Center, gave a status briefing on recent NASA missions, including the recent Tropical Rainforest Measurement Mission (TRMM). Marius assessed the students' initial knowledge of NASA and the space industry with an informal quiz, which he reviewed throughout his lecture. Students were introduced to the Space Operations Institute (SOI)—Capitol College's student-run NASA satellite control center—by the SOI student workers themselves.

Workshop 3: NASA Leadership (System Engineering). Tom Bagg, a Principle Systems Engineer at NASA Goddard Flight Center spoke on what a systems engineer does and what it takes to be a systems engineer. He gave real-life examples of the challenges and rewards that students would be likely to face in that field. Students then viewed a video lecture by Gentry Lee titled "So You Want to be a Systems Engineer," which illustrated the required characteristics to thrive in this dynamic field.

Workshop 4: STEM Workforce and Leadership Development. This workshop was held in conjunction with Capitol College's Career Fair, which gave the workshop series exposure with a larger pool of Capitol College students. At the Capitol College Career Day, students were provided with an opportunity to interact with representatives from local and regional businesses (e.g., aerospace, NASA, NSA, etc.) and learn about local careers available in STEM fields.

The participants then attended one or more of our three-part workshops on how to land a job in their field. Rick Sample, Director of Library Services, showed students how to properly research companies in which they are interested. Tony Miller, Director of Graduate Recruiting, spoke on how to conduct a productive job search. Ken Crockett, Director, Critical Infrastructures and Cyber Protection Center and internship liaison, concluded with tips on how to network and land internships.

Workshop 5: Project Management. Jason Copley, Associate Director of Development, spoke on the difference between leadership and management. Students were asked to complete an evaluation of the program. The important leadership traits that apply to engineering and project management responsibilities were stressed. Through lectures, exercises, and case studies, students gained a better understanding of their individual leadership abilities.

Workshop 6: Putting it All Together. The final event took place on May 2, 2011 from 4 to 5 p.m. at Capitol College in the Student Center. Certificates of completion were presented by Vice President Michael Gibbs, Ed.D., and Vice President Dianne Veenstra from Capitol College to the students who attended four of the five workshops.

3. Participants

Sixty students from Capitol College attended one or more workshop.

- Ten students received Certificates of Completion for attending four out of the five workshops.

- The participants' ages ranged from 18 to 42.

- The average participant age was 21.

- Females, 37%

- Males, 63%

4. Evaluation

When asked the questions below, participants answered as follows.

- Did this workshop provide you with information regarding career opportunities in STEM fields?

  - 90% indicated that they received information regarding career opportunities.

- Was this information of benefit to you?

  - 90% indicated that the information was beneficial.

- Would you recommend this program to others if it were offered again?

  - 100% would recommend this program to others.

- How would you rate the overall Emerging STEM Leaders Program workshop series? (This question was asked on a scale of 1 to 7, 7 being the highest score.)

  - 5.90 was the average score.

- How would you rate the overall presentations provided regarding career opportunities in a STEM field? (This question was asked on a scale of 1 to 7, 7 being the highest score.)

  - 5.7 was the average answer.

- How would you rate the overall presentations provided regarding leadership? (This question was asked on a scale of 1 to 7, 7 being the highest score.)

  - 6 was the average answer.

Acknowledgments. Capitol College and the Center for Space Science Education and Public Outreach thanks their sponsors: Maryland Space Grant Consortium, Lockheed Martin and Northrop Grumman. Without their financial support, this program would not be possible. The authors also thank Rachel Burns for her leadership in implementing the workshops during the 2010–2011 academic year.

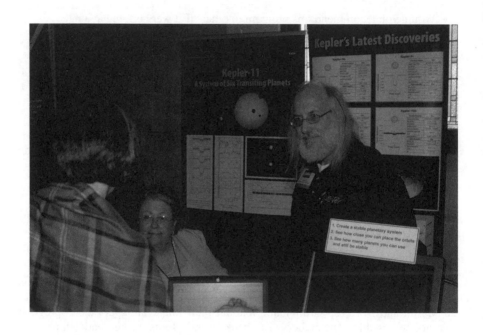

Edna DeVore and Alan Gould at the Kepler exhibit. Photo by Rich Berendsen.

Connecting People to Science
ASP Conference Series, Vol. 457
Joseph B. Jensen, James G. Manning, Michael G. Gibbs, and Doris Daou, eds.
© *2012 Astronomical Society of the Pacific*

Lifelines for High School Climate Change Education

Alan Gould

University of California, Berkeley, The Lawrence Hall of Science, 1 Centennial Drive #5200, Berkeley, California 94720-5200, USA

Abstract.   Lifelines for High School Climate Change Education is a project to establish a network of practicing high school teachers actively teaching climate change in their courses. The key aim of the project is creation of professional learning communities (PLCs) of teachers who meet mainly through teleconferences or webinar meetings to share best practices, strengthen knowledge, share resources, and promote effective teaching strategies. This is a NASA-funded project that incorporates analysis of NASA Earth observation data by students in classrooms. The project is exploring techniques to achieve the most effective teleconference meetings and workshops. This promotes not only teaching about minimizing environmental impacts of human activity, but minimizes environmental impacts of professional development—practicing what we preach. This poster summarizes project progress to date in this first year of a 3-year grant project. A number of PLCs are established and have ongoing meetings. There are openings for addition PLC Leaders to join and form PLCs in their regions.

1. Goals

Goals of Lifelines for High School Climate Change Education are:

- Establish Professional Learning Communities (PLCs) of high school teachers.

- The PLCs' goal: effective climate change education in existing courses.

- Identify the best resources for use in high school climate change education.

- Identify the best strategies for distance-meetings.

- Connect teachers with climate scientists.

2. Cascading leadership

We recruit leaders for PLCs who in turn recruit members for their PLCs (see Fig. 1). The idea is for project staff to recruit and organize 20 Professional Learning Community (PLC) Leaders, who in turn recruit PLC members (up to 15 members in each PLC). So there are 3 tiers: LHS organizers; PLC Leaders; PLC Members.

3. Partners

The Lifelines project partners are:

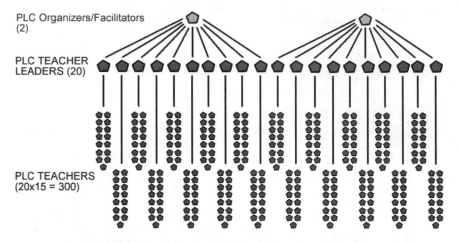

PLC Organizers/Facilitators (2)

PLC TEACHER LEADERS (20)

PLC TEACHERS (20x15 = 300)

Figure 1.   The project is identifying the optimum sizes for PLCs. Fifteen members seems about the upper limit for good communications and participation. Fewer than 5 and there can be faltering activity.

- Global Systems Science (GSS; LHS-UC Berkeley), providing core project staff.

- Great Explorations in Math and Science (GEMS), recruiting PLC members.

- Earth Science Information Partners (ESIP) Federation, providing connections with Earth Systems organizations, sources of data, and speakers.

- EOS-Webster (University of New Hampshire), providing satellite and Earth Systems data, scientist advisor and presenter.

- climatechangeeducation.org, advising on publicizing and dissemination of information.

4.   Distance communication tools

Various distance communication tools are being tested:

- Web sites, telecons with desktop sharing, e-mail lists.

- Telecon tools: Skype, DimDim, ReadyTalk

- Website tools: Google sites, Wikispaces, Wiggio

Lessons learned:

- Google sites is free and very easy to use as a repository/communication center (See Fig. 2).

- During real-time PLC telecons, notes can be taken on the PLC website, and sessions can be recorded for members to hear later.

- Skype, while free, can have problems in meetings with large groups. A paid service seems to work much better and may be worth the money for problem-free meetings.

- An online page of "Photo Name Tags" of each of the group members is helpful when hearing only the voices of people (see Fig. 3).

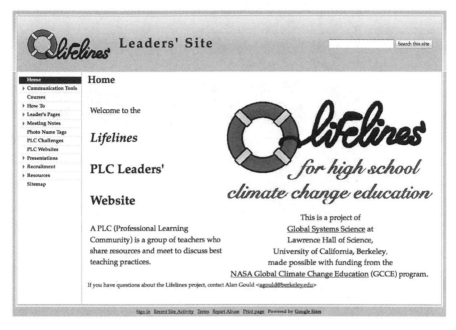

Figure 2.   This is the home page of the Lifelines leaders' website on Google Sites. The left navigation section has a number of useful features for the PLC leaders (See Fig. 4).

5. Recruitment

The project organizers/facilitators recruited 20 teacher leaders, but a few have dropped out, not being able to sustain their efforts. About 200 PLC members have been recruited, with varying levels of actual participation in the PLCs. Not all PLCs achieve 15 members.

A recruitment flier can be downloaded from the Lifelines website. We used a very nice feature of a Google Docs spreadsheet to create online forms for PLC applications.

6. Presentations

We have had presentations from several scientists and educators including Annette Schloss from the EOS-Webster project, Laura Tennenbaum from NASA's Global Climate Change website, and Matt Lappe from the Alliance for Climate Education (see Fig. 6).

Figure 3.   The "Photo Name Tag" page from the Lifelines leaders website.

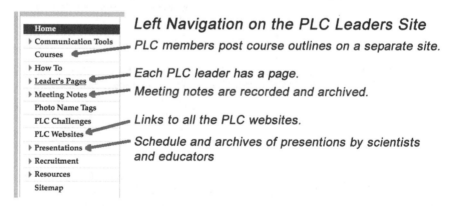

Figure 4.   This is a screenshot of the left navigation menu of the Lifelines leaders website, with explanatory comments. The "Resources" section was mostly moved to the "Course outlines" site (separate website) so that all PLC members from all PLCs could access, edit and add resources that they use in their courses.

We are experimenting with archiving webinar presentations with either straight recordings of the presentations or "talking slides" (audio of the presenter attached to each presentation slide). The "talking slides" concept seems very promising in that they are engaging and for the most part easier to download than movies.

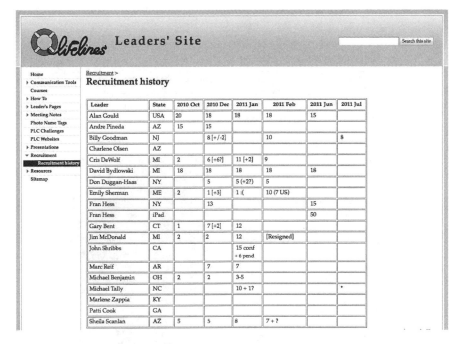

Figure 5.   The recruitment page on the Lifelines leaders website show recruitment progress in the project's first year.

Reference URLs

Lifelines for High School Climate Change Education websites:

- Main webpage: `http://www.globalsystemsscience.org/lifelines`

- Leaders site: `https://sites.google.com/a/berkeley.edu/lifelines`

- Course Outlines/Resources: `https://sites.google.com/a/berkeley.edu/lifelinescourses/`

Lifelines Presentations Archived

2010 Sep 22. Alan Gould, UC Berkeley, Lawrence Hall of Science, Director of Global Systems Science. **Overview** of the NASA Lifelines GCCE Project. For archive, see attachments at the bottom of <u>Alan's Leader page</u>.	**See also "<u>More Presentations</u>"** **(non-Lifelines archived** **presentations)**
2011 Feb 23. **EOS-Webster** (Accessing Satellite Data Pertinent to Climate Change) Annette Schloss, Univ of New Hampshire. Archive at <u>http://www.lawrencehallofscience.org</u> <u>/gss/lifelines/archive</u>	
2011 Mar 16. **The NASA Global Climate Change Website** Laura Faye Tenenbaum, education specialist for the climate communication team at JPL/NASA, and adjunct faculty at Glendale Community College. Archive at <u>http://www.lawrencehallofscience.org</u> <u>/gss/lifelines/archive</u>	
2011 Apr 6. **Alliance for Climate Education** - Sparking young mind to care about climate change Matt Lappé <<u>matt.lappe@climateeducation.org</u>>, Program officer for Alliance for Climate Education (ACE); 720-542-9815 Archive at <u>http://www.lawrencehallofscience.org</u> <u>/gss/lifelines/archive</u>	

Figure 6.   Presenters give telecons in which they share their desktops with all PLC members across the country—at least the ones who are available to attend. Presentations are recorded so those who can't attend "live" can still hear and see the presentation.

Connecting People to Science
ASP Conference Series, Vol. 457
Joseph B. Jensen, James G. Manning, Michael G. Gibbs, and Doris Daou, eds.
© 2012 Astronomical Society of the Pacific

A Teacher Professional Development Program on Comets

Mary Kay Hemenway,[1] Anita Cochran,[1] Judith Meyer,[2] Wade Green,[3] and Magdalena Rood[4]

[1] *University of Texas at Austin, 1 University Station C1400, Austin, Texas 78712, USA*

[2] *McDonald Observatory, 82 Mt. Locke Rd., McDonald Observatory, Texas 79734, USA*

[3] *Stony Point High School, 1801 Tiger Trail, Round Rock, Texas 78664, USA*

[4] *Third Coast Research and Development, 1108 S. 1ˢᵗ Street, Austin, Texas 78704, USA*

Abstract.   Various elements of a workshop centered on the EPOXI flyby of comet 103P/Hartley 2 were combined to meet several goals: participation in the Galileo Teacher Training Program, relation to a NASA mission, introduction to the Year of the Solar System, and continuing relations with teachers to pass current information on to their communities. The program elements include a residential workshop for teachers at McDonald Observatory, a short workshop at the Texas state science teachers' meeting, evaluation, and continuing electronic communication with participants. Evaluation results indicate the workshop successfully prepared teachers to implement activities and disseminate the new information with their students and their colleagues.

1. Introduction

From June 17–19, 2010, McDonald Observatory held an immersive experience workshop for fourteen secondary level teachers from Arizona, Florida, New Mexico, Oklahoma, and Texas. This workshop was the central element in a larger professional development program for these and other teachers. Other elements of the program included a shorter workshop and continuing electronic communication.

2. Program Elements

2.1. Summer Workshop

Although some of the planning for the summer workshop began with the original funding proposal, the majority occurred between October 2009 and May 2010. We chose an experienced high school astronomy teacher (who had participated in previous McDonald observatory programs) as a co-facilitator for the workshop. He assisted with the planning and preparation. His perspective as a classroom teacher provided an important point of view to the planning team, and eventually to the participants. Recruitment for the workshop occurred from November 2009 through mid-February 2010. We received

133 applications from which 15 were selected (this provided an extensive waiting list that was used to fill the program, as several of those originally selected canceled). An internal website was established as a communication tool for the participants. For our Texas Galileo Teacher Training program we wanted to include four elements: a product from the International Year of Astronomy, observing an object Galileo observed, using a computer-based application, and some basic astronomy concepts. The schedule shown in Table 1 incorporates these elements.

Workshop schedule:
June 17, 2010
Introductions/IRB documents/Pre-assesment
Activity: Scientific models survey
Activity: Optics and Galileoscope construction/use
Presentation: Dark Skies (light pollution)
Activity: Navigating the Night Sky; Celestial sphere
Reflection on relations to science/math teaching standards
Tour 2.7-m Harlan J. Smith telescope with Astronomer M. Williams
Activities: Formation of Solar System
Observing at Visitor Center with variety of telescopes
June 18, 2010
Review of observing session
Activity: Assembling and using a spectrometer
Tour: Hobby-Eberly Telescope (Herman Kriel, site manager)
Activities: Dry-ice comet video, Elliptical orbits, Comet Chronicles (Comet Bingo)
Reflection on relations to science/math teaching standards
Talk: Anita Cochran (science PI) on her comet research
Activities: Planisphere construction and use; Modeling the Night Sky (solar system motions), size and scale of solar system
(No observing due to cloudy weather)
June 19, 2010
Review of telescopes/instruments toured and used
Activity: Seasons; construct solar motion demonstrator
Activity: Concept map
Reflection on relation to science/math teaching standards
Post-assessment
Focus Group evaluation

2.2. Extension Workshop

In addition to the summer workshop, the instructional team was joined by two of the summer workshop participants at the Texas Science Teachers' Association annual conference in Houston on November 11, 2010, a week following the EPOXI fly-by of comet 103P/Hartley 2. Twenty-five teachers received materials about comets, performed an activity from the summer workshop, and were engaged with new science information from the fly-by.

2.3. Communication

Electronic mail was sent to the summer participants concerning astronomy events, opportunities, updates on the EPOXI mission and fly-by, and Year of the Solar System events. These teachers remain on our email list and continue to receive periodic updates on local and national programs.

3. Workshop Evaluation Procedure

A multi-phased evaluation was used to assess our success in meeting the goals of providing effective instructional materials for teachers to use in their classrooms, to enhance their knowledge base, to immerse them in the life of an astronomer, and increase their interactions with astronomers and other teachers. Evaluation methods were certified through approval by the University of Texas Institutional Review Board.

At the summer workshop, a pre/post self-evaluation instrument about telescopes was administered (locating celestial objects, basic physical principles, the telescope as a machine). Individual activities were assessed within the context of the activities. At the conclusion of the workshop, a focus group questionnaire was used to find out what the participants perceived they had learned, what connections to science standards they recognized, and what activities they planned to use in their instruction. Six months following the workshop, an on-line evaluation instrument was administered to assess the impact of the professional development workshop and use of the materials received. A telephone interview was conducted with a randomly selected participant. For the Texas Science Teachers' Association workshop, an informal survey of participants occurred at the conclusion of the workshop. Participants were asked to describe in writing what they had learned, what they planned to use from the workshop, and what they planned to share from the workshop.

4. Results

The evaluation results indicate minor differences between the summer workshop participants' replies at the conclusion of the workshop and those done six months later. At the conclusion of the workshop, the activities they reported as most likely to use were: dry-ice comet, Modeling the Night Sky (planetary motions), and seasons. Six months later they most frequently listed: dry-ice comet, Modeling the Night Sky, and telescopes. Although all teachers reported using NASA resources prior to the workshop, 70% of those completing the evaluation reported learning of additional NASA materials that were new to them. The following quotations from various evaluation instruments describe typical responses:

Science Standards: *"They are excellent for vertical alignment and build on the previous [standards] as the student goes up in grades. They are scaffolding."*

Teacher Needs: *"The activities were inexpensive, hands-on, reinforcement of content. There were diverse ways to add to instruction to hopefully add to student understanding. I found a lot of activities to reinforce process skills with my students."*

Most Useful Activities: *"The materials will better enhance the students' understanding of the ideas of space. More hands-on activities make it much easier to understand."*

Personal Interactions: (from focus group reports)

- Sharing with other teachers: *"The sharing with other teachers was a valuable part because we were able to see how other states have similar and different expectations, how other teachers use information in their classrooms, and how to connect elementary science topics to middle and high school topics."*

- With Astronomers: *"The interactions were a very valuable part of the workshop. The astronomers and staff were able to express their work and ideas in easy to understand ways..."*

5. Summary

Our concluding evaluation, from the survey six-months following the workshop, indicated that the primary benefits of the workshop were gains in knowledge, meeting researchers, the materials and resources they were given, and the activities ready to teach students. The workshop stimulated participants' interest in learning for themselves and sharing with colleagues. Teachers could immediately use the materials and experiences in the classroom. Within the fall semester following the summer workshop, most respondents had begun to integrate workshop materials into their lesson plans and were using the activities with students. Teachers noted that student outcomes included knowledge and a love of science activities.

Acknowledgments.   The National Aeronautics and Space Administration provides support for this project under an Education and Public Outreach supplement to Grant NNX08AO52G issued through the Office of Space Science. The support of Joe and Lucy Parsley for the Texas Galileo Teacher Training Program is also gratefully acknowledged.

References

Concord Consortium 2004, "Scientific Models Survey (NASA Universe Forum)", (Harvard-Smithsonian Center for Astrophysics), `http://www.cfa.harvard.edu/seuforum/mtu/`

National Academy of Science 1996, "National Science Education Standards", (National Academy Press)

Connecting People to Science
ASP Conference Series, Vol. 457
Joseph B. Jensen, James G. Manning, Michael G. Gibbs, and Doris Daou, eds.
© 2012 Astronomical Society of the Pacific

The Lowell Observatory Navajo-Hopi Astronomy Outreach Program

Kimberly A. Herrmann,[1] Deidre A. Hunter,[1] Amanda S. Bosh,[2]
Megan Johnson,[1] and Kevin Schindler[1]

[1]*Lowell Observatory, 1400 West Mars Hill Road, Flagstaff, Arizona 86001, USA*

[2]*Massachusetts Institute of Technology, 77 Massachusetts Avenue, Cambridge, Massachusetts 02139, USA*

Abstract. We present an overview of the Lowell Observatory Navajo-Hopi Astronomy Outreach Program, which is modeled after the ASP's Project ASTRO (Richter & Fraknoi 1994). Since 1996, our missions have been (1) to use the inherent excitement about the night sky to help teachers get Navajo and Hopi students excited about science and education, and (2) to help teachers of Navajo and Hopi students learn about astronomy and hands-on activities so that they will be better able to incorporate astronomy in their classrooms. Lowell astronomers pair up for a school year with an elementary or middle school (5th–8th grade) teacher and make numerous visits to their teachers' classes, partnering with the educators in leading discussions linked with hands-on activities. Lowell staff also work with educators and amateur astronomers to offer evening star parties that involve the family members of the students as well as the general community. Toward the end of the school year, teachers bring their classes to Lowell Observatory. The classes spend some time exploring the Steele Visitor Center and participating in tours and programs. They also voyage to Lowell's research facility in the evening to observe at two of Lowell's research telescopes. Furthermore, we offer biennial teacher workshops in Flagstaff to provide teachers with tools, curricula materials, and personalized training so that they are able to include astronomy in their classrooms. We also work with tribal educators to incorporate traditional astronomical knowledge. Funding for the program comes from many different sources.

A Riddle:

> Drive to the res, in hours, a few;
> Discuss Astronomy, old and new.
> Lead hands-on activities at the school;
> Hold a star party for all—oh, so cool!
> Bring the class to the hill named for Mars,
> Also a mesa—to unveil the stars!
> As part of what program do we teach?
> Lowell's Navajo-Hopi Astro Outreach!

The Partnerships: We typically work with 5th to 8th grade classes on the Navajo or Hopi Nations and with one teacher per school for only one school year. Roughly once per month we visit the class (often two to four hours travel time, each way) and leave all the materials at the school. We heavily use activities from "The Universe at Your Fingertips" and some of our favorite activities are: Moon Phases, 1000 yard Model of the Solar System, Venus Topography Boxes, Making & Using Galileoscopes, Make a

Comet, Making Craters, Reasons for the Seasons (see Fig. 1), Is it Alive?, and Constellations & Planispheres.

Figure 1.   Lowell pre-doc student, Christopher Crockett, demonstrating northern winter at the 2010 teacher workshop.

School Star Party: We believe it is important to involve the families of our participating students as well as their general community, so holding a star party at the school is an excellent opportunity. It can be as simple as collecting volunteers to run a few telescopes, bringing a few green laser pointers, and having the students themselves run a few Galileoscopes. Sometimes we also bring out a portable planetarium.

Class Visit to Lowell Observatory: We invite up to 35 students and several adult chaperones to come to Lowell Observatory. The covered expenses are typically one or two lunches, a dinner, and one night at a Flagstaff hotel (breakfast included). The field trip usually consists of (1) several special programs at Lowell Observatory's Mars Hill campus (often Cosmic Cart plus Clark and Pluto Telescope tours; for more specials, see our website[1]) and (2) a special visit to Lowell's research telescopes (Fig. 2) at Anderson Mesa (~30 min rotations among 4 to 6 stations which can include the 72- and 42-inch telescopes, a 8- to 10-inch portable telescope, a liquid nitrogen demonstration, Solar System Bingo and Snacks, and making a comet flipbook).

Teacher Workshops: Roughly every other year we hold a 1.5 day teacher workshop at Lowell Observatory for past and current teacher partners in the program. We invite each teacher to bring another teacher from his/her school. All expenses (travel, lodging, meals) are covered and each attendee leaves with materials (see Fig. 3) for all activities.

Cultural Sensitivity: Since we are working with two cultures with a rich traditional astronomical knowledge, we try to partner with Native American educators (see Fig. 4, left panel) to share traditional knowledge during our visits to the classes, star parties, the Lowell field trip, and/or teacher workshops. We are also very careful about cultural

---

[1]http://www2.lowell.edu/outreach/progref.php

Figure 2.   A student drawing of Anderson Mesa, her favorite part of the class trip.

Figure 3.   Lowell volunteer coordinator Mary DeMuth standing with the overflow materials she gathered for participants of the 2010 teacher workshop.

taboos (e.g., telescope viewing of the moon).

Past and Current Funding Sources: Though a NASA IDEAS grant was used to start the program, since then funding has come from a variety of sources, including: NASA EPO, the O. P. & W. E. Edwards Foundation, Honeywell, and The Bank of America. Over the past two years, through a partnership with the Hands-On Optics (HOO) team, we have been given 120 Galileoscopes, 34 tripods, and 30 HOO modules (~$19,540 worth of materials) funded by Science Foundation Arizona. Please see the website[2] for a complete list of our donors.

Program Evaluation: Since the start, we have had an external educator evaluate our program. This is facilitated by several questionnaires: (1) pre- and post-program questionnaires for the students and teachers, (2) post-program questionnaires for the participating astronomers, and (3) questionnaires for the teachers during the teacher workshops. Our external evaluator provides a report every 2 or 4 years.

A Success Story: The following e-mail was received from a partner teacher: *"Okay, if you ever had any doubt that your program is a life altering experience, this ought to set your mind at rest. [Jake], one of our fifth graders, was about the most unmotivated guy you ever saw at the beginning of the year. Poor work completion, 'this is boring' kind of guy. Well, last night, and of course following all the cool stuff you have shown him and the Lowell Trip, we had a benefit Bingo here at school. The fifth grade suggested that one of the [G]alileoscopes be used as a prize since it was a good cause and besides, they really wanted people to stay interested. To make a long story short, [Jake]'s mom won the scope and within 5 minutes [Jake] was down there and set up his own "mini Star Party" for the kids whose parents were here for the bingo. He focused on the moon and was telling about how craters are made. He sort of pointed where Saturn was the other night and said he thought it was around there, but that the moon was probably [too] bright to see it. He talked about constellations...it was a hoot! So, yes, this program has clearly been a life altering experience for [Jake] and I have [no/little] doubt it altered the others because they totally wanted it to be in the bingo so that someone could continue to enjoy using it. Just thought I'd share that story...you do good work! [Y]our program made a HUGE difference in [Jake]'s life and made him aware of options for his future that he never considered and which I think he may work hard to pursue and I think that is the beauty of this program."*

Figure 4.   Left: Verna Tallsalt, one of our partner Navajo Educators, giving a presentation at the 2006 teacher workshop. Right: Deidre Hunter, the leader of the Navajo-Hopi program.

For more information: Please see Hunter et al. (1999), the Outreach Program website,[2] or e-mail Deidre Hunter at dah@lowell.edu (Fig. 4).

Acknowledgments.   The authors would like to the thank the ASP and all the presenters for a stimulating meeting.

References

Hunter, D. A., Bosh, A. S., Stansberry, J. A., & Hunsberger, S. D. 1999, Mercury, 28, 18
Richter, J., & Fraknoi, A. 1994, Mercury, 23, 24

---

[2]http://www2.lowell.edu/users/outreach/outreach.html

Marni Berendsen and Anna Hurst Schmitt lead the "Finding Science in the Night Sky" workshop. Photo by Paul Deans.

Connecting People to Science
ASP Conference Series, Vol. 457
Joseph B. Jensen, James G. Manning, Michael G. Gibbs, and Doris Daou, eds.
© *2012 Astronomical Society of the Pacific*

Spatial Sense and Perspective: A 3-D Model of the Orion Constellation

Inge Heyer,[1] Timothy F. Slater,[1] and Stephanie J. Slater[2]

[1] *University of Wyoming, 1000 E University Ave., Laramie, Wyoming 82071, USA*

[2] *Center for Astronomy & Physics Education Research, 604 S 26th St., Suite C, Laramie, Wyoming 82070, USA*

Abstract. Building a scale model of the Orion constellation provides spatial perspective for students studying astronomy. For this activity, students read a passage from literature that refers to stars being strange when seen from a different point of view. From a data set of the seven major stars of Orion they construct a 3-D distance scale model. This involves the subject areas of astronomy, mathematics, literature and art, as well as the skill areas of perspective, relative distances, line-of-sight, and basic algebra. This model will appear from one side exactly the way we see it from Earth. But when looking at it from any other angle the familiar constellation will look very alien. Students are encouraged to come up with their own names and stories to go with these new constellations. This activity has been used for K–12 teacher professional development classes, and would be most suitable for grades 6–12.

1. Introduction

If we take a close look at people who are successful in their field of endeavor, we will likely find that not only were they generally creative and able to think outside their field box, but also that they were able to make connections—mental, conceptual, and physical—between things in ways that no one else has thought of before. Archimedes took a bath and made the connection to volume displacement, Galileo thought of using a spy glass for something other than spying, and a Persian mathematician thought that there should be a mathematical term for nothing, hence inventing the concept of zero, which allowed Newton to invent calculus. And it took a very creative engineer to think of using car air safety bags to land the Sojourner rover safely on Mars.

In the professional development activity described here, we have employed astronomy, literature, mathematics and art to create a 3-D scale model of the Orion constellation.

Science Fiction and Fantasy literature tends to contain a wide variety of references to astronomical phenomena, therefore this genre is well suited to make a connection between literature and astronomy. Recent books turned into movies such as *The Lord of the Rings*, *Harry Potter*, and *The Chronicles of Narnia* already have a wide following among students of all ages, and hence could be used as a good teaser to get them to think about astronomy issues, both fanciful and very real. "*The Lord of the Rings* (1955) draws heavily on Tolkien's academic interests in language, culture, mythology, and science. The works contain an impressive litany of astronomical knowledge (pertaining to such concepts as constellations, motions and phases of the moon, eclipses, and aurora) reflecting Tolkien's childhood interest in astronomy and other sciences" (Larsen & Bednarski 2008). Not only have ideas like this been used successfully in various observatory EPO informal science education activities, but they can also be applied in the formal class room.

2. Outcome Goals and Needed Supplies

This exercise was initially developed as one activity of a year-long professional development class taught in astronomy for K–12 teachers on the Big Island of Hawaii during the 2009 International Year of Astronomy (IYA). It is designed to be appreciated by students of varying interests, so even if they hate math and love art (or vice versa), there should be something of interest for most students. Furthermore, connections between these fields are made, and hopefully the students will pick up on all of the parts by working together. This activity is best for high school students (grades 9–12), but can also be used for grades 6–8 (with more scaffolding) and for college introductory astronomy classes (with less scaffolding). The activity should take about two hours.

Figure 1.   Hawaiian teachers building a scale model of the Orion constellation.

The outcome goals of this exercise are: (1) recognize science-related content in literature; (2) learn about perspective, line-of-sight, and relative distances; (3) apply basic algebra to determine the relative distances from actual data; and (4) apply artistic skills to construct and paint the parts of the Orion constellation model. The subject areas addressed are: astronomy, mathematics, literature, and art. The matching national science standards for grades 9–12 are: understanding and science inquiry, objects in the sky, science as a human endeavor, science in society, and the history of science.

This activity is best done in groups of three to six students. For each group the following supplies are needed:

- 2 square Styrofoam boards $12 \times 12 \times 0.5$ inches,

- 4 dowel rods 12 inches long,

- 16 beads 0.25 inches in diameter,

- 7 one-inch Styrofoam balls,

- 1 cotton ball

- 8 strings, thick thread, fishing line, or pipe cleaners 18 inches long

- Scotch tape,

- 2 sheets of 8.5 × 11 inch white paper,

- needle,

- ruler,

- 2 pencils,

- calculator, and

- student handout.

Figure 2.   The Orion constellation as seen from Earth (left) and from the side (right).

The literature paragraph is from J. R. R. Tolkien's *The Lord of the Rings* (The Fellowship of the Ring, Book 2, Ch. 2): "Aragorn smiled at him; then turned to Boromir again. 'For my part I forgive your doubt,' he said. 'Little do I resemble the figures of Elenil and Isildur as they stand carven in their majesty in the halls of Denethor. I am but the heir of Isildur, not Isildur himself. I have had a hard life and a long one; and the leagues that lie between here and Gondor are a small part in the count of my journeys. I have crossed many mountains and many rivers, and trodden many plains, even into the far countries of Rhun and Harad where the stars are strange.' "

Major Stars in the Orion Constellation: Betelgeuse (α Ori, at a distance of 427 ly), Rigel (β Ori, 772 ly), Bellatrix (γ Ori, 243 ly), Mintaka (δ Ori, 916 ly), Alnilam (ϵ Ori, 1342 ly), Alnitak (ζ Ori, 817 ly), Saiph (κ Ori, 721 ly).

3.   The Activity

1. Have one student read the literature paragraph. Ask the students to discuss in their groups what astronomical issues are being addressed ("the stars are strange"). Ask how and why stars would be "strange" in other places (here on Earth or elsewhere). Then have the groups share their ideas.

2. Give each group a copy of the Orion template. Point out the seven major stars and the Orion Nebula. Ask the students if they think Orion looks the same wherever they are. Why? Let them speculate, but don't give them the answer yet.

3. Hand out the rest of the supplies to the groups.

4. Take the first sheet of paper and mark off 0.5 inch from the top and bottom. Pick which side represents your view from Earth and mark it as such. Put the Orion template half on top of it, so you can trace the position of the eight objects (seven stars and one nebula) onto the "starting line" (the view from Earth). Be sure to mark your tick marks with the names of these objects.

5. From the table of stars pick the one furthest away (ϵ Ori, 1342 ly), and mark that on the far side from Earth (at 10 inches distance) exactly opposite the corresponding tick mark on the Earth side. Now we are ready to do some math. We need to figure out the relative distances of the other seven objects.

   Example: If 1342 ly corresponds to 10 inches, 1 ly corresponds to 10/1342 inches. Therefore, 427 ly (α Ori) corresponds to 3.18 inches. This requires proportional reasoning, which might present a challenge to students with lesser math abilities.

6. Do this for all the objects, and mark their positions on the paper exactly opposite their tick marks on the Earth side. Then put this paper onto one of the Styrofoam squares, put the other paper underneath, and the second Styrofoam square under that. Now take the needle, and punch one hole at each of the eight object positions through both papers and squares. Be sure to mark the orientation of the squares.

7. Other members of the group can meanwhile paint the stars and nebula appropriate colors. One could even paint the Styrofoam boards with a space background.

8. Someone else can at the same time knot one bead at the end of each string. Once we have made the holes, use the needle to put one thread through each hole of the bottom paper/Styrofoam combo. Then, put one painted Styrofoam ball onto each string for the stars, and the cotton ball for the nebula.

9. Have one person hold up the Orion template on the far side from Earth, so that they can determine how high up the objects should be. Determine the height by eye, then put a bit of scotch tape under each object to keep it in place. Next put the dowel rods in each corner. Be very careful of the orientation of the top paper/Styrofoam combo (the paper should be on top). Then use the needle to put the top of each thread through the appropriate hole in the top plate. Finish by tying a bead to each thread to keep things in place.

10. We now have a distance scale model of the Orion constellation. When looking at this model from the Earth side, we see the familiar constellation. But when looking at it from the side, it looks very different. The stars that appear next to each other in constellations are in reality nowhere near each other. So when viewed from anyplace but Earth, folks would see something very different. Refer the students back to the discussion at the beginning. Why does Orion look different from another point of view? (differing distances of stars).

11. Have each group design a new constellation from the Orion stars based on the different view point. Create a new name and write a new mythology for this new name, based on an invented culture. Here's where the creative writers get to shine, and the artists can contribute by painting the new constellation character (similar to the one of Orion the Hunter as seen from Earth).

12. Ask the groups to share their new constellation stories and pictures. This exercise can be done for other constellations as well; there are 88 to choose from.

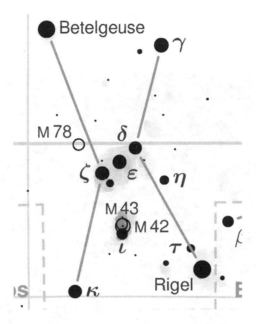

Figure 3. Orion template for the 2-D representation.

Acknowledgments. The inspiration for this activity came from a FINESSE workshop at the January 2009 AAS meeting.

References

Larsen & Bednarski 2008, "Muggles, Meteoritic Armor, and Menelmacar: Using Fantasy Series in Astronomy Education and Outreach," ASP Conference Series 400, 82

Poster and exhibit hall. Photo by Rich Berendsen.

Connecting People to Science
ASP Conference Series, Vol. 457
Joseph B. Jensen, James G. Manning, Michael G. Gibbs, and Doris Daou, eds.
©2012 Astronomical Society of the Pacific

A New Approach to Active Learning in the Planetarium

Tracy M. Hodge[1] and Jon C. Saderholm[2]

[1]*Physics Program, Berea College, Berea, Kentucky 40404, USA*

[2]*Education Studies Program, Berea College, Berea, Kentucky 40404, USA*

Abstract.   In a recent survey, Small & Plummer (2010) found that the goals of planetarium professionals are aligned with inquiry-based, active learning. However, most planetarium shows are designed as passive entertainment, with education as a secondary goal. In addition, there are very few research-based studies on the types of activities which promote greater learning within the planetarium environment, particularly at the post-secondary level. We report the results of the pilot test of a novel use of the planetarium to provide a simulated night sky, which students use to make longitudinal observations and measurements of planetary positions. In spite of several pragmatic limitations, the planetarium environment is well suited to student construction of both geocentric and heliocentric models of the solar system from direct observation. The curriculum we are developing addresses common misconceptions about the nature of science, in particular the use of modeling in the development of scientific knowledge.

1.   Introduction

The study of geocentric and heliocentric models of the solar system provides a particularly rich environment for addressing students' alternative conceptions of the motion of the Earth, the difference between observation and inference, and the use of models in science. There is ample evidence that many alternate conceptions on topics such the motion, relative size, and location of the Earth, Sun, Moon and stars persist in spite of specific instruction on the heliocentric model (Danaia & McKinnon 2007; Trumper 2001a; Trumper 2001b). Of particular interest, Zeilik has found that, while some alternative conceptions are relatively easy for college-age students to change with only limited corrective instruction, concepts related to frames of reference and relative motion are particularly resistant to change (Zeilik, Shau, & Mattern 1998; Zeilik & Morris 2003). Concepts such as these can be characterized as *entrenched presuppositions* that become organized into robust mental models which are difficult to adapt or replace (Vosniadou 1994).

When the question of heliocentric cosmology is addressed in the middle and high school curriculum, the geocentric model is often dismissed as "incorrect," depriving the students the opportunity to develop a deeper understanding of the motions that they observe every day, and leading to cognitive dissonance between what is perceived and what authority says is true (Shen & Confrey 2010). This leads students to incorrectly understand the role of scientists as creating authoritative objective knowledge, and undermines the importance of early historical models in the development of our current understanding of the universe (Desautels 1998). We argue that, in order to assemble a mental cosmological model that accurately reflects the natural world, students must be afforded the opportunity to analyze their existing explanations through an explicit metacognitive process (Vosniadou 1994).

The virtual environment created by a planetarium allows students to observe these celestial phenomena, free from the constraints of weather, time of day, or season. However, few planetarium facilities have been used as laboratory settings for astronomy students and there has been little research done on the effectiveness of inquiry-oriented, planetarium-based curricula (see Brazell & Espinoza (2009) for a meta-analysis of planetarium research). This may be the case because, rather than being conceived as active learning environments, planetaria are typically used as informal, passive learning environments that people visit perhaps once or twice (Small & Plummer 2010). We envision a planetarium-based laboratory curriculum which allows students to observe and measure celestial phenomena, through which they can consciously deconstruct their prior conceptions and build a more accurate model of celestial motion.

2. The Pilot Test

This instructional model was pilot-tested with 17 students enrolled in a 200-level science course designed for elementary education majors. Earlier questioning indicated that these students possessed little, if any, understanding of celestial motions, the nature of celestial objects, or the historical development of the modern heliocentric model of the solar system. For example, students could describe the sun as rising in the east and setting in the west, but did not know if the sun was ever directly overhead in the sky, the nature of lunar phases, or how to recognize a planet.

2.1. Observations

The students used a simple quadrant to measure the location of objects on the sky. The quadrants consisted of a shaft cut to 57.3 cm with a standard ruler fastened to one end. At a radius of 57.3 cm, an arc length of 1 cm will subtend an angle of one degree, providing a convenient scale for measuring angular separation to within a fraction of a degree. A protractor and plumb bob were attached to the shaft of the quadrant to measure vertical elevation (see Fig. 1).

Before any astronomical instruction was provided, students were introduced to the concepts of indirect measurement, altitude, and angular separation, and practiced using the quadrants to measure the height and angular size of a bell tower on campus. They then gathered in the planetarium where the current evening's sky was projected on the dome. The time was slewed forward, quickly causing the sun to set and the motion of the stars to become easily observable. After a short period of time, the motion was stopped and the students were asked to record what they had observed. The sky was reset to just after sunset, and the students were asked to measure the altitude of several stars in the western sky, and of Saturn (the only planet visible in the sky at sunset) using their quadrants. Azimuth was projected around the base of the dome using the existing mechanical projector. They also measured the angular separation of Saturn from Porrima (γ Vir), the nearest star, and recorded all their measurements in a chart provided to them. The daylight projector was turned on while the date was slewed forward one week. Again, the night sky was projected at sunset and students repeated the altitude, azimuth, and angular separation measurements. There was time to repeat

Figure 1.   Students built a simple quadrant for use in the planetarium.

this process one more time within the class period. Students were then asked to explain the changes they observed.

2.2.   Discussion

Students were able to observe diurnal motion of the Sun and stars. They were surprised by the existence of circumpolar stars, however, and had difficulty incorporating them into a model describing their observations. They did not autonomously describe the motions in terms of a celestial sphere rotating around the earth, instead relying on the authoritative heliocentric explanations they had learned previously. Only after a carefully scaffolded discussion, were students able to describe their observations of the diurnal motion in terms of a rotating celestial sphere.

Students were asked to describe their repeated observations of the sunset sky. They were quickly able to explain their observations as a westerly motion of stars into the sunset. Again, however, only after carefully scaffolded discussion, this time accompanied by video clips of examples of the relative motion of terrestrial objects, were they able to describe their celestial observations as originating from an eastwardly moving sun relative to the fixed stars. Because this instructional model was tested in a planetarium with mechanical projector there was not sufficient time to slew the date forward sufficiently for students to effectively observe any significant change in Saturn's position.

2.3. Limitations

The major difficulty in implementing the observations was due to the mechanical projector, which has a large footprint in the middle of the dome where student observations would ideally take place. The resulting parallax creates some variation in the angular separation between objects as well as azimuthal position. If students stay within several feet of the projector, however, these variations are manageable. It should also be emphasized that if they always view from the same location, changes in position are still easily measured. Altitude measurements are additionally affected by student height but can be corrected either mathematically, or by specifying the height from which students perform their measurements.

The most frustrating difficulty presented by the mechanical projector is the slow rate at which one can cycle through time. This presents pragmatic limitations to the time-frame of celestial motion which can realistically be observed in a single class period, and limits students' perception of subtle changes in celestial and planetary positions.

3. Future Work

Our long-term goal is to develop a set of a set of planetarium-based laboratory exercises which introduce students to the process of scientific inquiry through first-hand observation of celestial motion. The first module, which is the focus of this pilot test, will explore scientific observation and model building through the construction of first a geocentric and then a heliocentric model of the universe. In the next phase, students will be asked to make direct observations of retrograde motion, and modify their solar system models based on the new data. With a fully digital projector, students could then recreate Galileo's observations of the phases of Venus and the motions of the jovian satellites, leading into a discussion of Kepler's laws.

It should be emphasized that the purpose of this inquiry-based exercise is not primarily to present students with the concepts of either geocentrism or heliocentrism, but rather to engage them in the perceptual and cognitive processes resulting in the formation of those models. In this way, the features of scientific models and their relationship to the creation of scientific knowledge can become the subject of explicit instruction. Significantly, we believe these experiences will also prepare students to observe the natural night sky on their own.

References

Brazell, B. D. & Espinoza, S. 2009, "Meta-analysis of planetarium efficacy research," Astronomy Education Review, 8, 010108

Danaia, L. & McKinnon, D. H. 2007, "Common alternative astronomical conceptions encountered in junior secondary science classes: Why is this so?" Astronomy Education Review, 6, 32

Desautels, J. 1998, "Construction-in-action: Students examine their ideas of science," Constructivism in Education, Larochelle, N. B. & Garrison, J., (Cambridge University Press), Cambridge, 121

Shen, J. & Confrey, J. 2010, "Justifying alternative models in learning astronomy: A study of k-8 science teachers' understanding of frames of reference," International Journal of Science Education, 32, 1

Small, K. J. & Plummer, J. D. 2010, "Survey of the goals and beliefs of planetarium professionals regarding program design," Astronomy Education Review, 9, 010112

Trumper, R. 2001a, "Assessing students' basic astronomy conceptions from junior high school through university," Australian Science Teachers Journal, 47, 21

Trumper, R. 2001b, "A cross-college age study of science and nonscience students' conceptions of basic astronomy concepts in preservice training for high-school teachers," Journal of Science Education and Technology, 10, 189

Vosniadou, S. 1994, "Capturing and modeling the process of conceptual change," Learning and Instruction, 4, 45

Zeilik, M., Shau, C., & Mattern, N. 1998, "Misconceptions and their change in university-level astronomy courses," The Physics Teacher, 36, 104

Zeilil, M., & Morris, V. 2003, "An examination of misconceptions in an astronomy course for science, mathematics, and engineering majors," Astronomy Education Review, 1, 101

Marni Berendsen and Pablo Nelson at the ASP exhibit. Photo by Joseph Miller.

Connecting People to Science
ASP Conference Series, Vol. 457
Joseph B. Jensen, James G. Manning, Michael G. Gibbs, and Doris Daou, eds.
© *2012 Astronomical Society of the Pacific*

Astronomy Behind the Headlines: Developing a Series of Podcasts on Current Topics in Astronomy

Brian Kruse,[1] Carolyn Collins Petersen,[2] Mark C. Petersen,[2] Suzanne Gurton,[1] Anna Hurst Schmitt,[1] Andrew Fraknoi,[1] and Leslie Proudfit[1]

[1]*Astronomical Society of the Pacific, 390 Ashton, San Francisco, California 94066, USA*

[2]*Loch Ness Productions, Nederland, Colorado 80466, USA*

Abstract.   From 2009 to 2011, the Astronomical Society of the Pacific produced a series of podcasts for informal educators. Each episode gave a look behind the headlines in astronomy and space science, featuring interviews with leading astronomers involved in the research that made the news. Along with the interviews, links to related resources and activities were provided to help educators interpret the exciting topics for their audiences. We will describe the process of producing the podcasts and compiling the related resources, as well as the impact on podcast users.

1. Introduction

With support from an IDEAS grant from the Space Telescope Science Institute, the Astronomical Society of the Pacific produced a series of podcasts from 2009 to 2011. Called *Astronomy Behind the Headlines* (ABH), each podcast episode gives a more in depth look at current topics in astronomy and space science than is generally available in the mass media, interpreting the excitement of these discoveries for educators as well as the general public. Each episode is hosted on the ASP website[1] and also available on iTunes.[TM] In the summer of 2010, the first nine episodes were compiled onto a CD. Each episode on the CD includes, in addition to transcripts and resource guides, a featured activity for educators to use to engage learners in the topic. The CD was a featured resource at the fall 2010 Sky Rangers workshop in Yosemite National Park.
 Each episode consists of the following elements:

- five to seven minute interviews with leading researchers in astronomy and space science,

- interview transcripts,

- list of resources for further exploration,

- follow up discussions on the *Astronomy From the Ground Up* forum,

- follow up webchat discussions with featured scientists, and

---

[1]http://www.astrosociety.org/abh

- featured activities included on CD.

2. Podcasts

Eleven episodes were produced as a part of the *Astronomy Behind the Headlines* series of podcasts covering a wide range of topics from the news:

Episode 1: Astrobiology with Chris McKay

Episode 2: Cosmic Debris with Peter Jenniskens

Episode 3: Water on the Moon with Brian Day

Episode 4: Impact on Jupiter with Heidi Hammel

Episode 5: Measuring the Black Hole with Shep Doeleman

Episode 6: Kepler and the Sun-like Stars with Natalie Batalha

Episode 7: Gamma Ray Bursts with Dale Frail

Episode 8: Supernovae with Alex Filippenko

Episode 9: Cosmic Chemistry with David Grinspoon

Episode 10: Galaxies in the Early Universe with Rogier Windhorst

Episode 11: The Active Sun with Phil Erickson

3. Lessons Learned

By its very nature ABH is dependent on timely news for the heart of each program. At any given time, there are usually at least one or two stories that grab press interest. The ABH team members searched out topics they thought would interest the audiences, and tried to keep an open mind about the science results coming out. Selecting only *one* of many good ideas to use was, at times, problematic—every topic demanded an episode. As the team acquired more experience, the topic selection process was refined along with episode scripting, with some lessons learned.

- Topic selection evolved from a free-form approach in choosing without regard for any connections to other episodes to a more integrated approach in later episodes. This elevated the importance of connections between phenomena and the small set of basic scientific principles that explain the phenomena. Connecting episodes thematically can build a greater depth of basic scientific understanding in podcast listeners.

- Each episode was connected to some basic scientific background, even if only in the introduction. This served to help introduce the guest scientist, and took advantage of the opportunity to explain how basic science is used to explain the exciting phenomena appearing in the news.

- Not every scientist is a "pro" when "on mic." It is very helpful to pick scientists who have experience talking to the public and/or media.

- Scientists were involved in the scripting from the beginning. This gave them ownership of the final product, and invariably something unexpected came out of the interview.

- When developing the script, scientists were given the opportunity to explain the basic science. In the early episodes, the introduction was too comprehensive, and at times the scientist did not appear until almost halfway through the episode.

- In later episodes, not everything was scripted out in minute detail. The goal now is to have the interview sound like a conversation, not a reading.

- The script provided a framework for the scientists' explanation. Some scientists really needed this to keep them from going off on tangents.

Other lessons learned:

- In addition to the resource guides, we found that it was important to incorporate lessons and activities into each episode. This makes the podcasts more useful to formal educators.

- We need to expand the target audience. While there was success at reaching the target audience of informal educators with the podcasts, the formal education and amateur astronomer communities had little beyond the interview itself to use in their educational efforts.

4. Preliminary Evaluation Results

Preliminary results of an evaluative survey of ABH users with 40 respondents led us to conclude that:

- most users are associated with informal science education institutions;

- the most common audience ABH users interact with is the general public, followed by K–8 students;

- the most common venue where ABH users interact with their audience is at "star parties" or other public viewing events, school based events such as family science nights, and special events for the public such as scouting events and parties;

- the vast majority of ABH users are members of other astronomy education networks;

- ABH users listened primarily because they were curious about the subject matter and for their own personal use;

- about half of the ABH users indicated the use of the podcasts impacted their ability to share current astronomy news with their audience; and

- well over half of ABH users found the podcasts helpful in increasing their ability to answer questions from their audience about current astronomy news.

These results are encouraging as the ABH team looks towards future collaborations to produce future episodes of *Astronomy Behind the Headlines*.

5. More Information

For more information, go to: `http://www.astrosociety.org/abh`. If you are interested in partnering with us on producing podcasts, contact: Brian Kruse at 415-337-1100 extension 126 or at `bkruse@astrosociety.org`. To learn more visit the websites of the Astronomical Society of the Pacific (`http://www.astrosociety.org`) and Loch Ness Productions (`http://www.lochnessproductions.com`).

Acknowledgments.   The entire team at *Astronomy Behind the Headlines* would like to thank all the scientists for their willingness to share their insights on the wonders of the universe. A special thank you to Seth Shostak.

Connecting People to Science
ASP Conference Series, Vol. 457
Joseph B. Jensen, James G. Manning, Michael G. Gibbs, and Doris Daou, eds.
© *2012 Astronomical Society of the Pacific*

Using Wide-Field Meteor Cameras to Actively Engage Students in Science

David M. Kuehn[1] and Joy N. Scales[2]

[1] *Pittsburg State University, 1701 S. Broadway, Pittsburg, Kansas 66762, USA*

[2] *Blacksburg High School, 3109 Price's Fork Road, Blacksburg, Virginia 24060, USA*

Abstract.   Astronomy has always afforded teachers an excellent topic to develop students' interest in science. New technology allows the opportunity to inexpensively outfit local school districts with sensitive, wide-field video cameras that can detect and track brighter meteors and other objects. While the data-collection and analysis process can be mostly automated by software, there is substantial human involvement that is necessary in the rejection of spurious detections, in performing dynamics and orbital calculations, and the rare recovery and analysis of fallen meteorites. The continuous monitoring allowed by dedicated wide-field surveillance cameras can provide students with a better understanding of the behavior of the night sky including meteors and meteor showers, stellar motion, the motion of the Sun, Moon, and planets, phases of the Moon, meteorological phenomena, etc. Additionally, some students intrigued by the possibility of UFOs and "alien visitors" may find that actual monitoring data can help them develop methods for identifying "unknown" objects. We currently have two ultra-low light-level surveillance cameras coupled to fish-eye lenses that are actively obtaining data. We have developed curricula suitable for middle or high school students in astronomy and earth science courses and are in the process of testing and revising our materials.

1. Introduction

The wide availability of inexpensive, light-sensitive video cameras has opened the door of meteor tracking research to middle and high school students. The continuous monitoring allowed by dedicated wide-field surveillance cameras can provide students with a better understanding of the behavior of the night sky including meteors and meteor showers, stellar motion, the motion of the Sun, Moon, and planets, phases of the Moon, meteorological phenomena, etc. Additionally, some students intrigued by the possibility of UFOs and "alien visitors" may find actual monitoring data helpful for developing methods for identifying "unknown" objects. In this paper we describe a simple meteor camera system that is suitable and inexpensive, some of the associated software, and discuss some of the phenomena students can observe with the monitoring camera.

2. Materials

The core of our meteor camera is the Watec 902H Ultimate black-and-white video camera, which is a very low light level camera that senses a substantial amount of infrared light (in addition to visible light). Purchased new, these cost about $200, but we have found them on eBay for as low as $50. The lens we used was a Rainbow (model #L163VDC4P) 1.6–3.4 mm DC auto-iris lens. The camera/lens combination was mounted within a 3 inch SCH 40 PVC pipe using 1/4-20 bolts. There are other low-light-level cameras that work well, too. There are also a myriad of ways one could put together a camera housing by utilizing all the available PVC plumbing parts available at the local hardware store. Students could design a different one and make it a learning experience.

The camera is protected from the weather by a clear acrylic hemispheric dome ($12 from SurplusShed), set on a PVC toilet flange, and an outer 4-inch SCH 40 PVC pipe, in which the 12-Volt power supply for the camera is placed. Finally, the outside is covered with reflective bubble wrap for insulation purposes (not shown in Fig. 1).

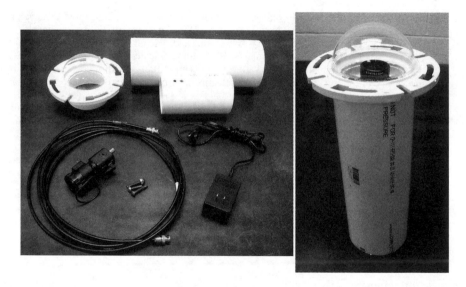

Figure 1.   Images of the disassembled and assembled meteor camera.

The video signal is digitized by a Hauppauge (Impact VCB model #64900) video digitizing card inside a standard Pentium 4-class computer. There are other hardware/software combinations that also work. For example, there are USB-based video digitization units that will collect video signals, allowing one to use a laptop computer.

Another important item is the software for triggering the video data collection. We use WSentinel from Sandia National Laboratory. It provides basic functionality for capturing meteors, but also takes video of other phenomena that might not be of interest (aircraft, birds, etc.). There are other software packages available, some of which work with more types of hardware, such as UFOCapture.[1] While it is expensive (about $250), you can get a free download of the fully-functioning software for testing purposes. It has many built-in features that aid meteor detection and tracking, filtering of non-

---

[1]http://sonotaco.com/e_index.html

meteor sources, and the company also provides other software for more automated analysis.

3.   What Can Students Learn?

Students can learn about geometry concepts (parallax, trigonometry) as well as physical science (displacement, velocity, optical distortion, time systems). Other phenomena such as cloud patterns and motion, lightning, stellar motion, and phases of the Moon contribute to students' understanding of Earth science. Finally, they can learn about how science is performed!

3.1.   Geometry

The camera's fish-eye lens provides a nearly hemispherical view of the sky, with the zenith (directly overhead) in the center of the image, and the horizon around the outer edge. Getting students to think of the geometry of this view might be challenging. Before attempting that, perhaps it would be easier to introduce the altitude-azimuth ("alt-az") coordinate system to them and have them practice pointing their arms in various alt-az directions given to them.

Then, place the camera in a fixed position on the floor of a classroom or gymnasium and take an image. Have the students measure the altitude and azimuth to various objects in the room (by pointing to and measuring angles with a protractor) and compare these with locations on the image. In this way, a rough calibration of pixel (x, y) and (altitude, azimuth) can be obtained.

Understanding parallax is also important because this is how the altitude of a suspected meteor can be measured. When using multiple cameras in locations separated by several miles, it is quite easy to detect the parallax shift relative to the background stars. There are many lessons available for teaching parallax. For a concept lesson, students can use their outstretched thumb at arm's length and close one eye at a time to see their thumb move relative to the background scene. Students can develop quantitative skills by measuring the distance to an object across the school grounds by measuring a baseline in the classroom windows and measuring the angle to the object using a protractor. A scale diagram and bit of trigonometry will yield the distance.[2]

3.2.   Other Phenomena

Besides the momentary second or two of meteor trails, there is also the chance for students to discover features of the night sky that they might not have thought about. Diurnal motion of stars, planets, and the Moon show up with just a cursory search of the night's captured video. By examining the time stamps on the videos, the angular rates of motion of "normal" objects will be quickly determined. Distant lightning storms show wonderful high-altitude phenomena ("sprites"). Longer term phenomena such as the changing phase of the Moon and the motion of planets relative to the background stars could be discerned with careful attention.

Many students are interested in detecting "unidentified" objects. There are certainly many fascinating phenomena in the sky, both man-made and otherwise. This is

---

[2]For example, see http://www-istp.gsfc.nasa.gov/stargaze/Lparalax.htm

Figure 2.   Image of a meteor obtained on 9 July 2011 at 03:14:20 UT.

an area where the investigational methods of science can really shine. While the optical resolution of most CCD cameras coupled to fish-eye lenses will not allow imaging of "flying saucers," if these objects did exist, their motions might be detected. Questions that students must address are flight speed, accelerations, and try to decide if humans have built such crafts. Some of the best video I have seen is the mating dances of fireflies caught with these cameras! To answer these questions, students must complete the entire data reduction procedure. By the way, it's fairly simple to identify fireflies since they will not show up on two cameras at the same time!

3.3.   Further Information and Material Sources

- NASA All-Sky Fireball Network: http://fireballs.ndc.nasa.gov/

- New Mexico State University All-Sky Network:
  http://skysentinel.nmsu.edu/allsky/

- SurplusShed: http://www.surplusshed.com/

- Cloudbait Observatory:
  http://www.cloudbait.com/projects/allskycamera.html

Connecting People to Science
ASP Conference Series, Vol. 457
Joseph B. Jensen, James G. Manning, Michael G. Gibbs, and Doris Daou, eds.
© *2012 Astronomical Society of the Pacific*

Bringing the Virtual Astronomical Observatory to the Education Community

Brandon Lawton,[1] Bonnie Eisenhamer,[1] Barbara J. Mattson,[2] and
M. Jordan Raddick[3]

[1]*Space Telescope Science Institute, 3700 San Martin Drive, Baltimore,
Maryland 21218, USA*

[2]*NASA/Goddard Space Flight Center and Adnet Systems, Inc., Greenbelt,
Maryland 20771, USA*

[3]*Johns Hopkins University, 3400 N. Charles Street, Baltimore, Maryland
21218, USA*

Abstract. The Virtual Observatory (VO) is an international effort to bring a large-scale electronic integration of astronomy data, tools, and services to the global community. The Virtual Astronomical Observatory (VAO) is the U.S. NSF- and NASA-funded VO effort that seeks to put efficient astronomical tools in the hands of U.S. astronomers, students, educators, and public outreach leaders. These tools will make use of data collected by the multitude of ground- and space-based missions over the previous decades. The Education and Public Outreach (EPO) program for the VAO will be led by the Space Telescope Science Institute in collaboration with the High Energy Astrophysics Science Archive Research Center (HEASARC) EPO program and Johns Hopkins University. VAO EPO efforts seek to bring technology, real-world astronomical data, and the story of the development and infrastructure of the VAO to the general public and education community. Our EPO efforts will be structured to provide uniform access to VAO information, enabling educational and research opportunities across multiple wavelengths and time-series data sets. The VAO team recognizes that the VO has already built many tools for EPO purposes, such as Microsoft's World Wide Telescope, SDSS Sky Server, Aladin, and a multitude of citizen-science tools available from Zooniverse. However, it is not enough to simply provide tools. Tools must meet the needs of the education community and address national education standards in order to be broadly utilized. To determine which tools the VAO will incorporate into the EPO program, needs assessments will be conducted with educators across the U.S.

1. What is the VAO?

The VAO is the U.S. virtual observatory project that provides access to data taken from a multitude of ground- and space-based missions. The VAO is funded by the NSF and NASA. It has a comprehensive EPO component that will provide general access to astronomical data via computer tools.

2. Purpose of the VAO EPO Program

Members of the VAO EPO program are in the process of introducing the education community to the VAO. We aim to provide standards-based activities that utilize the flexibility and power of the VAO. Along the way, We will identify those in the education community who want to work with the VAO EPO program and who will commit to a long-term partnership and the sharing of lessons learned.

3. Why Use the VAO?

The VAO is akin to the wealth of information-sharing brought about by social networking. Individual telescopes have their own custom-built data archives that store their images. The VAO connects all of these data archives with a common architecture. This will allow anyone in the U.S. to easily access and work with the data and do real computer and science lessons with that data. The benefits of using the VAO EPO resources include the following:

- the VAO EPO effort will provide the education community with a way to understand key science concepts via curriculum support tools that use real astronomical data and are aligned with national standards;

- the VAO has access to images from more than 30 telescopes, including NASA's Great Observatories (Hubble Space Telescope, Spitzer Space Telescope, Chandra X-ray Observatory, and the Compton Gamma Ray Observatory);

- the VAO incorporates a comprehensive blend of science, technology, engineering, and math (STEM), making it an appropriate tool for introducing STEM concepts into the classroom;

- the VAO provides students with opportunities to participate in real-world astronomical investigations, such as working with astronomical images, retrieving data from telescopes that observe across the electromagnetic spectrum, conducting image analysis via computers, and analyzing the chemical make-up of astronomical objects via spectra; and

- curriculum support tools will be available to support the studies of gravitation, dynamics, electromagnetism, and other basic physics principles.

4. Implementation Strategy

The VAO EPO team has adopted the following implementation strategy:

- introduce the VAO EPO program to the education community;

- conduct a needs assessment to ascertain the variety of ways the VAO can best be used by the community;

- work with VAO partners to deliver content that meets STEM education standards, including those of Project 2061;

- provide professional development opportunities to educators to show how to use VAO tools and integrate them into their programs;

- conduct impact studies to determine if and how the VAO tools are being used and if they need to be modified to better meet the needs of educators;

- utilize a flexible approach and develop new tools as necessary;

- make resources available on a dedicated VAO EPO website and through other commonly used astronomy EPO sites;

- provide curriculum support tools such as informational posters, activities, science content reading, and reflection questions; and

- explore new opportunities to connect with citizen-science projects and reach amateur astronomers.

5. VO Partners and Tools

We are currently exploring VAO EPO projects and collaborations with various VO groups, including the International Virtual Observatory Alliance (IVOA), Euro VO, Microsoft, Google, and several universities and NASA centers. The following VO tools are being explored by the VAO EPO project:

- World Wide Telescope (Microsoft),

- Google Sky (Google),

- Student Hera (HEASARC),

- Stellarium,

- SkyAlert,

- Zooniverse,

- PocketVO,

- Sloan Digital Sky Survey,

- Aladin, and

- DataScope.

Jason Tumlinson speaks during Science Night with STScI. Photo by Paul Deans.

Connecting People to Science
ASP Conference Series, Vol. 457
Joseph B. Jensen, James G. Manning, Michael G. Gibbs, and Doris Daou, eds.
© *2012 Astronomical Society of the Pacific*

Mobile Phone Application Development for the Classroom

Preston Lewis,[1] Daniel Oostra,[1] Sarah Crecelius,[1] and Lin H. Chambers[2]

[1]*Science Systems and Applications Inc., 1 Enterprise Parkway Suite 200, Hampton, Virginia 23666, USA*

[2]*NASA Langley Research Center Mail Stop 420, Hampton, Virginia 23681, USA*

Abstract.   With smartphone sales currently surpassing laptop sales, it is hard not to think that these devices will have a place in the classroom. More specifically, with little to no monetary investment, classroom-centric mobile applications have the ability to suit the needs of teachers. Previously, programming such an item was a daunting task to the classroom teacher. But now, through the use of online visual tools, anyone has the ability to generate a mobile application to suit individual classroom needs. The "MY NASA DATA" (MND) project has begun work on such an application. Using online tools that are directed at the non-programmer, the team has developed two usable mobile applications ("apps") that fit right into the science classroom. The two apps generated include a cloud dichotomous key for cloud identification in the field, and an atmospheric science glossary to help with standardized testing key vocabulary and classroom assignments. Through the use of free online tools, teachers and students now have the ability to customize mobile applications to meet their individual needs. As an extension of the mobile applications, the MND team is planning web-based application programming interfaces (API's) that will be generated from data that is currently included in the MND Live Access Server. This will allow teachers and students to choose data sets that they want to include in the mobile application without having to populate the API themselves. Through the use of easy to understand online mobile app tutorials and MND data sets, teachers will have the ability to generate unit-specific mobile applications to further engage and empower students in the science classroom.

Outcome

Generating customized mobile applications enables classroom teachers and other non-programmers to add a very powerful tool to their pocket in the form of a mobile app. After walking through the MND app generation system and completion of customized online tutorials, the classroom teacher will have the ability to utilize MND web API data sets to look at volumes of NASA satellite information. This will allow the teachers and students to pick and choose the data and locations that they study using authentic NASA data.

Poster displays. Photo by Paul Deans.

Connecting People to Science
ASP Conference Series, Vol. 457
Joseph B. Jensen, James G. Manning, Michael G. Gibbs, and Doris Daou, eds.
© *2012 Astronomical Society of the Pacific*

Astronomy Beat: A New Project to Record and Present the "Behind the Scenes" Story of Astronomical Projects and Programs

James Manning, Andrew Fraknoi, and Leslie Proudfit

Astronomical Society of the Pacific, 390 Ashton Avenue, San Francisco, California 94112, USA

Abstract.   We report on a relatively new project at the ASP that captures the spirit of astronomy research and astronomy outreach projects while the key players are still alive. Every two weeks, the Society publishes an "Astronomy Beat" column, explaining new developments and new ideas. At first, only members of the ASP can see them, but with time, more of the columns are being made available on the Web and through the educational programs of the Society.

For more than two years, the Astronomical Society of the Pacific has been experimenting with a new kind of web-based column that takes readers behind the scenes in astronomical research, education, and observing. In each 1000- to 1500-word edition, an "insider" profiles a project, program, or discovery in a first-hand way that the dry descriptions in professional journals just can't convey. For example:

- Bill Hartmann described how he and a colleague came up with the giant impact hypothesis for the origin of the Moon;

- Ray Weymann recalled how he and two fellow astronomers found the first gravitational lens;

- The webmasters of the "Astronomy Picture of the Day" explained where they find all the great images they feature;

- The Rev. Robert Evans, an amateur who held the world record for visual supernova discoveries, reminisced about his observing adventures;

- Pieces by Clyde Tombaugh and Edwin Hubble (written exclusively for the ASP, years ago) told the stories of their discoveries in everyday language;

- Michael Brown described how he finds and names dwarf planets beyond Pluto;

- Frank Drake told the full story of how he came up with the Drake equation.

You can see the titles and authors of the first 75 columns below.

At first, the columns are made available only to members and supporters of the ASP as a Society membership benefit. However, we have plans to make the columns more generally available in book format, and they are also used as appropriate in the many educational outreach programs and networks of the ASP.

The series was the idea of Jim Manning. When he took over as the Society's Executive Director, he wanted to take full advantage of the access the ASP has to astronomers, educators, and amateurs the world over. The idea is to continue the long tradition of helping our members "rub elbows" with the leaders of our field, but to do so in a way that fits our electronic age. Andrew Fraknoi edits the series and Leslie Proudfit designs and lays out the columns.

Authors receive a free year of membership in the Society and can, after a period of exclusive publication by the ASP, use the columns in their own educational work.

Some of the columns are also made public when they have particular educational function to play. David Morrison's debunking of the 2012 end of the world myth has been an especially popular outreach tool, gathering many tens of thousands of web hits. Clyde Tombaugh's story of the discovery of Pluto is a remarkable historical document and will be of interest as the New Horizons mission brings Pluto increasing public attention in the next few years.

Suggestions for new columns are most welcome. If you find that one of these columns is particularly relevant for your own educational program, you can feel free to approach the Society for permission to reprint. Over the years, we hope the series will form a useful archive of how astronomy and astronomy education projects were carried out during our times.

Astronomy Beat List of Columns (through June 2011)

Edited by Andrew Fraknoi

- *The Case Against Defining "Planet"*
  David Morrison, NASA Ames Research Center

- *Ceres: Not a Planet, Not a Plutoid (Or Is It)?*
  Mark Sykes, Planetary Science Institute

- *The Star of the Week* (website)
  James Kaler, University of Illinois

- *Exploring Earth-Approaching Asteroids with Radar*
  Steve Ostro, JPL

- *Observing Runs: Alone with the Night Sky*
  Michelle Thaller, Spitzer Science Center

- *Galileo's Glassworks: The Early Idea of the Telescope*
  Eileen Reeves, Princeton University

- *The Astronomy Picture of the Day Website: Behind the Scenes*
  Robert Nemiroff, Michigan Technological University

- *Astrophysics Faces the Millennium: How I Came Up with My Top Ten List for the Last Thousand Years*
  Virginia Trimble, (University of California Irvine

- *Where Science Meets Science Fiction*
  Mike Brotherton, University of Wyoming

- *Darkroom to Digital: A Photographic Journey*
  David Malin, Anglo-Australian Telescope

- *Observing with SOFIA, NASA's New Airborne Observatory*
  Dana Backman, SOFIA, SETI Institute, & the Astronomical Society of the Pacific

- *Prospects for Life on Jupiter's Ocean Moon: Tides are the Key*
  Richard Greenberg, (University of Arizona

- *Observing Mysterious Epsilon Aurigae*
  Jeff Hopkins (Hopkins Phoenix Observ.) & Robert Stencel (University of Denver)

- *Finding the First Clues to the Nature of Pluto and Its Distant Neighbors*
  Dale Cruikshank, NASA Ames Research Center

- *Celebrating the ASP's 120th Anniversary: Some Glimpses from Our Past*
  Andrew Fraknoi, Foothill College & the Astronomical Society of the Pacific

- *The Crowded Universe: The Search for Living Planets*
  Alan Boss, Carnegie Institution of Washington

- *What's In a Name? The Story of Makemake*
  Michael Brown, California Institute of Technology

- *The Navajo Sky–Down Under*
  Phil Sakimoto, University of Notre Dame

- *Galileo: Myths Versus Facts*
  Jim Lattis, University of Wisconsin, Madison

- *The Smell of the Oil and the Search for ET*
  Seth Shostak, SETI Institute

- *Satellite Revolution: Finding Small Moons of the Outer Planets*
  Scott Sheppard, Carnegie Institution of Washington

- *In Pursuit of Primeval Photons*
  Bruce Partridge, Haverford College

- *The Discovery of Pluto: Generally Unknown Aspects of the Story*
  Clyde Tombaugh

- *Planck Flies!*
  Bruce Partridge, Haverford College

- *Searching for Supernovae*
  Rev. Robert Evans

- *Doing Research on Astronomers*
  Erik Brogt, University of Arizona

- *The Universe in Living Color: The Secrets of the Hubble Images*
  Ray Villard, Space Telescope Science Institute

- *Spots on Neptune & Other Adventures in Planetary Astronomy*
  Heidi Hammel, Space Science Institute

- *Taking the Hubble Deep Field Image*
  Robert Williams, Space Telescope Science Institute

- *Earth Speaks: What Would You Say to ET*
  Douglas Vakoch, SETI Institute

- *Collecting Astronomy on Stamps*
  Ian Ridpath

- *Doomsday 2012, the Planet Nibiru, and Cosmophobia*
  David Morrison, NASA Ames Research Center

- *The Discovery of the First Gravitational Lenses*
  Ray Weymann, Carnegie Observatories

- *What if the Earth Had a Backwards-Orbiting Moon: The Art and Science of 'What If?' Questions*
  Neil Comins, University of Maine

- *The White House Star Party: Reports from the South Lawn*
  Stephen Pompea & Dara Norman, National Optical Astronomy Observatory

- *Preserving the Dark: Advocating Against Light Pollution at the American Medical Association*
  Mario Motta, MD

- *Heaven's Touch*
  James Kaler, University of Illinois

- *Finding the Naked s Process in Stardust*
  Donald Clayton, Clemson University

- *Dark Matter: A Talk with Vera Rubin*
  Douglas Isbell (Lawrence Berkeley Laboratory) & Stephen Strom (National Optical Astronomy Observatory)

- *Speaking for Astronomy: Harder than It Looks*
  Bruce Margon, University of California Santa Cruz

- *Exploring the Twitterverse*
  Peter Newbury, University of British Columbia

- *The Mojave Program: Exploring Powerful Jets from Supermassive Black Holes*
  Matt Lister, Purdue University

- *Discovering the Sodium Cloud Around Jupiter's Moon Io*
  Robert Brown, Space Telescope Science Institute

- *Uncovering the Secrets of Sirius*
  Jay Holberg, University of Arizona

- *My Sabbatical Year in Starry Sky National Park*
  Tyler Nordgren, University of Redlands

- *The Origin of the Drake Equation*
  Frank Drake (SETI Institute) and Dava Sobel

- *The Search for Molecules in the Universe*
  Phil Jewell, National Radio Astronomy Observatory

- *The Discovery of the Kuiper Belt*
  David Jewett, University of Hawaii Manoa

- *Celestial Harmony: Astronomical Music That Might Strike a Chord*
  Andrew Fraknoi, Foothill College & Astronomical Society of the Pacific

- *Adventures in Cosmology: Discovering the Expanding Universe*
  Edwin Hubble and Milton Humason, Mount Wilson Observatory

- *Explaining the Origin of the Moon*
  William Hartmann, Planetary Science Institute

- *The Formation of Enormous Black Holes*
  Mitchell Begelman (University of Colorado) & Martin Rees (University of Cambridge)

- *African Cultural Astronomy: The Conference, the Book, the Network*
  Jarita Holbrook, University of Arizona

- *Atoms as Historians: The History of A Course on History through Science*
  David Helfand, Columbia University

- *Why William Herschel is the Father of Modern Astronomy*
  Martin Griffiths, University of Glamorgan

- *Exploring Lakes on Titan*
  Alexander Hayes, California Institute of Technology

- *Dwarf Planets Yesterday and Today*
  Michael Brown, California Institute of Technology

- *How Kids Learn Astronomy: Insights from Research and Practice*
  Cary Sneider, Portland State University

- *Never-Ending Postcards from Mars*
  Jim Bell, Cornell University

- *Monster Stars: How Big Can They Get?*
  Paul Crowther, University of Sheffield

- *The Bruce Medalists: Makers of Modern Astronomy*
  Joseph Tenn, Sonoma State University

- *Encounter with Comet Hartley 2*
  James Green, NASA Headquarters

- *Astronomy in Popular Culture Quiz 2010*
  Andrew Fraknoi, Foothill College

- *Astronomy in Popular Culture Quiz 2010: Answers and Further Leads*
  Andrew Fraknoi, Foothill College

- *To Measure the Sky: Astronomy as an Intellectual Enterprise*
  Frederick Chromey, Vassar College

- *Stargazing Centaurs: The Astronomy of Harry Potter*
  Kristine Larson, Central Connecticut State University

- *Discovering a Cold Volcano on Titan*
  Rosaly Lopes, Jet Propulsion Laboratory

- *The Power of Stars: The Archaeoastronomy of Skyscrapers*
  Bryan Penprase, Pomona College

- *Rose O'Halloran: First Lady of the ASP*
  Rudi Lindner, University of Michigan

- *LCROSS: How We Crashed Into the Moon (On Purpose and for Science!)*
  Jennifer Heldmann, NASA Ames Research Center

- *Some Personal Reflections on the History of 20th Century Cosmology*
  Allan Sandage, Observatories of the Carnegie Institution

- *Comet Elenin: Cosmic Threat or Celestial Visitor*
  David Morrison, NASA Ames & SETI Institute

- *Nasty Astronomers: An Experiment in Prosopography*
  Thomas Hockey, University of Northern Iowa

- *Debating Pluto: Searching for the Classroom of the Future and Ending up in the Past*
  Tony Crider, Elon University

- *Starry Messages: Is Interstellar Archaeology Possible*
  Dick Carrigan, Fermilab

Connecting People to Science
ASP Conference Series, Vol. 457
Joseph B. Jensen, James G. Manning, Michael G. Gibbs, and Doris Daou, eds.
©*2012 Astronomical Society of the Pacific*

An Informed Approach to Improving Quantitative Literacy and Mitigating Math Anxiety in Undergraduates Through Introductory Science Courses

Katherine Follette and Donald McCarthy

University of Arizona, Steward Observatory, 933 N Cherry Ave, Tucson, Arizona 85721, USA

Abstract. Current trends in the teaching of high school and college science avoid numerical engagement because nearly all students lack basic arithmetic skills and experience anxiety when encountering numbers. Nevertheless, such skills are essential to science and vital to becoming savvy consumers, citizens capable of recognizing pseudoscience, and discerning interpreters of statistics in ever-present polls, studies, and surveys in which our society is awash. Can a general-education collegiate course motivate students to value numeracy and to improve their quantitative skills in what may well be their final opportunity in formal education? We present a tool to assess whether skills in numeracy/quantitative literacy can be fostered and improved in college students through the vehicle of non-major introductory courses in astronomy. Initial classroom applications define the magnitude of this problem and indicate that significant improvements are possible. Based on these initial results we offer this tool online and hope to collaborate with other educators, both formal and informal, to develop effective mechanisms for encouraging all students to value and improve their skills in basic numeracy.

1. Background

As technology, politics, medicine, and even ethics adapt to the challenges of modern life, educated people are faced with a continuous barrage of statistics, projections and graphical representations of trends intended to inform their decisions, and yet some estimate that as few as one-half of American adults have mastered the skills necessary to interpret such data. Innumeracy can have dire consequences for the American adult, as lower numeracy skills are postulated to result in "lower quality of life and more limited employment opportunities" (Kirsch et al. 1993, Charette and Meng 1998).

Over the past two decades, numeracy has been increasingly recognized as an essential component of literacy as a whole. It has been measured as one of three fundamental areas of literacy, along with prose and document literacy, on the Department of Education's National Assessment of Adult Literacy since 1985 (called the Young Adult Literacy Assessment in 1985). The 2003 results indicated that 55% of adults had either below basic or basic quantitative literacy skills, with only 13% performing at the level of proficient (U.S. Department of Education 2003). In order to correct this problem, both the pedagogy and the psychology of quantitative reasoning need to be carefully addressed.

2. Investigatory Study

We have begun a study to inform on whether the inclusion of numerical skills into the curricula of introductory science courses for non-majors improves students' comfort with and ability to manipulate and reason with numbers (i.e. their quantitative literacy). Essential skills identified by the National Council on Education and the Disciplines and the National Assessment of Adult Literacy include, but are not limited to: graph reading, proportionality, percentages, probability, and number sense. Each of these skills will be assessed in this study. We hope that the results will reveal whether introductory science courses for non-majors are a good forum for developing numerical skills and improving attitudes towards mathematics.

2.1. Study Motivation

While simple mathematical skills are essential to the understanding of the basics of virtually every scientific discipline, the increasing discomfort and lack of ability of college students to think and reason numerically has led many instructors to remove nearly all of the mathematical reasoning from their non-major science courses. Examples of mathematical misconceptions that we have encountered in our classroom (a very small subsample) are given in Table 1.

While math anxiety and quantitative illiteracy are well-documented phenomena, there is a decided lack of educational research into how to alleviate their effects at the college level outside of mathematics courses. This is the fundamental motivation for this study. We hope to answer some of the key questions about quantitative literacy in the science classroom. This includes assessing whether students can take the quantitative skills they may learn in Astronomy 101 and apply them to quantitative information they encounter in everyday life, as well as whether their attitudes towards mathematics can be improved. We will use these results to inform the development and conduction of a faculty development workshop series to support science faculty and graduate students in learning about evidence-based mathematics teaching practices that can be incorporated into the introductory science classroom.

Table 1.   Common mathematical misconceptions encountered frequently in our classrooms.

Operation	Common Incorrect Answer
$1 \div 5$	0.5
$0.5 =$	5%
How many seconds in an hour?	60 sec/min + 60 min/hr = 120 sec
$10^2 =$	20
$4.3 \times 10^6 =$	4.3000000

Obtaining such data is essential to inform this debate. This research will be important in providing a first quantitative look at whether and where basic quantitative illiteracies exist among college students enrolled in general education astronomy courses and whether they can be addressed through these courses.

2.2. Regarding Important "At Risk" and Influential Populations

Another key aspect of this study will be to inform on how certain pedagogical approaches affect members of populations that are particularly at risk for math anxiety and quantitative illiteracy. Several of these "at risk" groups have been identified in the literature, most notably/comprehensively by the results of the National Assessment of Adult Literacy (NALS), which surveyed 26,000 American adults in the areas of prose, document, and quantitative literacy in 1992, with a follow-up in 2003. This NALS defined quantitative literacy as "the knowledge and skills required to apply arithmetic operations, either alone or sequentially, using numbers embedded in printed materials; for example, balancing a checkbook, figuring out a tip, completing an order form, or determining the amount of interest from a loan advertisement," and included among the key results were the following conclusions:

1. More than 20% of American adults performed at the lowest level of quantitative proficiency, and an additional 25% of adults demonstrated the next lowest degree of proficiency. Both levels indicate an inability to perform more than a single operation or use more than one number from a document to determine information.

2. The quantitative proficiency of younger adults was lower than that of older adults, suggesting that the general numerical proficiency of the U.S. population is declining with time.

3. Individuals with the highest numerical proficiencies reported weekly wages two to three times higher than individuals in the lowest levels and were more likely to be employed.

4. Black, American Indian/Alaskan Native, Hispanic, and Asian/Pacific Islander adults were more likely to perform in the lowest two levels than white adults.

5. The quantitative proficiencies of men were somewhat higher than those of women.

6. Adults with physical, mental or other health conditions were more likely to perform in the lowest two levels of proficiency.

3. Conclusion

The dual problems of math anxiety and quantitative illiteracy are widely acknowledged and lamented in the educational literature, and a resounding call for action has been issued. We propose to rise to this call and begin studying how numeracy issues can be resolved or mitigated using introductory science courses, which we feel are uniquely suited to this purpose. Introductory science courses for non-majors, as they are often a student's last opportunity to see applied mathematics in a formal educational setting, are an essential forum for promoting quantitative literacy.

Results from the literature further suggest that neither math anxiety nor innumeracy are correlated with math ability (Ashcraft 2002). Both are strongly correlated, however, with math avoidance. This means that both innumeracy and math anxiety are significant sources of leaks in the STEM (science, technology, engineering, and mathematics) pipeline. In fact, the relationship between low self-efficacy expectations in mathematics and a tendency to avoid science-based college majors was revealed more than 25 years ago (Betz 1983), yet very few studies have since investigated how to alleviate it, and none through the vehicle of college-level introductory science. We believe that our study is an important first step in revealing whether introductory science courses for non-majors can be used to reverse negative attitudes towards mathematics, alleviate math anxiety, and develop the numerical skills that are essential for our students' success in life.

4. How *You* Can Help

Give the survey to your class! We're still recruiting instructors for fall 2011 and spring 2012! We'd also love your feedback, especially in the following areas:

- Do you feel there are any necessary quantitative literacy skills missing from our list? Which do you feel are the most important?

- We'd love to see how you emphasize quantitative skills in your labs, lectures, homework assignments, etc. and will be happy to keep any materials you provide confidential or to add them to our resource database for the workshop series.

- How do you connect quantitative skills to science, astronomy, and real life in your classroom?

- Is a one day workshop on Quantitative Literacy in the Astronomy Classroom something that you'd be interested in attending at a future ASP, AAS or AGU conference?

Please contact us with questions, comments or for more information. We'd love to hear from you! The authors can be contacted at `kfollette@as.arizona.edu` and `dmccarthy@as.arizona.edu`.

References

Ashcraft, M. H. 2002, Current Directions in Psychological Science, 11, 181
Betz, N. E. & Hackett, G. 1983, Journal of Vocational Behavior, 23, 329
Charette, M. F. & Meng, R. 1998, The Canadian Journal of Economics, 31, 495
Henderson, C. 2008, American Journal of Physics 76, 179
Kirsch, I. S. et al. 1993, National Center for Education Statistics
Madison, B. L. & Steen, L. A., 2003, "Quantitative Literacy: Why Numeracy Matters for Schools and Colleges," National Council on Education and the Disciplines
Meece, J. L., Wigfield, A. & Eccles, J. S. 1990, Journal of Educational Psychology, 82, 60
National Adult Literacy Survey 1992
National Assessment of Adult Literacy 2003, U.S. Department of Education
Peters, E., Hibbard, J., Slovic, P., & Dieckmann, N. 2007, Health Affairs, 26, 741
Resnick, H., Viehe, J., & Segal, S. 1982, Journal of Counseling Psychology, 29, 39
Steen, L. A. 1999, Educational Leadership, 57, 8
Steen, L. A. 2001, "Mathematics and Democracy: The Case for Quantitative Literacy," The National Council on Education and the Disciplines
Wigfield, A. & Meece, J. L. 1988, Journal of Educational Psychology, 80, 210

Connecting People to Science
ASP Conference Series, Vol. 457
Joseph B. Jensen, James G. Manning, Michael G. Gibbs, and Doris Daou, eds.
© *2012 Astronomical Society of the Pacific*

Engaging in Online Group Discussions Using Facebook to Enhance Social Presence

Scott T. Miller

Department of Physics, Sam Houston State University, Farrington Building, Suite 204, Box 2267, Huntsville, Texas 77341-2267, USA

Abstract.   A comparison study between two different methods of conducting online discussions in an introductory astronomy course was performed to determine if the use of Facebook as an online discussion tool has an impact on student participation as well as student response time. This study shows that students using Facebook for their online discussions participated more frequently and responded more quickly than students using a traditional online discussion forum.

1.   Introduction

While students within a traditional course are provided with multiple opportunities to interact with their peers, students within online courses find it more difficult to do so. Many course management tools provide discussion boards for faculty and students to communicate with one another, but students do not take full advantage of these resources. Discussion boards, in particular, are not highly effective at recreating the natural interaction between students in a classroom environment. In order for a student to participate in an online discussion, the student must log into the course, navigate to the discussion activity and choose to participate. Once a student leaves a discussion, he has no way of knowing if further discussion has taken place without returning to the course and to the discussion activity.

I used Facebook within two online astronomy courses as a tool for fostering a social presence among the students. A majority of students already use Facebook to interact with their friends and feel comfortable within its environment, facilitating their use of Facebook for course discussions. One advantage that Facebook has over course management discussion tools is the fact that when one student posts a topic to Facebook and another student responds, Facebook emails everyone who contributed to the discussion, notifying each student of the update. Another advantage of Facebook is that, in order to participate in a discussion topic, students don't need to log in to the course and navigate to the discussion activity. Instead, they simply read the postings on their Facebook status page, and when they see a post from the course, they can comment on it right then and there. This helps facilitate participation in the discussion activities and foster a sense of community within the course.

2. Data and Analysis

2.1. Comparison of Total Responses Per Semester

Figure 1 presents a histogram of the total number of posts plus responses per student based on the discussion forum used. During a typical semester, students were required to post a minimum of four articles (once per unit) and respond to other posted articles once per week (over twelve weeks) for a total of sixteen posts and responses. The vertical black line represents the minimum number of posts plus responses required for full participation.

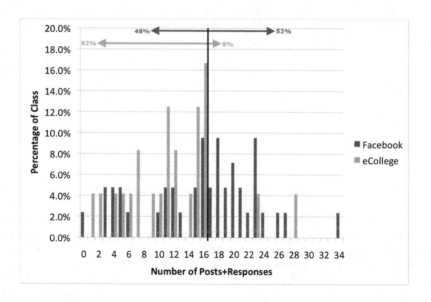

Figure 1.   A histogram of the total number of posts and responses in an online discussion forum over two semesters, based on the tool used to host the discussions.

Looking at the total number of posts plus responses, it is clear that students in the Facebook section were much more likely to post above the minimum requirement than the non-Facebook students. Over 50% of the Facebook students participated more than the minimum required, while only 8% of the non-Facebook students did. Looking at the median response rate for both classes, the median value for the non-Facebook students was 11.5 posts throughout the semester, while the median value for the Facebook students was 17 posts, or 48% more often.

2.2. Comparison of Total Responses Per Week

Figure 2 displays the number of responses posted per week as a function of the total number of students per class. Values over 100% indicate that there were more posts during that week than students in the class.

Every week, the Facebook section of the course participated more frequently than the non-Facebook section. During many weeks, the participation by the Facebook students was almost twice as frequent as that by the non-Facebook students. For at least the first half of the semester, participation from the Facebook students often exceeded 100%, while participation by the non-Facebook students never reached more than 75%. The difference between the two classes, though, diminishes over time. By week 8, the Facebook students are posting only 10%–20% more frequently than the non-Facebook students.

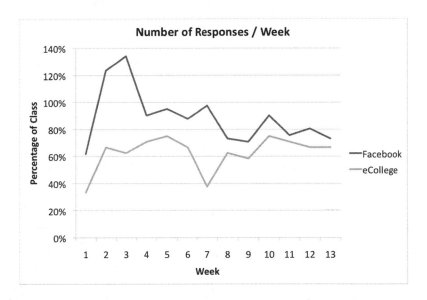

Figure 2.   The weekly number of responses to news articles for both sections. A value of 100% corresponds to a number of responses equal to the number of participating students.

2.3. Comparison of Response Times

Figure 3 presents a quartile plot of how quickly students responded to posts based on the day of the week it was posted. Each bar spans the range over which 25%–75% of the students first responded to an article. The median response time is indicated by a color change in the bar.

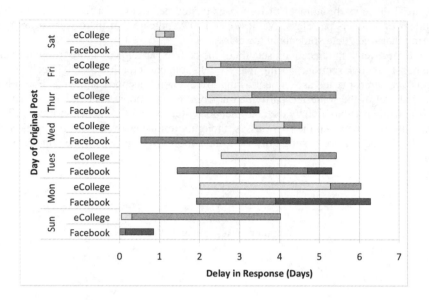

Figure 3.   Quartile plots of the lag time between when an article was posted and when students responded, as a function of the day on which the article was posted.

Articles were posted by the instructor roughly every day of the week while the weekly assignment deadline was Sunday at midnight. This accounts for the quick response rate to Sunday posts. For most days, the fastest 25% of Facebook student responses were almost a full day sooner than the fastest 25% of non-Facebook student responses, the exceptions being on Mondays (the day after the assignment deadline, when students were less prone to respond quickly) and Thursdays. Looking at the median values (the time it took for 50% of the students to respond), the difference isn't as pronounced. With the exception of Mondays and Wednesdays, the differences in the median response times are less than half of a day.

2.4.   Frequency of Multiple Responses Per Week

We can also look at individuals in both courses to see if students in one section were more likely to post multiple times per week than students in the other. Figure 4 graphs the percentage of students in each section who posted more than one response per week. Students in the non-Facebook section rarely responded to more than one post per week. For most of these students, they only completed the bare minimum assignment each week, if even that much. Clearly most of them were only doing what they had to in order to complete their assignment and nothing more! The same is not true for the Facebook students. Almost every week at least 10% of the students (or just under) responded to multiple posts, and many weeks the multiple response rate was closer to 20%–25%.

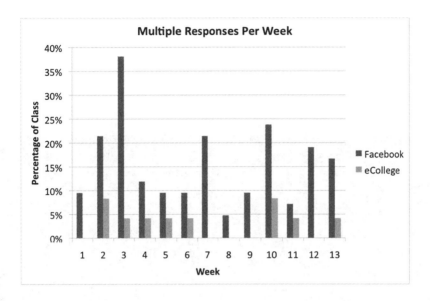

Figure 4.   A histogram of the percentage of students in each section who posted more than one response, by week.

3. Conclusions

Facebook can be an effective tool for facilitating online discussion among students, who are already comfortable using this medium. In comparing the two sections, Facebook students not only responded more frequently to articles posted for discussion, but they also responded more quickly to posted articles. While Facebook students typically posted more frequently, it appears as if the rate of responses dropped off over the course of the semester, almost down to the level of the non-Facebook students. It could be that the novelty of using Facebook for online discussions wore off around half way through the semester, but other factors, such as an increased workload in other courses, could also play a role.

Nick Veronico at the SOFIA exhibit. Photo by Rich Berendsen.

Connecting People to Science
ASP Conference Series, Vol. 457
Joseph B. Jensen, James G. Manning, Michael G. Gibbs, and Doris Daou, eds.
© *2012 Astronomical Society of the Pacific*

"Lunar Phases Project" as a Foundation for the Development of Innovative Inquiry Based ASTRO 101 Activities Utilizing Existing Concept Inventories as Assessment Tools

Manuel J. Mon and Angela Osterman Meyer

Evelyn L. Egan Observatory and the Whitaker Center for STEM Education, College of Arts and Sciences, Florida Gulf Coast University, Fort Myers, Florida 33965-6565, USA

Abstract. The cause and process of the lunar phases are difficult concepts for undergraduates and non-science majors to grasp. At Florida Gulf Coast University (FGCU) we have combined an inquiry-based instructional method (Mental Model Building) which can be more effective increasing students' conceptual understanding of the lunar phase cycle, together with the students' own observations. Undergraduate and non-science major students completed a hands-on project designed to integrate real observations, application of the scientific method, and Mental Model Building to connect the students' own observations to the Earth-Sun-Moon orientations responsible for their findings. Students' learning was assessed by administering the Lunar Phases Concept Inventory (developed by Rebecca S. Lindell and James P. Olsen, Southern Illinois University) before and after students completed the project, with positive results. We describe the methodology and activities utilized in our Lunar Phases Project, and propose their expansion to a variety of astronomical topics for undergraduate non-science majors and pre-service teachers. We emphasize developing and implementing new instructional strategies through the expansion of the Mental Model Building and similar pedagogical methodologies to develop innovative inquiry-based projects and activities in a variety of astronomical topics for undergraduate non-science majors and pre-service teachers. In order to meaningfully assess the new curriculum tools, we recommend utilizing already existing research-validated concept inventories specific to the astronomy content in the curriculum tools. These inventories can be analyzed to determine the conceptual learning gains achieved by the participating students and with further analysis can be used to refine portions of the activity under study.

1. Introduction

The growing field of research into astronomy education provides an excellent opportunity to improve upon the teaching of specific concepts such as the teaching of the lunar phases. *In this project, we propose to improve the teaching of astronomical topics to undergraduate non-science majors by using already proven assessment tools. In addition, we will assess the efficacy of our Lunar Phases Project using an established concept inventory so that we can directly link our efforts to student learning gains.*

2. Implementation of the Lunar Phases Project (LPP):

Before engaging in any discussion of the lunar phases or starting the project, we administer the Lunar Phases Concept Inventory (LPCI) (Lindell 2001, Lindell & Olsen 2002) as a pre-test of students' baseline knowledge. After that, the project materials and basic background information are provided. The students gather observational data through a full lunar phase cycle, plus one week. A table of Moon rise and set times provides a guide for when to record an observation on any given day. Students use a simple sextant they make themselves to measure and record their individual daily observations of the Moon. Once a week they confirm and share their observations in small groups from which they progressively connect the Moon's appearance to the corresponding position in its orbit. After completing the observations, a prediction and a last observation are made to confirm the students' understanding. Each student hands in a "portfolio" consisting of their weekly observations and plots, their predictions/observations, and a final model of the lunar phases summarized in one drawing along with a written extrapolation for the actual cause of the phases of the Moon. Finally, students are given the LPCI again as a standard tool to comparatively gauge their pre- and post-project comprehension.

3. Assessment of LPP

The Lunar Phases Project was implemented in a pilot study in 2009 and has continued since then in introductory astronomy and physical science courses. To date, over 270 FGCU undergraduate students have completed the Lunar Phases Project and the LPCI administered both as a pre- and post-test to gauge the efficacy of the Lunar Phases Project. To assess the students' improvement on the LPCI, we chose to use the normalized gain

$$g = (\%posttest - \%pretest)/(100 - \%pretest)$$

The LPCI average score went from about 41% to 56% across all students. The average g varied greatly from class to class; the lowest was 0.06, the highest 0.47, with a mean of 0.23. A g of at least 0.30 indicates that the project is effective at improving many students' performance on the chosen concept inventory (Hake 1998). The pre- and post-project scores on the LPCI combined with the pre- and post-project written explanations indicate significant improvement in students' understanding of the lunar phase concepts after completing the Lunar Phases Project.

4. Proposed Assessment Implementation

The LPP represents a way in which to develop an effective means of incorporating research into learning and education. Here are some suggestions for general application:

1. Search for validated assessment instruments related to the topics covered by the method you wish to evaluate (Internet, journals, authors of existing concept inventories, etc.)

2. Consult experts in the field related to the topic covered by your method. They can help to determine the information the students need to know regarding the topics included in your method for assessment.

3. Remember that existing assessment tools (such as existing concept inventories) provide you with an already validated tool for assessment as long as they are made for the topics of the method you wish to evaluate.

4. Validated concept inventories and assessment tools are available for a handful of specific topics. Analysis of the questions related these specific topics may provide you with guidance on what part of your teaching method needs refining.

5. Don't be afraid to partner with an education expert that can help you further analyze your results.

A short list of available and validated concept inventories include:

- Lunar Phases Concept Inventory (LPCI),

- Light and Spectroscopy Concept Inventory (LSCI),

- Greenhouse Effect Concept Inventory (GHECI), and

- Star Properties Concept Inventory (SPCI).

These instruments and contact information for their authors is available on the Internet.

References

Lindell, R. S. 2001, "Enhancing college students' understanding of lunar phases," Ph.D. Dissertation, University of Nebraska Lincoln (Lincoln, Nebraska)
Lindell, R. S. & Olsen, J. P. 2002, "Developing the Lunar Phases Concept Inventory," Proceedings of the 2002 Physics Education Research Conference

Michael Gibbs, William Gutsch, Gordon Meyers, and James Manning at the Awards Banquet. Photo by Paul Deans.

Connecting People to Science
ASP Conference Series, Vol. 457
Joseph B. Jensen, James G. Manning, Michael G. Gibbs, and Doris Daou, eds.
© 2012 Astronomical Society of the Pacific

NASA Heliophysics Education and Public Outreach Forum Product Analysis Effort

Lindsay Bartolone,[1] Michelle Nichols,[1] Heather Withnell,[1] Nancy Alima Ali,[2] Bryan Mendez,[2] Ruth Paglierani,[2] and Laura Peticolas[2]

[1] *The Adler Planetarium, 1300 S. Lake Shore Dr., Chicago, Illinois 60605, USA*

[2] *Center for Science Education at Space Sciences Laboratory, University of California at Berkeley, 7 Gauss Way, MC 7450, Berkeley, California 94720, USA*

Abstract. The NASA Science Education and Public Outreach Forums (SEPOF) have as two of their goals to improve characterization of the contents of the Science Mission Directorate Education and Public Outreach (SMD EPO) portfolio (Objective 2.1) and assist SMD in addressing gaps in the portfolio of SMD EPO products and project activities (Objective 2.2). The Heliophysics Forum has undertaken the task of product analysis for the entire heliophysics portfolio of K–12, informal, and outreach products and programs. The product analysis process determines the alignment of the portfolio to the AAAS Benchmarks. Along with the work of the other forums, the goal of this project is to conduct a "gap analysis" to ascertain how many products and programs align to the Benchmarks and to discover where the "gaps" are located, i.e. those benchmarks that are not covered sufficiently (or at all) by the portfolio. Two anticipated outcomes of this work are to inform the creation of the future online "one-stop-shop" catalogue of NASA Science Mission Directorate (SMD) educational materials and help inform NASA EPO professionals of the need for portfolio materials to "bridge the gaps" that would address the needs of our audiences. The goal of product analysis is not to redo the work of the NASA Review processes, only to analyze the degree of alignment of these approved products to the AAAS benchmarks and a characterization of the instructional strategies and assessment tools (as well as other details) that would inform the future online catalogue. This paper highlights the product analysis process, the Heliophysics Forum team's progress, and the results of Heliophysics Forum product analysis to date.

1. Preparation For Product Analysis

Between October 2009 and September 2010, the cross-forum product analysis team accomplished the following tasks regarding the process of product analysis:

- determination of the overall questions that product analysis was designed to answer, especially with regard to the future online catalogue of NASA Science Mission Directorate products and programs;

- creation of the tools by which all materials would be analyzed;

- training of analysts in all four forums, and

- development of a system for quality assurance (i.e. "inter-rater reliability") for analysis.

All analysts were trained by October 2010. From October to November 2010, the Heliophysics product analysis team established inter-rater reliability for the analysis process. Each team member independently analyzed five Heliophysics Forum products, and the results were compiled to show the degree of agreement between all analysts for the various sections in the product analysis tool. Product analysis for all Heliophysics Forum K–12 products that had passed NASA Education Review prior to June 2010 began in November 2010 after inter-rater agreement had been established. *Analysis of informal, outreach, and new K–12 materials will take place from late 2011 to mid-2014.*

2.   Heliophysics Forum Product Analysis Progress To Date (as of July 2011)

At the beginning of the product analysis effort, seventy-three Heliophysics K–12 products were available for analysis. The Heliophysics K–12 portfolio, as a whole, includes many different material formats, including lesson plans, educator guides, brochures, web-based materials, mathematics guides, posters, video products, and more. The scope of these products ranges from one to as many as thirty-five individual lessons or sections, and from as small as a one-page brochure to a massive 165-page curriculum unit. Team members analyze each lesson in a product separately, and the average amount of time to properly analyze a single lesson is about two hours. As of the writing of this paper, fifty-eight products have been analyzed by one Heliophysics product analysis team member, and eight products have been analyzed by two team members, providing a way for the team to periodically check inter-rater reliability throughout the analysis process.

The Heliophysics team uses the materials developed by the cross-forum team, including the analysis tool spreadsheet, the full list of AAAS Benchmarks, and a vocabulary list that covers all of the instructional strategies and assessment methods included in the analysis spreadsheet tool. Each product is assigned to one or two specific team members to analyze, and the number of products assigned is based on the total amount of time each analyst has been assigned to this project. The team meets by phone and WebEx each month to discuss progress to date, any challenges that need resolution, and the results of the monthly cross-forum tag-up calls. In between the phone calls, team members discuss issues in person or via email, try to come to consensus with their analyses, and utilize the Analysis Cataloguing and Implementation Team area of the SMD EPO workspace for further cross-forum discussions.

2.1.   The Use of AAAS Benchmarks for Product Analysis

The AAAS Project 2061 Benchmarks are a set of K–12 learning goals that are widely used by education professionals nationwide for curriculum development and program planning. Benchmarks are a familiar resource for various audiences, making it an excellent choice to which we can align the SMD K–12, informal, and outreach materials.

There is one specific drawback that must be addressed by the Heliophysics team, however. The Benchmarks contain very few heliophysics topics. While the alignment to the Benchmarks is necessary and continuing so that the cross-forum community has a coherent set of results to consider, the Heliophysics team has also chosen to align products in the portfolio using one additional set of tools. In Spring 2011, the Heliophysics forum team began to create a heliophysics conceptual framework using a variety of source materials, including the concept maps contained in the Great Explorations in Math and Science (GEMS) Space Science Sequence curriculum materials, thematic concept maps created by members of the Heliophysics EPO community at a recent retreat, and the National Science Education Standards. This rounds out the information available to Heliophysics product developers, and the framework will go beyond the content available in the AAAS benchmarks, while still keeping the content focused on heliophysics concepts. Additional work will be needed by the team to reanalyze each product using this conceptual framework.

The flexibility of the product analysis process allows for this additional work so each forum can compile a set of information that is most useful for the overall gap analysis, for their own identification of forum-specific gaps in the portfolio, and for the work related to the future online catalogue.

2.2. Difficulties

The product analysis process has proceeded smoothly, though the team has had to work together to resolve a few minor issues. The first issue came about just as the team was beginning its work in late 2010. The cross-forum product analysis team made the decision to only allow analysts to list up to three AAAS benchmarks per product, a product being defined as a single lesson or activity. The Heliophysics product analysis team developed a hierarchy process for benchmark analysis to help team members narrow down the list of benchmarks. The steps in this process are as follows:

1. Consider any standards and/or benchmarks that are listed in the lesson.

2. Consider any objectives written by the developer.

3. Consider the assessments. What are the students expected to demonstrate that they have learned?

4. consider the content in the entire scope of the lesson. Look at the lesson plan, any worksheets that the students need to fill out, notes they need to take, lectures, etc. If there is background material that is included for the teacher, only consider it if it is listed as "present this to the students." Look for assessments elsewhere in the lesson, in the lesson packet, online, etc.

5. Align benchmarks that have the highest level of focus in the lesson.

This list has helped the team tremendously, especially when working on a particularly long lesson or large product.

2.3. Results

At the beginning of the product analysis process, the Heliophysics Forum team was concerned that heliophysics products would be difficult to align to the AAAS benchmarks due to the lack of relevant content in the AAAS benchmarks as a whole. For the most part, this is largely the case. The space science-themed benchmarks center around Chapter 4 of the AAAS Benchmarks, "The Physical Setting." 4G – Forces of Nature is the section most often used to describe the heliophysics content in the products. When content benchmarks are not relevant, though, the team must look elsewhere for relevant benchmarks to characterize a product. Many of the heliophysics products are aligned to benchmarks in Chapter 12 – "Habits of Mind." Chapter 12 covers many skills, habits, and values that students should develop, and quite a few heliophysics products align to 12D – Communication Skills. This is interesting to note, as similar skills are highlighted in the new Framework for K–12 Science Education, Dimension 1: Scientific and Engineering Practices.[1]

When looking at instructional strategies, the list is more varied, though "non-linguistic representations" (creating non-verbal renderings such as graphs, maps, models, pictures, and kinesthetic activities), "discussions," and "hands-on learning" (in that order) are the three strategies used by developers most often in their products. As for assessment strategies, heliophysics product developers turn to "alternative assessments," any type of assessment that is not a test.

3. Conclusions

The team is very impressed with the wide range of heliophysics-themed products available for the K–12 audience, and the heliophysics EPO community should be commended for tackling what can sometimes be very high-level concepts for teachers and students to understand. The Heliophysics Forum team continues to analyze the remaining K–12 products in this first set of materials, and work on Informal and Outreach materials will begin in the near future. We look forward to contributing information to be used for the construction of the online catalogue.

References

American Association for the Advancement of Science 2009, "Project 2061, Benchmarks for Science Literacy," (New York, Oxford University Press),
    http://www.project2061.org/publications/bsl/online/index.php
American Association for the Advancement of Science 2007, "Project 2061, Atlas of Science Literacy, Volumes 1 and 2," (Washington, D.C., American Association for the Advancement of Science and National Science Teachers Association),
    http://www.project2061.org/publications/atlas/default.htm

---

[1]A graphical representation of the alignment of Heliophysics products to 12D is available at
http://cse.ssl.berkeley.edu/forum/asp/Helio_PA_poster.ppt

Connecting People to Science
ASP Conference Series, Vol. 457
Joseph B. Jensen, James G. Manning, Michael G. Gibbs, and Doris Daou, eds.
©2012 Astronomical Society of the Pacific

Women in Planetary Science: Career Resources and e-Mentoring on Blogs, Twitter, Facebook, Google+, and Pinterest

Susan M. Niebur,[1] Kelsi Singer,[2] and Kathryn Gardner-Vandy[3]

[1]*Niebur Consulting, 418 Hillsboro Drive, Silver Spring, Maryland 20902, USA*

[2]*Washington University, One Brookings Drive, St. Louis, Missouri 63130, USA*

[3]*Lunar and Planetary Laboratory, University of Arizona, 1629 East University Boulevard, Tucson, Arizona 85721, USA*

Abstract. Fifty-one interviews with women in planetary science are now available as an e-mentoring and teaching resource on WomeninPlanetaryScience.com. Each scientist was nominated and interviewed by a fellow member of the planetary science community, and each gladly shared her advice for advancement in the field. Women in Planetary Science was founded in 2008 to connect communities of current and prospective scientists, to promote proposal and award opportunities, and to stimulate discussion in the planetary science community at large. Regular articles, or posts, by nearly a dozen collaborators highlight a range of current issues for women in this field. These articles are promoted by collaborators on Twitter, Facebook, and Google+ and shared again by the collaborators' contacts, reaching a significantly wider audience. The group's latest project, on Pinterest, is a crowd-sourced photo gallery of more than 350 inspiring women in planetary science; each photo links to the scientist's CV. The interviews, the essays, and the photo gallery are available online as resources for prospective scientists, planetary scientists, parents, and educators.

1. Women in Science

Although almost as many women as men are working towards bachelor's degrees in science and engineering today, there is still significant underrepresentation of women at the graduate level in each of the disciplines that make up the field of planetary science. The National Science Foundation (NSF) reports that 19% of new physics Ph.D.s are women, 24% of new astronomy Ph.D.s are women, and 37% of new earth sciences Ph.D.s are women. The representation of women on the faculty at U.S. universities is even smaller. Only 17% of tenured or tenure-track faculty in the physical sciences overall are women (NSF 2011), with the numbers the same or smaller in both astronomy and physics (Ivies 2008).

Of even greater concern is the number of physical science departments nationwide without a single woman on the faculty. An American Institute of Physics (AIP) survey of all physics and astronomy departments nationwide (with a 96% response rate) showed that 43% of all physics departments in the U.S. have no women at all on the faculty (Ivies 2008). The 2011 Planetary Science Workforce Survey of 48 departments thought to include faculty involved in planetary science research found that 16 of the 48 departments (33%) had only male planetary science faculty. The same survey showed that only 13% of planetary science faculty are women (Bagenal 2011). It is therefore quite possible for a student to complete her bachelor's degree, master's degree, and Ph.D. dissertation having never collaborated nor learned from a female faculty member in her field of choice.

2.   WomeninPlanetaryScience.com

In 2008, one of us called together the women attending the 38th Lunar and Planetary Science Conference (LPSC) for an informal networking breakfast. Held in a basement room at 7 a.m., tables were set for several dozen attendees, but the room quickly filled with over 100 women eager to air and solve problems. Table discussions were lively with talk of career advancement, flight mission work, collaboration, advisors, students, and retention of women at the department level. It was clear that these women, used to sitting in conference rooms filled with mostly older, mostly male scientists, were eager to talk with other women in their field. Attendees decided to continue communication throughout the year, so an email list and website were created and opened to the community, with the address `WomeninPlanetaryScience.com`.

The website was set up as a blog in order to facilitate community participation in a Web 3.0 model, soliciting comments, discussion, community contributions, followup posts, and interaction on the associated email list, which was also used to highlight significant posts on the blog quarterly or more often. Posts were promoted through collaborators' accounts on Twitter.com, Facebook.com, and Google+. A blogroll linked individual and department blogs in the geological and space sciences, and existing career resources were highlighted. Collaborators participated in discussions on other blogs, linking back to `WomeninPlanetaryScience.com`.

With an email list of just 100 subscribers, the site has received over 100,000 hits from 136 countries. The major topics of discussion include: collaboration; demographic studies; flexible work; announcement of new proposal opportunities; proposal tips for scholarships, fellowships, and grants; specific questions from site users; congratulating new faculty members on their appointments and groups of scientists on mission successes; posting of paired job announcements, where both members of a couple may apply to the same institution; essays on the academic job search; career advice; and preparation for major conferences in the field.

A significant portion of the articles or posts have been directed at sharpening skills and proposing research to NASA and NSF grant programs. Participating Scientist Programs for NASA flight missions were highlighted when opportunities became available for new scientists to join flight mission teams. Annual research solicitations were announced and potential proposers encouraged to read NASA's guidance, site-specific tips from tenure-track professors, museum scientists, civil servants, postdoctoral fellows, and students. Advice came from people working at large research universities, small teaching colleges, nonprofit research institutions, NASA Headquarters, and NASA Centers. Those who had received research grants and fellowships were able to pass on both knowledge and encouragement. Senior collaborators posted proposal tips for NASA research solicitations and service on NASA missions, graduate students advocated strategies for finding and applying for fellowships, and undergraduate students posted and used tips on selecting an advisor or area of research. Reader questions and community announcements were encouraged.

3. Original Interviews with 51 Women in Planetary Science

A recent study by the American Astronomical Society (Marvel 2009) showed that while women do not win professional awards at the same rate as men, they both nominate and are nominated much less often than men. To encourage women to nominate their colleagues, articles frequently appeared on the site increasing awareness of such awards, reminding scientists of approaching deadlines, and noting whether women won the most recent awards. After the discussion was well underway and the premise accepted that for women to win more awards, more must be nominated, a project was initiated whereby readers would go through the nominating process for each other. To work, the project needed to be merit-based, have a low barrier to entry, and possess clearly stated goals separable from the nomination procedure.

To meet these objectives and raise the profile of women for future awards, the project concept was simple: interview outstanding women in the field. Recognizing that success in planetary science has more than one definition, the two basic criteria for nomination were 1) a peer-reviewed publication in the last two years or 2) current service on a NASA mission (which often precludes conducting significant scientific research during mission development). The first criterion emphasized the importance of publishing to a career in science. The second criterion raised the profile of women working on missions; at the time, only two women (and 75 men) had been selected to lead NASA planetary science missions in the previous thirty years (Niebur 2009). A few exceptions were made for inspirational mentors and researchers; all non-mission scientists were nominated by users of the site. To date, 51 interviews have been conducted by the authors, edited, and posted, forming a permanent archive publicly available at WomeninPlanetaryScience.com.

4. Pinterest Gallery of 351 Women in Planetary Science

The set of 51 interviews is a new resource illuminating career pathways and choices of women employed in a number of different institutions and at all career levels, from undergraduates with their first journal publication to senior scientists leading instrument developments. "Crowd-sourcing" these interviews—asking community members to nominate their peers—resulted in more interview candidates than anticipated. At the same time, a new social media platform was launched at Pinterest.com. Pinterest seemed to be a picture gallery of tempting new designs, fashions, and edible treats. As an experiment, one of us began "pinning" images of female planetary scientists at work or in the field, linking the earlier interviews to this new virtual bulletin board.

The collaborators "tweeted" and "Facebooked" the gallery of earnest, smiling scientists, sharing the diversity of women in planetary science with online and offline communities, and asking for additional nominations. New names poured in, and soon the gallery included over 350 women and links to their CVs. By crowd-sourcing the Pinterest content, other members of the community became invested in the gallery. The project has been recognized by an editor of Scientific American, prominent and international scientists, teachers; mothers who sat fascinated with their daughters and then tweeted us about it, and by Pinterest itself. This gallery is now a resource for students seeking role models, for award nominees, for inspiring girls and young women, and as an answer to people who question whether there are any women going into planetary science.

5. Social Media Evolution

While the specifics of our social media strategy change yearly (or more often), the main approach can be stated simply: go where the scientists and students are. Promote the presence of underrepresented minorities such as women. Increase awareness of the important actions for career advancement. Work together, and enable more senior women to mentor younger women at every level; as senior scientists mentor early career scientists, postdoctoral fellows can mentor graduate students, and undergraduates can encourage and inform freshmen and high school students.

To support our core audience, such as the women who attend the annual LPSC (Fig. 1), WomeninPlanetaryScience.com will continue to expand with the next wave of innovation in social media and bring our resources and support wherever the students, scientists, and curious public go. The reader is invited to join our journey and to use our interviews and gallery to encourage others to follow their own dreams of exploring the planets and beyond.

Figure 1.   Women in Planetary Science Networking Breakfast at LPSC, 2011. Photo courtesy Heather Dalton, Lunar and Planetary Institute

Acknowledgments.   We thank Dr. Jim Green, Director of the Planetary Science Division at NASA, for his support of the annual networking breakfast at LPSC. One of us (Niebur) acknowledges NASA support from an outreach supplement to contract NNH08CC65C. All of us thank the collaborators and contributors to WomeninPlanetaryScience.com and the 51 featured women in planetary science.

References

Bagenal, F. 2011, "Summary of Results of Survey of U.S. Academic Departments that Include Planetary Science,"
        http://lasp.colorado.edu/mop/resources/links/
        PlanetaryScienceWorkForceSurvey2011/
Ivie, R. 2008, "Women in Physics & Astronomy," AIP Statistical Research Center,
        http://www.aip.org/statistics/trends/highlite/women3/faculty.htm
Marvel, K. 2009, "The Ongoing Demographic Shift in the AAS, Status: A Report on Women in Astronomy," January 2009, 1
National Science Foundation 2011, "Women, Minorities, and Persons with Disabilities in Science and Engineering," NSF 11-309.
Niebur, S. M. 2009, "Women and Mission Leadership," Space Policy, 26, 257

Connecting People to Science
ASP Conference Series, Vol. 457
Joseph B. Jensen, James G. Manning, Michael G. Gibbs, and Doris Daou, eds.
©*2012 Astronomical Society of the Pacific*

Innovative Low Cost Science Education Technology Tools: Increasing Access to Science for All

Jacob Noel-Storr[1] and Brandon N. Cole[1,2]

[1] *Chester F. Carlson Center for Imaging Science, Rochester Institute of Technology, Rochester, New York 14623, USA*

[2] *Gollisano College of Computing and Information Systems, Rochester Institute of Technology, Rochester, New York 14623, USA*

Abstract.   We present three low cost educational technology tools that have been developed by the Rochester Institute of Technology Insight Lab. Our technology tools are designed with cost and "user-tinkerability" in mind, to increase the potential for technology-rich access to scientific data to be in the hands of a much larger slice of the population. The three tools presented are the *Planeterrainium*—A digital interactive floor projection system allowing users to explore the planets in 3D; the *Digital Solar Explorer*—a 5 foot inflatable sphere designed to allow for the exploration of solar imagery; and the *Scube* – a digital immersive tentware system. We describe projects that involve both undergraduate and high school students in the development of content for these systems, encouraging the growth of both scientific and technological literacy in the process.

1. Introduction

Owing to the demonstrated need for low-cost, "user-tinkerable" (easily modified or configured by the users) science education technology tools (Office of Educational Technology 2010), the Insight Lab at the Rochester Institute of Technology operates a Science Education Technology Key Working Area, involving faculty, students, and local high school teachers and students in the development and implementation of novel low cost science education technology tools (Noel-Storr, Kruchoski, & Cole 2010). Our projects are developed with key benefits for the end user in mind in order to maximize their potential impact in increasing access to science education technology across audiences.

- **Cost Effectiveness**: It is critical to provide cost effective solutions to presenting digital visual information (Interagency Working Group on Digital Data 2009). Many commercial projection technologies can be too expensive for use in most schools and other small institutions (Dede 1998). It is for this reason that we construct our technologies with common and readably available materials. For example, we often use PVC pipes and fittings in our projects because of the availability and relatively low cost of the materials.

- **Portability**: Science education technologies must be robust and portable so that they can be used by different teachers on a site, or shared across sites (Fabry and

Higgs 1997). Our projects are designed with portability as a key component. Each project can fit within three suitcases or less, and are robust enough to withstand the harsh treatment from airline baggage handlers (as we have confirmed!). They are also compact in nature, which allows them to fit into the space available in a typical classroom, after-school program location, or small science center.

- **Easy Implementation**: To encourage the next generation of STEM professionals, it is essential that science education technologies are "tinkerable" by teachers and students alike (Office of Educational Technology 2010). A benefit we strive to build into all of our projects is that it is easy to change the content being presented. Any typical computer user can create content for these tools to cater to the subject they wish to discuss. As teachers or students experiment more they are able to develop new ways of using these technologies.

- **Immersive and Interactive**: Today's media-savy youth (and public) expect media rich-experiences (Lenhart et al. 2001), which, while ubiquitous, are often expensive and cumbersome to develop and deliver. The immersive and interactive nature of our projects captivates the attention of audiences more effectively than conventional learning techniques, and allows teachers and other facilitators of science education programs to easily move beyond the PowerPoint presentation or interactive whiteboard to provide richer and more meaningful media experiences in the science classroom.

2.   Projects Developed

Thus far we have developed and are testing three low cost education technology tools, which we implement in classrooms, museums, planetaria, and other out-of-school settings.

- **The *Planeterrainium*** (Fig. 1) was built with the goal of effectively presenting stereoscopic 3D imagery of planetary surfaces. To do this, we built a custom projector stand that allows the projector to display imagery on the ground in front of viewers. Developing content for this project is simple. The system can present PowerPoint slides, Flash animations, static images, videos, or any other type of digital content. If connected to an interactive device (such as the Nintendo WiiMote^{TM} or the Microsoft Kinect^{TM}), this system can present content for users to interact with. The *Planeterrainium* is best suited to display any type of imagery that can be viewed more effectively on a horizontal surface than on a vertical wall. The main focus thus far has been planetary surfaces, but this system can easily be adapted for any other types of surfaces.

- **The *Digital Solar Explorer*,** or *DSE* (Fig. 2), allows viewers to visualize the Sun in an innovative way. The *DSE* consists of a standard projector on a tripod, aimed at a large 5-foot inflatable sphere. This system currently utilizes a Nintendo WiiMote^{TM} so that users can interact with the content being displayed. The *DSE* is designed to present spherical content. Thus far, our team has developed extensive solar content so that users can explore properties such as magnetic field strength, solar flares, temperature, and brightness of the Sun. However, this educational tool is perfect to display anything that is spherical in shape, such as other planets or moons.

Figure 1.   The *Planeterrainium* shown being demonstrated by Brandon Cole at the semi-annual RIT/NASA AstroZone event, held Summer 2011 at the Boston Museum of Science. Viewers wear 3D red/cyan glasses to visualize planetary surfaces.

Figure 2.   The *Digital Solar Explorer* shown demonstrated by Insight Lab Assistant Director Drew C. Lierheimer at the American Astronomical Society Solar Physics Division Conference (2011) in Las Cruces, New Mexico.

- **The *Scube*,** or the Science Cube (Fig. 3), provides a way to immerse the audience in educational content that is displayed all around them. The *Scube* is made up of four walls and four external projectors, each projector aimed at a wall. A wide variety of immersive and educational presentations have been developed by many different groups. One of these presentations guides the viewers through the universe and allows them to follow historic missions such as the *Voyager*, *Galileo*, and *Cassini* space missions. Another popular presentation takes viewers through a digital replica of a monkey's brain. Content for the *Scube* has been designed by high school groups through college seniors. This tool allows educators to immerse an audience into any kind of 360 degree environment.

Figure 3.   A group of elementary age students immersed in the Universe inside the four-walled *Scube* projection system.

Acknowledgments.   Funding for this work was provided in part by education supplements to NASA contracts NNX08AO03G and NNX07AM68G.

References

Dede, C. 1998, "Six Challenges for Educational Technology," in the 1998 ASCD YearBook (Alexandria, Virginia: ASCD)

Fabry, D. L. & Higgs, J. R. 1997, Journal of Educational Computing Research, 17, 385

Interagency Working Group on Digital Data (IWGDD) 2009, "Harnessing the Power of Digital Data for Science and Society," (Washington, D.C.: Office of Science and Technology Policy)

Lenhart, A., Rainie, L., & Lewis, O. 2001, "Teenage Life Online," (Washington, D.C.: Pew Internet Life Project)

Noel-Storr, J., Kruchoski, Z. S., & Cole, B. 2010, Science Education Technology Key Working Area, http://insight.rit.edu/node/37

Office of Educational Technology 2010, "Transforming American Education: Learning Powered by Technology," (Alexandria, Virginia: U.S. Department of Education), xvi

Connecting People to Science
ASP Conference Series, Vol. 457
Joseph B. Jensen, James G. Manning, Michael G. Gibbs, and Doris Daou, eds.
© 2012 Astronomical Society of the Pacific

A Place of Transformation: Lessons from the Cosmic Serpent Informal Science Education Professional Development Project

Laura Peticolas,[1] Nancy Maryboy,[2] David Begay,[2] Jill Stein,[3] Shelly Valdez,[4] and Ruth Paglierani[1]

[1] *Space Sciences Laboratory, University of California, Berkeley, 7 Gauss Way, Berkeley, California 94720, USA*

[2] *Indigenous Education Institute, PO Box 898, Friday Harbor, Washington 98250, USA*

[3] *Institute for Learning Innovation, 3168 Braverton Street, Suite 280, Edgewater, Maryland 21037, USA*

[4] *Native Pathways, P.O. Box 248, New Laguna, New Mexico 87038, USA*

Abstract.   A cultural disconnect exists between Western scientists and educators and Native communities in terms of scientific worldviews and Indigenous ways of knowing. This cultural disconnect manifests itself in the lack of participation of Native Americans in Western science and a lack of appreciation by Western scientists of Native science. Our NSF-Funded project "Cosmic Serpent: Bridging Native and Western Learning in Museum Settings" set out to provide a way for informal science education practitioners and tribal museum practitioners to learn about these two worldviews in such a way as to inform their educational practice around these concepts. We began with a pilot workshop in year one of this four-year project. We then provided two week-long professional development workshops in three regions within the Western U.S., and culminated with a final conference for all participants. In total, the workshops served 162 participants, including 115 practitioners from 19 tribal museums and 41 science, natural history, and cultural museums; 23 tribal community members; and 24 "bridge people" with knowledge of both Indigenous and Western science. For this article, we focus on the professional and personal transformations around culture, knowledge, science, and worldviews that occurred as a part of this project. We evaluated the collaborative aspects of this grant between the Indigenous Education Institute; the Center for Science Education at the University of California, Berkeley; the Institute for Learning Innovation; Native Pathways; Association for Science and Technology Centers; and the National Museum of the American Indian. Using evaluation results, as well as our personal reflections, we share our learnings from a place of transformation. We provide lessons we learned with this project, which we hope others will find relevant to their own science education work.

1.   Transformation

For the purposes of this paper, the idea of transformation stems from a Native worldview as an ongoing, self-driven, and emergent process that includes many steps along the way, and many possible pathways. This kind of transformation can include new realizations, deeper cultural understanding, shifts in how one thinks about his or her work, new personal and emotional awareness, and building relationships to support authentic collaboration.

Considering transformation in individuals as a process within a larger system is similar to seeking to understand the change of individual molecules that occurs when ice changes to water. Using this analogy, it may be helpful to imagine the Cosmic Serpent project as changing from one state to another, much as ice changes to an ice-liquid mixture and then over time, becomes water when warmed. In both processes, it is possible to understand the changes in individual participants or molecules within the context of a larger system. What type of changes people experience in the Cosmic Serpent project depended very clearly on the state they were in when the project started—just as the change a water molecule undergoes when ice is warmed depends on where it is at the start of the process of warming up—whether it was buried deep in the ice or closer to the surface of the ice.

Figure 1.   Project Goals organized by the Diné Cosmic Model (Maryboy & Begay copyright 2003, 2007, 2009).

2.   Project Goals Organized by the Diné Cosmic Model

The project goals were organized along the four directions, according to the Diné Cosmic Model as shown in Figure 1 (Maryboy and Begay 2003). The goals for the professional audience of the Cosmic Serpent project—primarily museum practitioners—did not explicitly aim for one particular type of transformation.

3. Collaboration with Integrity: the Cosmic Serpent Team

The leadership team's own growth had a direct impact on the approaches used in the workshops. In order to get to the place of transformation, they had to invest the time to experience relationship building and a deeper understanding of the Diné model. They also needed to factor in an awareness of where the participants were in their understanding and perceptions of worldviews of Western science and Native Ways of Knowing.

The team came in with some background on Native Ways of Knowing. The degree of this background knowledge was dependent on the individual's learning experiences; because of this everyone brought a different strength to the collaborative effort and organically, it morphed into a strong collaborative partnership based on trust and consensus of team members. During this process, there were individuals on the team with strong shifts in understandings about Western science and Native Ways of Knowing. The team members experienced collective leadership from a Native worldview. In this case, the collaboration wasn't about the individual benefit, but about becoming a vehicle for accentuating the community's voice and active engagement, while creating environments for this model to unfold.

4. What We Heard from Cosmic Serpent Participants

Many participants described becoming aware of new ideas or having new realizations about worldviews, their own way of thinking, their own cultural assumptions, use of language, and cultural protocols. The following quote from a participant helps illustrate this point: "As a Native person I still have a lot of questions about Western science and where it is taking us. At the same time I am more aware of the areas that Indigenous knowledge and Western knowledge (science) can overlap and knowledge can be shared, in some cases."

Throughout the project, participants (mostly Western, in this case) shared how they gained a deeper understanding of Indigenous knowledge and/or Native worldviews, such as increased awareness and heightened sensitivity towards challenges faced by Native communities, increased understanding of how Native communities have experienced or view Western science, the importance of relationship building and trust; the importance of deep listening; and a stronger framework for thinking about Indigenous knowledge. One participant stated, "I would say that I have a better understanding of what is important to various tribes when it comes to Indigenous knowledge. It is not (just) a cultural sensitivity, but beyond the typical 'let's all just get along'." While learning about the Native worldview naturally emerged mostly from the Western participants, there were certainly examples of Native participants who came to view the culture of science and scientists with more trust, especially given a long history of Western cultural domination and oppression of Native communities in the U.S.

On a professional level, participants shared examples of becoming more reflective practitioners. Some felt the project has made them think more deeply about their work, the language that they use, and the complexity of cross-cultural collaboration.

The theme of building relationships was strong for many Cosmic Serpent participants, and relationship can be seen as the foundation for all transformational change. Participants talked about feeling that they had made deep personal connections, opened up to others, and made contacts with others that could support their work. There was a sense of growing openness and trust among participants in the three regions. As described by one participant, "face to face collaboration is not something you can substitute."

The Cosmic Serpent experience was a personal and emotional one for many participants. Some described learning more about themselves, their own assumptions and stereotypes through participation in the project, suggesting that what they learned had broader application to their personal lives, in addition to being relevant to their profession, "the more I participate the more I am realizing how little I know, how complex these issues are, and more than anything, humbling."

5. Lessons Learned

Through the shared Cosmic Serpent experience, which can be thought of as the "story" of this project, and reflecting on the lessons learned, it becomes clear that future collaborations need to be considerate of the time and space needed for creating relationships, and purposefully build them into the program environment. Professional relationships developed and maintained with respect and integrity are what stimulate and shape the pathways of productive and genuine collaborative work. Another area of learning was that all project activities need to be considerate of balance of voice, creating environments where all voices are heard, shared, and experienced. With this said, sometimes there may need to be more time spent on orienting the audience to Indigenous worldviews. Consideration of where the participants are in terms of knowledge levels and an effort to start with common ground is recommended. It is incredibly helpful to include people who work in both Western science and Indigenous knowledge in such cross-cultural collaborative projects.

Challenges have included: reducing defensiveness about one's own worldview; not knowing where to start or, alternatively, being overwhelmed when wanting to focus interests or develop collaborations; getting institutional buy-in or getting support from organizational leadership; and funding.

Acknowledgments. The authors express much gratitude for the hard work and respect brought to the project by the entire Cosmic Serpent team, the Cosmic Serpent participants, and the "Bridge People;" and to Karin Hauck for her work on Figure 1.

References

Maryboy, N.C., and D. Begay., 2003, A Cosmic Planning Model for the World Hope Foundation. Friday Harbor, WA, Indigenous Education Institute

Connecting People to Science
ASP Conference Series, Vol. 457
Joseph B. Jensen, James G. Manning, Michael G. Gibbs, and Doris Daou, eds.
© *2012 Astronomical Society of the Pacific*

Preservice Teachers' First Experiences Teaching Astronomy: Challenges in Designing and Implementing Inquiry-Based Astronomy Instruction for Elementary Students in After School Programs

Julia D. Plummer

Department of Curriculum and Instruction, Pennsylvania State University, University Park, Pennsylvania 16802, USA /tt jdp17@psu.edu

Abstract. This study examined preservice teachers' pedagogical content knowledge (PCK) in designing astronomy investigations for children. Fifteen pairs of preservice teachers taught groups of children in after school programs once a week for five weeks; their assignment was to guide the children in a multi-day inquiry investigation about astronomy. Pre- and post-program content assessments and five lesson plans from each pair were analyzed using a mixed-methods approach to understand the successes and challenges in developing PCK by new teachers in this domain. Findings suggest that while most preservice teachers were able to implement inquiry investigations in elementary astronomy topics, many also struggled to successfully connect a scientific question to an explanation based on evidence. A correlation between the teachers' content knowledge and the sophistication of their investigation plan was found.

1. Introduction

State and national standards suggest that students in elementary school should be learning astronomy while also engaging in the practices of science (e.g., National Research Council 2011). Research is needed to uncover the ways elementary teachers design instruction that applies scientific inquiry to the domain of astronomy as well as methods of supporting their pedagogical knowledge through professional development and curricula. A pedagogical content knowledge (PCK) framework was used to interpret dimensions of preservice teachers' development during an elementary science methods course. PCK describes teachers' understanding of how to design the learning environment to support children's development of specific content and practices. This paper focuses primarily on their knowledge of instructional strategies and representations for teaching science (Magnusson, Krajcik, & Borko, 1999). Though some research on teachers' PCK in astronomy exists (Henze et al. 2008; Plummer & Zahm 2010), questions remain regarding specific features of elementary teachers' PCK in astronomy.

Elementary teachers, who are often non-science majors, have particular challenges in developing PCK in astronomy. First, new teachers may have limited understanding of how to tailor the learning environment to their students (e.g., Grossman 1991). Second, many teachers do not have scientific understanding of elementary astronomy topics (e.g., Plummer, Zahm, & Rice 2010). Third, research has shown that teachers, especially elementary teachers who are prepared as generalists, often do not understand scientific practices or the nature of science (e.g., Lederman et al. 2002). These challenges led to the following research questions:

1. To what extent are preservice teachers able to design inquiry-based, extended investigations in elementary astronomy?

2. How did their knowledge of elementary astronomy impact the sophistication of the investigations developed?

2. Methodology

Participants were preservice teachers in a 15-week elementary science methods course taught by the author (and co-taught by an instructor not involved in the study) at a small, private suburban university. All preservice teachers in each of three sections participated in the study (30 teachers total). Participants were primarily female (29) and Caucasian (23). Two sections were designated for undergraduates (18) while the third was for a post-baccalaureate certification program (12); however, the instruction and assignments for all three courses were identical. While most participants had no prior teaching experience, seven had prior experience teaching pre-K children, one had taught at the high school level, and one at the elementary level. Only three participants had studied astronomy in high school or beyond, and many could not recall ever having studied astronomy.

The following experiences define the first five weeks of the course: in-class guided inquiry instruction on astronomy concepts, discussion of reform-based readings on science in the classroom (e.g., National Research Council 2000, 2007), analysis of a commercially available curriculum (Full Option Science System 2007), and feedback from their professors as they developed lesson plans for fieldwork. During the second five weeks of the class, pairs of teachers designed and implemented lessons once a week in an afterschool program. Each pair was assigned to teach a group of four to ten children from grades K–6 across the five weeks of the program; children were grouped by grade level. The preservice teachers' assignment was to design and implement an extended inquiry investigation on elementary-level astronomy concepts.

A coding scheme was developed to analyze the lesson plans by identifying 1) specific inquiry practices, and 2) connections between practices and across lessons. The science practices described in the Inquiry in the National Science Education Standards (INSES; National Research Council 2000) were used in determining the level of sophistication in the extended inquiry presented by the teachers. Each lesson was coded individually as well as across lessons from each pair of students.

The second research question was addressed quantitatively. The codes developed to describe the use of inquiry practices were organized into categories for different types of investigations (looking across lesson-sets), as shown in Table 1. This was used to create an ordinal, ranked scale for the sophistication level of each lesson-set. The Spearman's ρ (or r_s) correlation was computed between the investigation scale and both pre- and post-content assessment scores using SPSS.

3. Findings

Research question 1: Each pair's five lesson plans were categorized using the criteria in Table 1; each pair of preservice teachers were assigned the level of the most sophisticated set of lessons created. Table 2 shows the frequency of pairs at each level as

Table 1.   Levels of sophistication for investigations using *INSES* criteria

Level 1	Investigation question must lead to examining data for evidence towards answering the question. An explanation must be constructed in response to the investigation question and must explicitly use evidence as support.
Level 2	Investigation uses a question that leads to interrogating data for evidence but leave the connection between evidence and explanation *implied*.
Level 3	Lessons allow children to engage with data in response to a question but do not attempt to construct an explanation in response to the question.
Level 4	Uses scientific practices but does not meet above criteria.

well as the number of lessons that the pair's investigation extended. Level 1 is most sophisticated while Level 4 included zero lessons classified as an investigation.

Table 2.   Frequency (number of teacher pairs) and length (number of lesson) for each level of investigation

Group	Level 1	Level 2	Level 3	Level 4
Undergraduate	4 [2, 2, 2.5, 4.5 lessons]	1 [1 lesson]	1 [1 lesson]	3
Non-traditional	3 [5, 5, 5 lessons]	2 [3, 4 lessons]	0	1

Seven pairs (47%) developed extended investigations extending from two to five lessons. Investigations were typified by the following structure: the teachers posed an investigation question that concerns a pattern of observations (such as the Sun's path or the phases of the Moon). Observations were made and recorded to determine this pattern. A preliminary explanation was offered that used evidence to construct a representation of the Earth-based observable pattern. The investigation then continued as the teacher engaged children to think about why the observational pattern exists through a psychomotor and/or kinesthetic modeling activity. An explanation was then constructed that drew upon the observational evidence to design or evaluate the space-based model of motion or observing orientation.

Some pairs were able to designate a *scientific question* that lead to collecting data (4 pairs; 27%) but fell short when it came to *connecting evidence to an explanation*. Teachers followed data collection and analysis with a broad summarizing question but no support or explicit guidance for constructing an evidence-based explanation. These were categorized as an *implied* use of evidence because of the temporal proximity of this summarizing event to the data collection phase. Other teachers did not make explicit connections between the initial observational pattern and the modeling activity; thus, the explanation from a space-based perspective lacked a clear evidence base. Finally, four pairs (27%) did not create extended inquiry investigations. These pairs either lacked a clear investigation question, the teachers answered the investigation question before collecting and analyzing data, or investigation questions lead to rote activities, such as completing worksheets.

Research question 2: Initially, the preservice teachers had a low level of knowledge of elementary astronomy, with a mean score of 8.8 (standard deviation = 2.9) out of 24 on the content assessment. The mean score on the post-test was 13.8 (standard deviation = 3.2), a statistically significant improvement ($t = 9.437$, $p < 0.001$). Though no correlation was found with the pre-test scores ($r_s = 0.332$, $p > 0.05$), a significant correlation between investigation-scale level and the post-test score was found ($r_s = 0.552$, $p < 0.01$). Teachers who developed an overall greater understanding of astronomy during the course showed greater sophistication in the investigation level they designed.

4. Conclusions

Prior research has demonstrated that preservice elementary teachers can adapt curriculum to be more inquiry based (Forbes 2011). This study demonstrates that preservice elementary teachers are also able to develop multi-lesson arcs of inquiry-based investigations that engage students in constructing explanations about observational astronomy concepts. Why were these preservice teachers successful? First, many drew from experiences from the methods course where they participated in guided inquiry investigations about astronomy topics. They also adapted aspects of a commercial curriculum read during the course and found new resources online. Teachers who achieved greater content knowledge during the course were also more successful at creating inquiry-based investigations. Finally, the investigation topic may have been a factor: six of the seven pairs that chose to investigate the sun's path (the simplest topic) were classified as developing an inquiry-based investigation (levels 1 and 2).

Most of the preservice teacher pairs that were able to construct a scientific investigation question were also able to use this to engage children with collecting and analyzing data and constructing some form of explanation (even if that was primarily through teacher-led discussion). Groups that were not classified in the more sophisticated levels of inquiry-based investigation plans were not able to connect a clear investigation question with engagement in data collection. This suggests that, in addition to improving their content knowledge, more explicit support on the types of investigation questions that are feasible for elementary astronomy, how to connect that to appropriate data collection methods, and why this is important for learning the practices of science, would be useful in professional development and curricula to support those teachers.

References

Forbes, C. 2011, Science Education, 95, 1
Full Option Science System 2007, Sun, Moon, and Stars, Lawrence Hall of Science, Berkeley, California
Grossman, P. L. 1991, Teaching and Teacher Education, 7, 345
Henze, I., van Driel, J., & Verloop, N. 2009, Journal of Teacher Education, 60, 184
Lederman, N. G., Abd-El-Khalick, F., Bell, R.L., & Schwartz, R.S. 2002, Journal of Research in Science Teaching, 39, 497
Magnusson, S., Krajcik, J., & Borko, H. 1999, in the Yearbook Of The AETS, J. Gess-Newsome & N. G. Lederman, eds. (Boston: Kluwer), 95
National Research Council 2001, "Inquiry in the National Science Education Standards," (Washington, D.C.: National Academies Press)
National Research Council 2011, "A Framework for K–12 Science Education," (Washington, D.C.: National Academies Press)
Plummer, J. & Zahm, V. 2010, Astronomy Education Review, 9
Plummer, J., Zahm, V., & Rice, R. 2010, Journal of Science Teacher Education, 21, 471

Connecting People to Science
ASP Conference Series, Vol. 457
Joseph B. Jensen, James G. Manning, Michael G. Gibbs, and Doris Daou, eds.
© 2012 Astronomical Society of the Pacific

Authentic Astronomy Research Experiences for Teachers: The NASA/IPAC Teacher Archive Research Program (NITARP)

L. M. Rebull,[1] V. Gorjian,[1,2] G. Squires,[1] and the NITARP Team

[1] *Spitzer Science Center, MS 220-6, 1200 E. California Blvd., Pasadena, California 91125, USA*

[2] *Jet Propulsion Laboratory, 4800 Oak Grove Drive, Pasadena, California 91109, USA*

Abstract.   How many times have you gotten a question from the general public, or read a news story, and concluded that "they just don't understand how real science works?" One really good way to get the word out about how science works is to have more people experience the process of scientific research. Since 2004, the way we have chosen to do this is to provide authentic research experiences for teachers using real data (the program used to be called the Spitzer Teacher Program for Teachers and Students, which in 2009 was rechristened the NASA/IPAC Teacher Archive Research Program, or NITARP). We partner small groups of teachers with a mentor astronomer, they do research as a team, write up a poster, and present it at an American Astronomical Society (AAS) meeting. The teachers incorporate this experience into their classroom, and their experiences color their teaching for years to come, influencing hundreds of students per teacher. This program differs from other similar programs in several important ways. First, each team works on an original, unique project. There are no canned labs here! Second, each team presents their results in posters at the AAS, in science sessions (not outreach sessions). The posters are distributed throughout the meeting, in amongst other researchers' work; the participants are not "given a free pass" because they are teachers. Finally, the "product" of this project is the scientific result, not any sort of curriculum packet. The teachers adapt their project to their classroom environment, and we change the way they think about science and scientists.

1.   What is NITARP?

The NASA/IPAC Teacher Archive Research Program (NITARP) partners small groups of teachers with a mentor astronomer, they do original research as a team, write up a poster, and present it at an American Astronomical Society (AAS) meeting. The teachers incorporate this experience into their classroom, and their experiences color their teaching for years to come, influencing hundreds of students per teacher. This program was originally operated as the Spitzer Teacher Program for Teachers and Students, and in 2009 was rechristened NITARP.

Our goal is to give teachers an authentic research experience, using real astronomical data on a current astronomical topic. We use real astronomical data from archives housed at the Infrared Processing and Analysis Center (IPAC), which includes (but is not limited to) Spitzer, WISE, other Infrared Science Archive (IRSA) holdings, the NASA Extragalactic Database (NED), the NASA Stars and Exoplanets Database (NStED), etc. Each team does a new, original project.

We select teachers from a national competitive application process; teachers must already be familiar with the basics of astronomy (e.g., what magnitude means) and astronomical data (e.g., what a FITS file is). Most of the educators are high school teachers, but 8th grade and community college, as well as non-classroom educators, may also participate.

2. How Does NITARP Work?

The nucleus of each team is the mentor astronomer. Each team usually includes four educators, one of whom has been through the program before and acts as a mentor teacher—essentially the scientist's deputy.

The educators who have been through the program have typically been high school teachers, but we have also had 8th grade and community college educators participate and benefit. Very recently, we have started to expand to more non-traditional educators—people who are not currently in the classroom, but who are still doing education, e.g., observatory education coordinators, regional coordinators, museum educators, or amateur astronomers doing public outreach.

The program runs roughly January to January—it starts at an American Astronomical Society (AAS) meeting with a kickoff workshop for the new NITARP participants on the Sunday before the meeting. The AAS meetings are the largest astronomy meetings in the world, and often represent a significant fraction of the professional astronomy community in the U.S. Part of the NITARP experience is immersion in the culture of professional astronomy. In addition to starting to learn the culture of astronomy, and in addition to their kickoff workshop, the teams start work on their project at the meeting. They learn about how to present science and education posters, what makes a good or bad poster, etc. After the AAS meeting, the teams go home and work remotely. They write a proposal, due in February or March, and it is reviewed. They do reviews of the published astronomical literature in the spring. They come out to visit the scientist at Caltech in Pasadena for three to four days in the summer to get started in earnest on their project; most teams save intensive data work for the summer visit. They go home and continue to work through the fall. As a team, they write two AAS poster abstracts by the appropriate deadlines (one science, one education), and then go to the AAS and present their results. The program pays for all (reasonable) travel expenses. Most of the work is conducted remotely.

After the AAS meeting (and in some cases, even beforehand), the teacher participants conduct at least 30 hours of professional development in their schools, districts, counties, regions, and states; many have conducted workshops on the national level. The teachers involve students in their classes during their intensive NITARP year, and then emerge with tools that can continue to be used in future years for future projects. The alumni, who have gone through the program serve as mentors (formal or informal) to the rest of the NITARP community of teachers and students. Almost without exception, teachers who have been through the program don't want to stop after just a year; they want to do more!

3. How does NITARP pick participants?

The educators are selected from an annual nation-wide application process. The applications are available annually in May and due in September.

Since one of the purposes of this program is to spread the word about how science works and expose more people to the process of science, it explicitly spells out in the application that if an applicant already has a Ph.D. in the physical sciences, that applicant probably already knows how science works, and thus is at a competitive disadvantage compared to people who don't already know how science works. We have had a few participants with doctorate degrees in other fields who are now teaching full time. Generally, we don't accept educators who already have astronomy Ph.D.s, because evidently they already know how astronomy research works.

The astronomy mentors are largely volunteers, and most have some affiliation with IPAC and/or Caltech. They are all active researchers, and represent a wide variety of sub-fields within astronomy.

4. What has NITARP Accomplished?

The list below includes those participants in the previous Spitzer program (starting in 2004) and NITARP, and are current as of July 2011:

- 56 educators trained (or training) in real astronomy research,

- 47 science or education posters presented,

- 4 research articles published in major refereed astronomical journals,

- 117 students (high school, middle school, college) visited IPAC and/or attended AAS meetings,

- more than 1200 students used data through the program,

- more than 100 students report that the program has influenced them to pursue careers in science or related fields,

- teachers and students have delivered at least 200 presentations, reaching over 14,000 people,

- at least 100 newspaper, radio, and TV reports (plus numerous Internet articles) discussing various aspects of teacher and student involvement,

- at least 43 high school students using their experiences in this program have received regional and international science awards.

Here are some selected quotes from recent participants (teachers and students); many more are on our website.[1]

---

[1] http://nitarp.ipac.caltech.edu

"...it invigorated me to become part of the greater message, which is the story of space and ground based observatories. Never in the history of this great science has so much data...been available to not just the scientific community but the general public as well. All one has to do is just ask!"

"I always thought just from programs on TV and in the classroom that astronomy was, more or less, completely figured out. Learning that it isn't is pretty exciting."

"Being there with my students was the most amazingly cool experience. I saw [them] explode in their willingness to ask questions and express an opinion. I was totally amazed by how their attendance made them reflective about the year and enthusiastic about science."

"EVERYTHING had a different flavor this year. I experienced everything through the lens of the research project of the past year. The entire experience was in context. When I look at how the intellectual process changed over the last year, I imagine it going from a diffuse look at research and the entire conference experience to the extreme focus on our own project during the year, and finally reaching outward again in Seattle to incorporate new information and understandings. Returning to AAS made the experience complete."

"I cannot say enough positives about the NITARP experience for the participating students. They have had the opportunity to learn and grow and see science applied in authentic research projects, while working with some of the coolest scientists around! It has allowed me to grow as a teacher and researcher, and be able to share my insight and newfound knowledge with students and peers."

"The best thing about the trip was the real world experience. Just like a real scientist, we worked with others to accomplish our goal by using the data and making graphs and calculations to find what we needed. We helped each other out, compared our answers, and learned from our findings and mistakes."

"This experience definitely changed the way I thought about astronomy and astronomers. I didn't realize that some of the calculations and applications were as accessible as they were. I also didn't realize how collaborative of a job it is...[and it's made up of many components]."

"[Real astronomy is] handling huge chunks of data and learning how to mine this information from sets so large that it is simply mind boggling. Many people are not aware of this, notably teachers in the trenches. They are teaching the science not as a process, but as a set of background material that acts as a starting point for conversation. The actual DOING of the science is a foreign thing to most teachers. This project is exactly why we are doing what we are doing! We want to convey what science is..."

"[The best thing about the trip was that] there are only a handful of people at home that I can speak to about astronomy, physics, and the like, so I was absolutely thrilled by the people I got to meet. Being surrounded by people at least as intelligent and, oftentimes, far more so is quite the exhilarating experience."

5. Come Play With Us!

Applications for NITARP educators are available annually in the spring and due in September. Scientists interested in mentoring a team (or subsidizing a team or teams) should contact us. The program runs from January to January. Please see our website[1] for more information, or email nitarp@ipac.caltech.edu with any questions.

Acknowledgments. We gratefully acknowledge funding from the NASA Astrophysics Data Program and the IPAC Archives.

2011 ASP Awards recipients, from left to right: Lonnie Puterbaugh, Kevin Apps, Doug Duncan, Jeremiah Ostriker, Paul Davies, and Mark Reed. Photo by Paul Deans.

Connecting People to Science
ASP Conference Series, Vol. 457
Joseph B. Jensen, James G. Manning, Michael G. Gibbs, and Doris Daou, eds.
© *2012 Astronomical Society of the Pacific*

Engaging Teachers and Students in the Rio Grande Valley in Earth and Space Science: Chapter II

Judit Györgyey Ries,[1] Margaret Baguio,[1] and Susana Ramirez[2]

[1]*The University of Texas at Austin, Center for Space Research*

[2]*The Rio Grande Valley Science Association of Texas*

Abstract.   In the summer of 2010, we received a NASA Science, Technology, Engineering, and Mathematics (STEM) education Cooperative Agreement Notice to prepare teachers in the Rio Grande Valley to become certified to teach the new fourth year capstone courses in astronomy and earth and space science. During the 2010 ASP conference, we reported on the earth and space science resources provided, guidance in curriculum development, and training in classroom activities. This two-year project began with the two 2010 summer workshops that concentrated on earth and space sciences, and were then followed up with two weekend training sessions, on-line training, and a Family Science Night during the school year. An important requirement of the new fourth year courses is a field investigation conducted by students. We offered minigrants for proposing teachers to support a field investigation. Here we highlight the outcomes of these follow-up programs and the two weeklong astronomy workshops in June 2011 in Edinburg, Texas.

1. Background

The U.S.-Mexico borderlands in south Texas is an area of rapid population growth where residents are younger and poorer than the U.S. population as a whole. Of the students in this region, 96% are Hispanic, 2% White, 1% Asian/Pacific Islander, and all other races make up the final 1%. Reducing of the dropout rate and attracting more students to scientific fields requires a workforce of educated, empowered, and enthusiastic teachers to generate interest in STEM education, and to nurture this interest from elementary through secondary school and on to college. To achieve this goal, in June of 2010 we conducted two week-long earth and space science workshops in Edinburg with a total of 45 teachers. We now summarize our activities and our accomplishments for the second year of the project.

2. Implementation Highlights

The following are updates and additions to the program described at last year's ASP meeting (Ries & Baguio 2011). Summer workshop participants requested additional activities to cover topographic maps, erosion processes, and volcanism for the November 2010 Science Saturday. We added an art component to the unit on planetary properties by designing travel brochures to different solar system objects (37 teachers enrolled). In March 2010, in preparation for the upcoming Texas Essential Knowledge and Skills test, we provided training on solar system scale models and the Earth-Moon system. Michael McGlone, Aerospace Education Specialist at NASA Johnson Space Center, led a session on how to estimate the size of the Sun and conducted training so the participants could be certified to check out the lunar rocks for classroom use (35 teachers participated).

An important requirement of the new fourth-year courses is a field investigation conducted by students. We funded one teacher proposal, "Destination Earth: Inside Out." On March 11, 2011, students from the Pharr-San Juan-Alamo Independent School District traveled to the San Marcos/San Antonio area to visit the Natural Bridge Caverns and investigate rock formations.

To encourage parent involvement in STEM education, on April 14, 2011 in San Juan, Texas, fifteen high school students, with the help of eight Rio Grande Valley Science Association teachers and students from the University of Texas Pan American (UTPA) organized an evening of fun science activities for area students and their families. Participants made sundials, built solar cameras to safely view the Sun, created Water Cycle Bracelets, and received Foldable Pocket Solar System guides for quick reference. The college students set up the mobile UTPA Star Lab planetarium in the school gymnasium and conducted presentations about the night sky. Fifty-one students and 49 adults attended the event.

3. Astronomy Workshops

June 13–24, 2011 Edinburg, Texas

3.1. Workshop Themes

This year our subject was astronomy, and we chose activities addressing topics from the Texas Essential Knowledge and Skills state education standards in all grade levels. Just as in 2010, UTPA hosted our two workshops on their campus, and helped to provide subject-matter support. In addition to classroom activities we had two night sky viewing sessions.

Day 1: Daily and Yearly motion of the Sun and Stars, The Sun, Electromagnetic Spectrum, HR Diagram, Life Cycle of a Star, Night sky viewing.

Day 2: Solar System Scales, Planetary System Formation, Small Bodies, Earth/Moon Comparison, and Phases of the Moon.

Day 3: Historical Perspective on the Solar System, Gravity, Tides, Black Holes, Our Place in the Milky Way, Galaxies, Big Bang.

Day 4: Night Sky Navigation, Constellations, Astronomical Technology, Making a Star Map, GalileoScope assembly (only for week 2), Night Sky Viewing with Binoculars/GalileoScopes.

Day 5: Exploring New Worlds, History of Space Travel, and Tribute to the Space Shuttle Program.

3.2. Survey Highlights

After completion of the workshop, participants completed an anonymous online survey about their background and answered self-reflective questions.

The majority of the 41 participants were women (70%). The ethnic composition was 70% Hispanic, 20% White, and 9% Asian, Pacific Islander or Black. Of the teachers, 34% had more than ten years experience, and an additional 31% had between six and ten years of teaching experience. Our statistics revealed that 69% of these educators, who are science teachers, never had a college level astronomy course.

Table 1. Level of perceived preparedness before and after the workshop

Subject Material	% Before	% After
Viewing the Night Sky	15	52
Phases of the Moon	60	95
Solar System Inventory	39	87
Sun and Stars	37	80
Galaxies	29	77
Electromagnetic Spectrum	30	68

3.3. Content Assessment

This year, in addition to the self-reflection, we introduced pre- and post-workshop content assessment. We chose ten questions to address common misconceptions and to apply knowledge by reasoning. The average score was 54% before the workshop. The post-workshop score reached 67%.

Three of the questions were chosen from the "Astronomy Diagnostic Test 2.0." They addressed the length of a shadow at noon, the daily motion of the Sun, and the concept of free fall. They had the lowest scores both for pre- and post-tests, but they also showed the second and third largest gains.

We covered a large quantity of material in five days, and it is not surprising that participants did not internalize some of the content. Questions relating to simple facts had higher scores on the post-test, and in the case of black holes, the teachers were eager to learn in anticipation of student questions relating to those objects. The largest post-test gain, 44%, was on a simple true/false question on black holes. We need to return to the concept questions in the follow-up workshops and encourage teachers to carry out long term projects, such as measuring the length of the shadows during the day for a whole semester and use planetarium software on their own. These additional activities will help teachers to better navigate the night sky. Also, we need to put significant effort into insisting that the media reporting on science news use correct expressions, and stop perpetuating misconceptions such as "zero gravity" in earth orbit.

4. Summary

Evaluation results at the close of the first year of NASA funding document the high level of need for professional development initiatives for science teachers in the Rio Grande Valley. The results also show that our project is making significant progress toward achieving its goals and objectives for engaging teachers and students in astronomy and earth and space science. While it is extremely difficult to sufficiently absorb the material during a workshop, we have significantly increased the motivation and comfort level for teaching astronomy and, given the hands-on nature of the activities, we expect the teachers will reach a deeper understanding as they prepare for the classroom. We have seen the dedication of these teachers. They are eager to help their kids learn and want to grasp difficult concepts so they can better assist their students. Survey responses at follow-up workshops in November 2010 and March 2011 showed more than half (57%) of the teachers were using one or more of the hands-on/inquiry guided activities from the June workshops with their students during school year. They also indicated that this program reached at least 450 elementary students, 860 middle school students, and 415 high school students in the Rio Grande Valley during the 2010–2011 school year. We have already reached half the STEM education goal stated in the original proposal. We are confident that, after completing our workshops, they will try even harder and will be able to motivate at least some of their students to pursue STEM fields as their career choices.

Acknowledgments.   Our project would not have been successful without team members Ms. Janie Leal (Rio Grande Valley Science Association), Dr. Nicholas Pereyra, Juan Gonzales and Hyun Chul Lee at UT Pan American at Edinburg Department of Physics and Geology. We also wish to thank our evaluator, Cindy Roberts-Gray (Third Coast R&D Inc.), Miss Suzanne Ramirez, Ms. Laura Espinoza and Mr. Efren Rodrigues for providing photography, and to Mr. Leo Ramirez for helping during the workshops. This work is supported by NASA Grant NNXI0AD31A.

References

Astronomy Diagnostic Test 2.0 `http://solar.physics.montana.edu/aae/adt/`
Ries, J. G. & Baguio, M. R. 2011, "Engaging Teachers and Students in the Rio Grand Valley in Earth and Space Science," in Earth and Space Science: Making Connections in Education and Public Outreach, J. B. Jensen, J. G. Manning, & M. G. Gibbs, eds., ASP Conference Series, 443, 304

Connecting People to Science
ASP Conference Series, Vol. 457
Joseph B. Jensen, James G. Manning, Michael G. Gibbs, and Doris Daou, eds.
© *2012 Astronomical Society of the Pacific*

Las Cumbres Observatory Global Telescope Network: Keeping Citizen Scientists in the Dark

Rachel J. Ross

Las Cumbres Observatory Global Telescope Network, 6740 Cortina Dr Suite 102, Goleta, California 93117, USA

Abstract. Las Cumbres Observatory Global Telescope Network (LCOGT) is creating a network of telescopes at excellent sites around the world providing 24/7 all sky coverage for astronomical observations. The network of telescopes, ranging in size from 0.4 m to 2.0 m, will be available for both scientific and education users.

The LCOGT telescopes are being built quickly and will be deployed soon. The two 2.0 m Faulkes Telescopes, one on Haleakala, Maui (FTN), the other at Siding Spring Observatory, Australia (FTS), are currently in operation. There is also a 0.8 m telescope in the Santa Ynez Valley, California (BOS), which is being used for commissioning and for many local outreach programs. The first 1.0 m telescopes will be heading to Chile and South Africa in 2011 and will each be accompanied by a 0.4 m telescope. Other sites, including Tenerife (Canary Islands, Spain), McDonald Observatory (Texas), Siding Spring (Australia), and Haleakala (Hawaii) will follow, with the possibility of up to two additional sites yet to be selected.

The LCOGT education and public outreach effort is transforming into a "Citizen Science" program. Several projects will encompass taking observations through the network, analyzing the data, and sharing the results with other citizen scientists from around the world. The first of these projects, "Agent Exoplanet," will be launched in mid-2011, and will involve analyzing brand-new data to create a light curve of an exoplanet. As the network is not yet complete, this test project will not include actual observing as future ones will.

More information about LCOGT and its Citizen Science program can be found online (http://www.lcogt.net). In addition to material to get started in the Citizen Science program, the website also includes resources and content for more hands-on activities using archived data, general astronomy pages, network information, complete access to the public data archive, current news, and recent publications. And don't forget to register for the LCOGT monthly newsletter.

1. The Telescope Network

The Las Cumbres Observatory Global Telescope Network (LCOGT) will have a total of 40 to 44 telescopes ranging in size from 0.4 m to 2.0 m at six to eight sites around the world providing 24/7 coverage of both the northern and southern hemispheres. The telescopes will be available for both scientific and educational uses, including our new citizen science program.

The sites include Haleakala, Maui and Siding Spring, Australia, where our two 2.0 m Faulkes telescopes (Faulkes Telescope North, FTN, and Faulkes Telescope South, FTS) reside. The other sites in the network will include Cerro Tololo in Chile, Tiede on Tenerife, Canary Islands, Sutherland in South Africa, and McDonald Observatory in Texas. Another potential site may be in China. Each of these sites will have two to three 1.0 m telescopes and two to four 0.4 m telescopes, for a network total of two 2.0 m telescopes, fifteen to eighteen 1.0 m telescopes, and about twenty-four 0.4 m telescopes.

There will also be dedicated weather stations, all-sky cameras, and webcams at each site that will be visible to the public. Each telescope will have high quality instrumentation and may include CCD imagers, high speed cameras, and spectrographs (not all instruments will be available on all telescopes, but there will be access to all at each site).

2.   Educational and Public Outreach with LCOGT

The LCOGT education effort is now primarily a Citizen Science program, which is open and available to learners of all ages and skill levels.

2.1.   Citizen Science

We are currently in the process of releasing our first project, "Agent Exoplanet," where users will be able to analyze data for transiting exoplanets. The data is analyzed individually and then is combined with other users to create a "master" light curve, averaging the photometry values for each image from each user.

As the telescope network comes online, users will have the added capabilities of taking their own observations to add to what will be a vast database. At the moment, the project uses new data that was taken specifically for this project using FTN, FTS, and BOS.

Agent Exoplanet, like all future projects, is entirely web-based, including photometry measurement tools and plotting light curve models. Observing with the network in the future will also be web-based. There is no need to download and install additional software!

All of our Citizen Science Programs are designed so that users get a better idea of what professional astronomers do and how they can participate and contribute to real, ongoing scientific research. This will include planning and making observations, analyzing the data, working with others from around the world, and determining what information the data can reveal. Many skills will be practiced or acquired, including critical thinking, planning and working through a project, participation in a large group, and more. Other citizen science projects are already in the works that are consistent with the main science goals of LCOGT.

Everyone is welcome to participate in Agent Exoplanet. To begin, register at our new Citizen Science Portal,[1] and then select an exoplanet to access brand new data taken specifically for this project. Analyze as many or as few images as you want using our online photometry tool (or use a photometry package such as SalsaJ and manually enter the values for the stars). Determine if your chosen comparison stars are good by comparing individual lightcurves; you can always go back and change them or add another. The good comparison stars will be used, and the data added to the combined data set. Tutorials will be available to all users. Help and discussion forums will also be available online, which all users are encouraged to utilize.

---

[1]http://portal.lcogt.net

2.2. Other Educational Uses of the Network

LCOGT telescope time will also be available for other educational projects, not just Citizen Science. Create your own observing program or use our vast data archive; there are countless projects that can be done.

2.3. More Educational Tools

LCOGT has other tools as well that may be useful. The public archive contains several years of data; it is available to all and can be explored to find the newest, most popular, and trending observations. Download and use as much data as you want.[2] *Virtual Sky* is a customizable, interactive, HTML5 based planetarium program that requires no software to download or browser plug-ins, and can be embedded in most webpages.[3] *Star In A Box* is an online tool to learn about the life cycle of a star.[4] *Space Book* is an online "text book" and includes general information about astronomy and telescopes.[5]

3. More Information

Check out the LCOGT website[6] for more information about everything LCOGT, including the telescope network, science, Citizen Science, and more.

To find out more about or to participate in Agent Exoplanet, check out the Citizen Science Portal.[1]

There are three main ways to keep up-to-date on current LCOGT news, including subscribing to the newsletter,[7] adding the RSS news feed,[8] and by finding us on Facebook and Twitter (links can be found on front page of website).
Rachel J. Ross: rross@lcogt.net

Acknowledgments. Thank you to everyone who is helping make LCOGT become a reality.

---

[2]http://lcogt.net/en/observations

[3]http://lcogt.net/en/virtualsky

[4]http://lcogt.net/siab/

[5]http://lcogt.net/spacebook

[6]http://lcogt.net/

[7]http://lcogt.net/newsletter/signup

[8]http://lcogt.net/en/blog

Andrew Fraknoi, James Manning, and Michael Gibbs. Photo courtesy of Andrew Fraknoi.

Connecting People to Science
ASP Conference Series, Vol. 457
Joseph B. Jensen, James G. Manning, Michael G. Gibbs, and Doris Daou, eds.
©2012 *Astronomical Society of the Pacific*

Are We Teaching Students to Think Like Scientists?

Louis J. Rubbo and J. Christopher Moore

Department of Chemistry and Physics, Coastal Carolina University, Conway, South Carolina 29528, USA

Abstract. University courses in conceptual physics and astronomy typically serve as the terminal science experience for non-science majors. Significant work has gone into developing research-verified pedagogical methods for the algebra- and calculus-based physics courses typically populated by natural and physical science majors; however, there is significantly less in the literature concerning the non-science population. This is quickly changing, and large, repeatable gains on concept tests are being reported. However, we may be losing sight of what is arguably the most important goal of such a course: development of scientific reasoning. Are we teaching this population of students to think like scientists?

University courses in conceptual physics and astronomy typically serve as the terminal science experience for non-science majors. Significant work has gone into developing research-verified pedagogical methods for the algebra- and calculus-based physics courses typically populated by natural and physical science majors; however, there is significantly less in the literature concerning the non-science population. This is quickly changing, and large, repeatable gains on concept tests are being reported. However, we may be losing sight of what is arguably the most important goal of such a course; development of scientific reasoning. Are we teaching this population of students to think like scientists?

A recent discussion in *The Physics Teacher* (Sobel 2009; Lasry, Finkelstein, & Mazur 2009) has led us to examine the central focus of our courses for non-science majors. Like Lasry et al. (2009), we certainly do not believe that this population of students is "too dumb" for physics, or that physics is in a "different category," accessible only to certain students such as science majors. However, there are very real differences in the two populations, especially when considering interest level, formal preparation, and prior development of scientific reasoning skills. It is the latter that we address in this article, since we believe development of scientific reasoning should be a central goal for these types of courses.

In particular, reasoning and metacognition development are essential if we hope to elevate non-science students to "expert-like" status with respect to problem solving, understanding and applying abstract concepts, and shifting between multiple representations (White & Frederiksen 1998; Etkina & Mestre 2004). However, non-science majors enter the classroom with a disadvantage not necessarily shared by their self-selecting science major peers. Non-scientists struggle with basic scientific reasoning patterns, which can hinder their growth in the course.

We have found that students in our conceptual physics and astronomy courses score significantly lower on Lawson's Classroom Test of Scientific Reasoning (LCTSR) compared to students enrolled in courses typically populated with science majors. The

LCTSR assesses reasoning patterns such as proportional reasoning, control of variables, probability reasoning, correlation reasoning, and hypothetico-deductive reasoning (Lawson 1978). Figure 1 shows average LCTSR pre-instruction scores ($N = 1061$, average = 74.2%) for freshman science and engineering majors enrolled in a calculus-based introductory physics course, as reported by Bao et al. (2009). The LCTSR was also administered to students taking a physics or astronomy course with one of the authors during the past three years. As shown in Figure 1, this population of students scores significantly lower ($N = 68$, average = 54%) than their scientist counterparts. Of particular interest, scores on LCTSR questions designed to test application of hypothetico-deductive reasoning, which can arguably be called the "scientific method," average an abysmal 38% in the non-scientist classes.

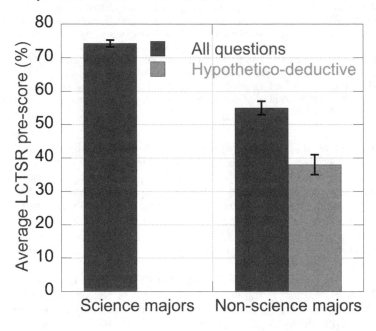

Figure 1.   Average scores on the LCTSR before instruction for science ($N = 1061$) and non-science ($N = 68$) majors. Hypothetico-deductive scores were obtained from LCTSR questions 21–24.

Acknowledgment of this dramatic difference in reasoning ability is important for development of good pedagogy, considering scientific reasoning has been linked to student gains in conceptual knowledge for both non-scientist and scientist populations. Coletta & Phillips (2005) observed a strong correlation between normalized gain on the Force Concept Inventory (FCI) and pre-instruction LCTSR scores. As shown in Figure 2, during assessment for our conceptual physics courses over the past two years, we have observed similar strong correlations between pre-instruction LCTSR scores and normalized gain on two concept inventories, the Determining and Interpreting Resistive Electric circuits Concept Test (DIRECT; Engelharda & Beichner 2004), and the Test for Understanding Graphs–Kinematics (TUG-K; Beichner 1994).

Strong correlations are seen for content requiring higher-order and more abstract reasoning. With a linear fit slope of 0.64 and $r = 0.59$, the correlation between TUG-K

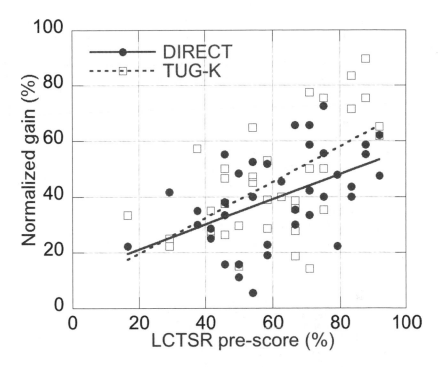

Figure 2. Normalized gain on DIRECT (filled circles) and TUG-K (hollow squares) versus LCTSR pre-instruction scores for non-science majors taking a conceptual physics course. DIRECT: slope = 0.45 and r = 0.50; TUG-K: slope = 0.64 and r = 0.59.

normalized gain and LCTSR score is similar to that seen for the FCI, and stronger than the correlation observed for the DIRECT assessment (slope = 0.45 and r = 0.50). This is not surprising, considering the TUG-K tests a student's ability to move between multiple representations, which rely on higher-order and more abstract thinking (Kohl & Finkelstein 2008). The FCI assesses a student's knowledge and application of the abstract concept of force. Lawson would classify these concepts as hypothetical (motion) and theoretical (force), requiring advanced reasoning development to achieve success (Lawson et al. 2000). A weaker correlation between DIRECT gains and LCTSR scores could be because strong scores are possible on DIRECT via good observation and retention from well designed inquiry based activities; many of the questions are descriptive requiring proficiency only in descriptive level reasoning. This suggests that if we wish to push our non-science students past the lower three levels of Bloom's Taxonomy of Educational Objectives (Anderson et al. 2000), then we need our courses to focus explicitly on scientific reasoning early and often.

Even with significant disadvantages, substantial gains in content knowledge can still be obtained in conceptual physics and astronomy courses, especially when those courses are designed around a research-verified, active engagement curriculum. For the conceptual physics and astronomy course, respectively, the authors use a large-enrollment implementation of *Physics by Inquiry* (PbI; Scherr 2003) and *Lecture Tuto-*

rials for Introductory Astronomy (Prather et al. 2004). Shown in Figure 3 are average normalized learning gains on DIRECT, TUG-K and the Star Properties Concept Inventory (SPCI; Bailey 2007) for students enrolled in our courses over the past three years. Although lower than reported for students completing some active-engagement algebra and calculus-based courses, these gains are still significant.

Even though we have been relatively successful with content, we have failed to improve reasoning ability. As seen in Figure 3, average normalized gains on the LCTSR for both physics and astronomy students are essentially equivalent to zero. This is particularly surprising for the conceptual physics course, which via PbI is completely designed around the process of scientific inquiry. Of course, this is not to suggest that gains in reasoning are unachievable. The content-specific education literature in other disciplines suggests that explicit intervention is necessary to improve reasoning (White & Frederiksen 1998; Etkina & Mestre 2004; Lawson 2000). In fact, we are beginning to see significantly larger gains in scientific reasoning via explicit instruction during our most recent courses, though these observations are preliminary.

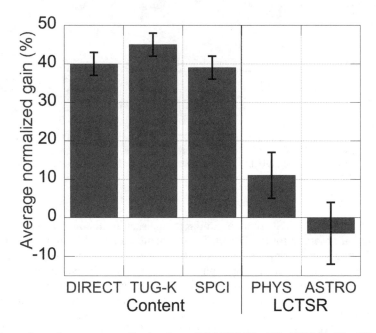

Figure 3.   Average normalized gain on DIRECT ($N = 39$), TUG-K ($N = 40$) and SPCI ($N = 28$) for non-science majors taking either a conceptual physics or astronomy course. Average normalized gains on the LCTSR during these courses are also shown.

Development of scientific reasoning is not only a necessary means to an end (making their thinking more scientific so that they can better grasp the content), but it is also a justifiable end in and of itself. We should expect our courses to affect our students beyond the classroom. Particularly for non-scientists, a broader approach should be expected since these types of courses are typically their terminal experience in formal science education. Are students in our courses learning to think like scientists? Do we care? Our purpose in writing this paper is to continue the discussion about how we should go about designing our courses for the non-science major. Specifically, is development of scientific reasoning an important goal? If yes, then are we currently achieving that goal? At least in the case of the authors, the answer to that question is "no." We are working on that, and we hope others will join us.

References

Anderson, L. W., Krathwohl, D. R., Airasian, P. W., Cruikshank, K. A., Mayer, R. E., Pintrich, P. R., Raths, J., & Wittrock, M. C. 2000, A Taxonomy for Learning, Teaching, and Assessing: A Revision of Bloom's Taxonomy of Educational Objectives, 2nd Edition, (Allyn and Bacon)

Bailey, J. M. 2007, "Development of a Concept Inventory to Assess Students' Understanding and Reasoning Difficulties about the Properties and Formation of Stars," Astronomy Education Review, 6, 2, 133

Bao, L., Cai, T., Koenig, K., Fang, K., Han, J., Wang, J., & Wu, N. 2009, "Leaning and Scientific Reasoning," Science, 323, 5914, 586

Beichner, R. J. 1994, "Testing Student Interpretation of Kinematics Graphs," American Journal of Physics, August, 62, 8, 750

Coletta, V. P., & Phillips, J. A. 2005, "Interpreting FCI Scores: Normalized Gain, Preinstruction Scpores, and Scientific Reasoning Ability," American Journal of Physics, December, 73, 12, 1182

Engelharda, P. V., & Beichner, R. J. 2004, "Students' Understanding of Direct Current Resistive Electrical Circuits," American Journal of Physics, January, 72, 1, 98

Etkina, E., & Mestre, J. P. 2004, "Implications of Learning Research for Teaching Science to Non-science Majors," (Spencer: Harrisburg)

Kohl, P. B., & Finkelstein, N. D. 2008, "Patterns of Multiple Representation Use by Experts and Novices during Physics Problem Solving," Physical Review Special Topics: Physics Education Research, June, 4, 1, 010111

Lasry, N., Finkelstein, N., & Mazur, E. 2009, "Are Most People Too Dumb For Physics?" The Physics Teacher, October, 47, 7, 418

Lawson, A. E. 1978, "The Development and Validation of a Classroom Test of Formal Reasoning," Journal of Reasearch in Science Teaching, January, 15, 1, 11

Lawson, A. E. 2000, "The Generality of Hypothetico-Deductive Reasoning: Making Scientific Thinking Explicit," American Biiology Teacher, September, 62, 7, 482

Lawson, A. E., Alkhoury, S., Benford, R., Clark, B. R., & Falconer, K. A. 2000, "What Kinds of Scientific Concepts Exist? Concept Construction and Intellectual Development in College Biology," Journal of Research in Science Teaching, November, 37, 9, 996

Prather, E. E., Slater, T. F., Adams, J. P., Bailey, J. M., Jones, L. V., & Dostal, J. A. 2004, "Research on a Lecture-Tutorial Approach to Teaching Introductory Astronomy for Non-Science Majors," Astronomy Education Review, October, 3, 2, 122

Scherr, R. E. 2003, "An Implementation of Physics by Inquiry in a Large-Enrollment Class," The Physics Teacher, January, 41, 2, 113

Sobel, M. 2009, "Physics for the Non-Scientist: A Middle Way," The Physics Teacher, September, 47, 6, 346

White, B. Y., & Frederiksen, J. R. 1998, "Inquiry, Modeling, and Metacognition: Making Science Accessible to all Students," Cognition and Instruction, 16, 1, 3

Closing reception. Photo by Paul Deans.

Connecting People to Science
ASP Conference Series, Vol. 457
Joseph B. Jensen, James G. Manning, Michael G. Gibbs, and Doris Daou, eds.
© *2012 Astronomical Society of the Pacific*

Anatomy of an App: Why Design and Content are Essential for Outreach

Jessica Suzette Santascoy[1] and Joe Cieplinski[2]

[1]*Astronomical Society of the Pacific, 390 Ashton Avenue, San Francisco, California 94112, USA*

[2]*Graphic Designer, San Francisco, California 94112, USA*

Abstract.   Design is a major component of outreach, but few organizations recognize the urgency of having good design, along with its sister, good content. Content includes anything that is produced, such as writing, audio, video, and photography. Good content is essential for good design and vice-versa. You need both to be successful, but we'll focus more on design in this paper. We cover the basics of app design, and many of these principles will work with regard to websites, print pieces, and posters. This paper is intended for outreach professionals interested in developing an app and/or interested in attracting a wider audience.

1. Introduction

Design is a major component of outreach, but few organizations recognize the urgency of having good design, along with its sister, good content. Content includes anything that is produced, such as writing, audio, video, and photography. Good content is essential for good design and vice-versa. You need both to be successful, but we'll focus more on design in this paper. We cover the basics of app design, and many of these principles will work with regard to websites, print pieces, and posters. This paper is intended for outreach professionals interested in developing an app and/or interested in attracting a wider audience.

Good design is critical for outreach. Design is a way to persuade people to listen and opens the door for them to absorb information. You have limited time to communicate why your project is important and relevant to your audience. If something is poorly designed, people mentally walk away. They may not even know what was off-putting and you might not either. For example, the choice and size of font is something a designer considers carefully. Designers know that font, size, and spacing of the lines in between text can make the difference in whether someone continues to read or not (Müller-Brockmann 1996). A good designer knows how to prevent eye fatigue. A designer respects users and knows how to engage them, even at detailed levels. If something is designed and written well, you have a higher chance of people staying, listening, and learning.

If you're an outreach professional, it's very helpful to have a basic understanding of design. If you don't have these skills, hire someone who has them and listen to him or her. Even better, hire a creative director to manage the processes between the designer and the content writer. Don't make the mistake of thinking that design or content is easy and you can do it yourself, or that every writer and designer have all the necessary skills.

You can have a great message with superb research and science, but if the content and design are bad, the project may fail to reach people. You want to ensure your outreach project truly makes a difference, and good design and content are two tools that a smart organization would never do without.

Now we'll discuss some basic ways to approach design by using apps as our entry point, so that you can work with designers constructively.

2.   What is Good App User Interface?

User interface (UI) is a term that describes the way a user responds and interacts with a product and/or an interface. An app has to have easy navigation that you understand in seconds. There should be no question as to where you're going next or what you're doing. The app should be simple to use for all ages. Grandparents, parents, and children should be able to use the app easily.

Why does an app have to be easy to use? Mobile devices, in particular an iPhone or Android smartphone, is something we use when we're bored, in-between tasks, or for a particular purpose. By design, a mobile device is an on-the-go device, whereas a laptop or even a tablet is most often used for longer tasks.

An app should also be stable and speedy. There's nothing worse than needing information quickly and waiting for the information. A large part of this is in the programmers' hands, but good UI design helps the user move through the app easily and increases the perception that the app is fast.

3.   How to Evaluate an App

People have short attention spans and so you've got to make sure your app is worth their time and attention. Josh Clark, author of *Tapworthy*, developed the "Glance Test." Hold your device at arms' length and see if you can still get the information you need. If you can get and use the information, then there's a pretty good chance that your app is user friendly.

Generally speaking, if you follow the rule of "less is more," with both content and design, you'll be on your way to reaching and engaging more people.

4.   Mobile vs. Website UI

Go StarGaze, the iPhone app that helps people find stargazing events and astronomy clubs, is an outreach app that was developed after the NASA Night Sky Network website. The website offers more features, such as downloadable astronomy activities and star charts. The app is meant to increase attendance at astronomy club events by putting star party and club information in an easily accessible form.

The key to deciding what would be included in the app was to think about what the user wanted and needed and not offer everything that's available on the website. We wanted people to be able to find a star party or astronomy club easily, at any time, when most convenient for the user. We limited the data to finding astronomy events and clubs.

Originally, data was included for up to 6 months in the app and the lengthy search time made it slow. Most users are using apps on the go, not to plan out their stargazing activities for the long term. The speed of the app was improved by limiting the search to two months. The original mistake was equating an app with a website, and they are different experiences. It's also important to understand and trust that the user will go to the website for further information.

Due to limited funds, we've been unable to make further changes to the app. An app needs to be updated and improved, and nonprofit organizations would do well to plan for additional funding for subsequent upgrades. Apps are dynamic, just like websites, and you'll want to be prepared to meet the next level of user engagement.

5. Conclusion

Design and content are important, and you can test the effects of them on yourself. Imagine a dingy café, where the staff is mildly nice and the food is mediocre. Then imagine this same café, renovated, with lots of natural light, tasty food, and a new staff that is happy to see you. In this example, design is the external environment of the café and the content is the staff and food. Which café do you prefer?

Jessica Santascoy is the Astronomy Outreach Coordinator at the Astronomical Society of the Pacific. She's the voice of the NASA Night Sky Network on Facebook and Twitter. She's also a member of the San Francisco Amateur Astronomers. Jessica is passionate about the outreach that astronomy clubs do, and thinks that everyone should experience the sky through a telescope.

Joe Cieplinski is a graphic designer, sound engineer, and musician. He designed *Go StarGaze*, the iPhone app for the NASA Night Sky Network that helps you find stargazing events and astronomy clubs. Joe also designs music for apps.

References

A List Apart, "A List Apart Magazine explores the design, development, and meaning of web content, with a special focus on web standards and best practices," http://www.alistapart.com/

Clark, J. 2010, Tapworthy: Designing Great iPhone Apps (O'Reilly)

Lidwell, W., Holden, K. & Butler, J. 2010, "Universal Principles of Design, Revised and Updated: 125 Ways to Enhance Usability, Influence Perception, Increase Appeal, Make Better Design Decisions, and Teach through Design," Second edition, (Rockport Publishers)

Müller-Brockmann, J. 1996, "Grid Systems in Graphic Design/Raster Systeme Fur Die Visuele Gestaltung," (Ram Publications; Bilingual edition)

Connecting People to Science
ASP Conference Series, Vol. 457
Joseph B. Jensen, James G. Manning, Michael G. Gibbs, and Doris Daou, eds.
© *2012 Astronomical Society of the Pacific*

Interactive Spherical Projection Presentations Teach Students about the Moon and Mercury

Sarah B. Sherman,[1] Brian James,[1] Collin Au,[1] Korie Lum,[2] and Jeffrey J. Gillis-Davis[1]

[1]*Hawaii Institute of Geophysics and Planetology, University of Hawaii, 1680 East-West Rd., Honolulu, Hawaii 96822, USA*

[2]*Moanalua High School, Honolulu, Hawaii 96818, USA*

Abstract. Using spherical displays like the Magic Planet and Science on a Sphere, this project aims to create interactive and engaging multimedia presentations for schools and the general public. These innovative multimedia devices offer a unique and stimulating way to display NASA data sets to a broad audience. Presentations will highlight data of the Moon from Clementine, Lunar Orbiter, Lunar Prospector, and the Lunar Reconnaissance Orbiter mission as well as data from the MESSENGER mission to Mercury. A key aspect of the project is the incorporation of clickers into the presentation in order to encourage audience participation and to promote interest.

1. Introduction

Spherical displays for digital media, such as the Science on a Sphere (SOS) and Magic Planet (MP), have outdated images of Mercury and the Moon and do not convey the robust geology of these planetary objects. We are in the process of creating updated images of Mercury with the new images from the MEcury Surface, Space Environment, GEochemistry, and Ranging (MESSENGER) satellite. MESSENGER's panchromatic, high-resolution, narrow angle camera accentuates surface features, while images from the multi-spectral wide angle camera provides mineralogical distribution information of Mercury.

Lunar data on SOS and the MP consist of digital airbrush maps showing shaded relief and a Clementine true-color image of the Moon with Apollo landing sites. We are also updating the lunar images to include the latest data from the Lunar Reconnaissance Orbiter (LRO), which has a camera with 50 cm resolution, and have created an improved topography map of the Moon (Fig. 1). Using data from Clementine, Lunar Orbiter, Lunar Prospector, as well as the LRO mission, we are creating multimedia applications for the MP and SOS for the Moon. Presenting the data on this innovative and stimulating medium captures the interest, stimulates curiosity, and inspires scientific learning in children, as well as general audiences.

2.   Why the Moon and Mercury?

With missions currently underway and new, more detailed data sets being released, the Moon and Mercury are both topical to current NASA research and of interest to the target audience. While very little is known about Mercury, it's characteristics as a small, rocky planet offer a link to understanding terrestrial planet evolution. The MESSENGER mission will greatly expand the amount of available data on the planet.

Figure 1.   Lunar Reconnaissance Orbiter data of the Moon projected onto the Magic Planet.

As the Earth's closest relative, the Moon is of similar interest. The Moon and Mercury have the most extreme surface thermal environments and both may have water ice. Further exploration of the Moon will require detailed knowledge of potential areas of resources. The current SOS projections that are available for Mercury and the Moon are outdated and do not adequately portray the geology of these planetary objects. Assimilating the newest data sets into MP and SOS compatible formats offers a unique display and exploration option. They are also a valuable visual teaching tool.

3.   What are Spherical Displays?

The MP and SOS are spherical projection screens ideally suited to display data of planets and spherical objects with minimal distortion (Figs. 1 and 2). They range in size from twelve inches to six feet in diameter. Linked to a computer, the screens are able to display seamless images and animations. Being both unique and innovative, the displays are a stimulating medium that captures audience interest, enables the visualization of complex three dimensional processes, and inspires scientific insight and learning.

4.   Interactive Presentations

The use of clickers to encourage audience participation is an effective way to maximize the learning potential of the SOS and MP. Clickers allow the audience to guide and adapt the presentation in real-time. This allows each presentation to accommodate a wide variety of knowledge levels. Throughout the presentation, the audience will be asked to make predictions, answer questions, and vote on where the presentation will go next. After displaying relevant data such as lunar surface, temperature, and TiO_2 distribution maps, the audience is asked to select the "best" lunar landing site for future exploration and a site for a permanent lunar base. This flexibility within the presentation and the increased involvement of the audience leads to greater interest and potentially retention of information.

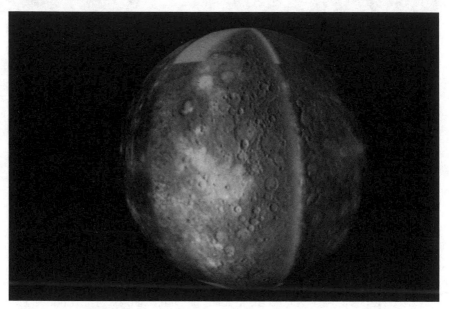

Figure 2.   Mercury images from Mariner and MESSENGER combined and projected onto the Magic Planet.

5.   Conclusion

The primary goals of this project are to 1) increase the general public's understanding of planetary science and awareness of NASA missions by engaging audiences with displays using the latest NASA data sets for Mercury and the Moon onto the high-tech, stimulating SOS and MP, and 2) promote interest in science, engineering, and/or technology careers through exposure to the current MESSENGER and LRO missions and their scientific findings. When finished, the movie presentations will be uploaded onto the SOS website for anyone to download.

Acknowledgments.   The authors would like to thank James Potemra for his generosity in granting access to his Magic Planet.

Connecting People to Science
ASP Conference Series, Vol. 457
Joseph B. Jensen, James G. Manning, Michael G. Gibbs, and Doris Daou, eds.
© 2012 Astronomical Society of the Pacific

Working Together Through NASA's Astrophysics Forum: Collaborations Across the Spectrum

Denise A. Smith,[1] Mangala Sharma,[1] Mitch Watkins,[1] Brandon Lawton,[1]
Bonnie Eisenhamer,[1] Dan McCallister,[1] Lindsay Bartolone,[2]
Michelle Nichols,[2] Heather Withnell,[2] Greg Schultz,[3] Andrew Fraknoi,[3]
James G. Manning,[3] William Blair,[4] and Luciana Bianchi[4]

[1]Space Telescope Science Institute, 3700 San Martin Dr, Baltimore, Maryland
21218, USA

[2]The Adler Planetarium, 1300 S. Lake Shore Dr., Chicago, Illinois 60605, USA

[3]Astronomical Society of the Pacific, 390 Ashton Ave., San Francisco,
California 94066, USA

[4]Johns Hopkins University, 3701 San Martin Dr., Baltimore, Maryland 21218,
USA

Abstract.   The NASA Science Mission Directorate (SMD) Science Education and
Public Outreach Forums are teams of scientists and educators that work closely with
NASA SMD and its education and public outreach (EPO) community to organize individual SMD-funded EPO activities into a coordinated effort. One Forum exists for
each of SMD's four science divisions: Astrophysics, Earth Science, Heliophysics, and
Planetary Science. Through the Forums, NASA SMD and the members of its EPO community work together to interconnect a wide spectrum of SMD-funded EPO activities,
resources, and expertise and to enhance the efficiency and effectiveness of the resulting ensemble of efforts. The Forums also support members of the broader science and
education communities by identifying opportunities and strategies for scientists and educators to participate in and make use of SMD EPO activities and resources according
to their needs. This article highlights examples of this work within NASA Astrophysics
EPO, where the Forum and community members are extending the impact of existing EPO resources through collaborative online professional development activities for
classroom educators, activities to engage girls in STEM in library settings, resources
to support the Astronomy 101 instructional community, and practical tips for scientists
interested in EPO.

1.   Introduction

The four NASA SMD Science Education and Public Outreach Forums provide resources and activities to facilitate communication, collaboration, and sharing of best
practices between SMD-funded mission EPO teams, instrument teams, grant awardees,
and SMD EPO leadership. These include regular conference calls (TagUps), informal networking events and brief meetings held in conjunction with major conferences
(meetings of opportunity), annual retreats, an online SMD EPO community workspace
(`smdepo.org`), professional development webinars, and sessions at conferences such

as "Connecting People to Science." These activities help SMD EPO community members to become more aware of each other's EPO products, activities, and areas of expertise, and to work together to carry out their EPO activities in a way that makes more efficient use of their collective resources. The Forum teams also identify opportunities and strategies for scientists and educators to participate in SMD EPO.

To assist members of the NASA Astrophysics EPO community in identifying priority areas for collaboration, the Astrophysics Forum has carried out a community survey and incorporated a variety of group brainstorming, prioritization, and consensus-building activities into community meetings and retreats. Emerging community priorities for collaborative efforts fall into the following categories: conducting professional development (coordinating professional development for informal science educators, creating an online professional development institute for classroom educators); developing key themes and resources (multi-wavelength resources, investigations/citizen science opportunities using mission/research data, continuing the best of the International Year of Astronomy); broadening EPO audiences (activities focusing on girls and young women); community EPO planning and publication (helping SMD to develop an EPO roadmap; creating community EPO publications); scientist engagement in EPO (partnership best practices and examples; menu of opportunities, strategies, and products); and higher education (EPO products and activities for the undergraduate classroom). Community members examined existing Astrophysics EPO resources and audience needs to refine these priorities into the initial set of community collaborations described below. Efforts to facilitate the involvement of scientists in EPO were initiated via a community discussion at the winter 2011 meeting of the American Astronomical Society, and are also described below.

2.   NASA's Multiwavelength Universe: Online Professional Development for Classroom Educators

NASA's Multiwavelength Universe is a pilot online course for educators that has been developed by several community members representing many different NASA Astrophysics missions and projects (Fermi, HST, Kepler, NuSTAR, Observing with NASA, SOFIA, Swift, WMAP, WISE, and XMM-Newton) and coordinated by the Astrophysics Forum. It was offered in July 2011 for academic credit or for continuing education units through Sonoma State University. During the two-week course, 22 middle and high school teachers explored the properties of the electromagnetic spectrum and learned about the different ways that NASA Astrophysics missions make use of the electromagnetic spectrum to observe the Universe. They also discussed common misconceptions that students may have as well as ways that NASA resources can help teachers address them. There were four live sessions and several homework assignments with online discussion components that allowed educators to experience multiple resources in a unified way; this combination of sessions and assignments provided context, access to science and education experts, and time for in-depth exploration of content and resources.

This community collaboration emerged from discussions of the well-known need for in-depth quality professional development for classroom educators on core science topics and a desire to reach classroom educators who may not travel to a professional development conference or workshop. Piloting an online course of this nature also allowed community members to gain experience with distance-learning technologies, online course management systems, and each other's EPO materials and expertise. The modular format of the professional development experience also extends the "life" of some mission resources while allowing flexibility to include other mission resources as the course evolves.

3.   Astro4Girls: Engaging Girls in STEM via Public Libraries

The "Astro4Girls" community collaboration aims to use existing NASA Astrophysics EPO resources in library settings to empower girls to see that they can and are doing science. The collaboration is designed to support community members and NASA in broadening audiences for NASA SMD EPO and to respond to needs for science programming expressed by public libraries (Smith et al; this volume) and research on engaging girls in STEM (Sharma et al; this volume). Community members share information about challenges and best practices in engaging girls in STEM, as well as the needs and opportunities of public libraries, through activities coordinated by the Astrophysics Forum. These include small group discussions at community retreats and meetings, discussion threads on the online SMD EPO community workspace, and professional development sessions at conferences (see the related articles by Bleacher et al., Sharma et al., and Smith et al. in this volume). Community members and Forum team members are also identifying existing SMD Astrophysics EPO resources and programs that can be utilized in library settings and adapted to be more girl-friendly. A pilot effort is planned for March 2012 in conjunction with Women's History Month. Participating community members represent a range of Astrophysics missions and EPO programs, including Fermi, NuSTAR, Swift, and XMM-Newton (Sonoma State University), the Hubble and James Webb Space Telescopes (STScI), Kepler and SOFIA (SETI Institute), WISE, Afterschool Universe, Observing with NASA, and Visions of the Universe. External partners include the American Library Association and National Girls Collaborative Project.

4.   Cosmology Resource Guide: Supporting Resources for Undergraduate Instructors

To assist SMD EPO and the NASA Astrophysics EPO community in meeting the needs of higher education faculty, we are also exploring how existing NASA Astrophysics EPO materials and programs can be utilized in undergraduate settings. Astrophysics Forum team members organized a Special Interest Group discussion on this topic at the 2010 Astronomical Society of the Pacific conference (Sharma et al. 2011) and are also carrying out informal conversations with undergraduate instructors. Based on emerging audience needs, the initial focus is to gather resources that support Astronomy 101 instructors in teaching cosmology into a "Cosmology Resource Guide." Cosmology is an area where rapid research progress is being made and concepts are challenging for both students and instructors. Conversations with Astronomy 101 instructors also reveal that instructors are seeking new approaches for teaching cosmology, simulations and animations that convey "big picture ideas," and updates on new results (Schultz et al. 2012). A single entry point to resources and training on existing materials are also desired. Collecting existing resources into a Cosmology Resource Guide is aimed at addressing these needs, providing a pathway for Astronomy 101 instructors to connect with NASA SMD EPO, and extending the impact of existing NASA Astrophysics EPO resources.

5.   Scientist Engagement in Education and Public Outreach

Astrophysics Forum and community efforts to engage scientists in EPO are aimed at providing scientists with easy access to information about SMD EPO resources and opportunities. We are compiling needs, challenges, and best practices regarding scientist participation in EPO from discussions with Astrophysics mission EPO teams (who actively involve scientists in their EPO efforts), informal discussions with scientists, and the literature. Based on this information, we are working together to craft materials that can assist scientists and outreach coordinators in connecting more efficiently with SMD EPO resources. For example, we are piloting a "sampler" brochure that outlines a subset of NASA SMD Astrophysics mission and program EPO resource materials, where to obtain the materials, and how to use them in EPO activities. Forum team members are also working with NASA Astrophysics EPO community members to create a "menu of opportunities" for scientists to participate in SMD EPO based on their time, interests, and expertise. We are also exploring connecting EPO professionals, scientists, and outreach coordinators to form an "EPO Resource Partner Network" that can provide practical tips for scientists wishing to become engaged in EPO.

Acknowledgments.   The Astrophysics Science Education and Public Outreach Forum is supported by NASA under Cooperative Agreement NNX09AQ11A through the NASA Science Mission Directorate. The Astrophysics Forum is a partnership between the Space Telescope Science Institute, the Adler Planetarium, the Astronomical Society of the Pacific, and Johns Hopkins University.

References

Bleacher, L. V., Peterson, K. A., Sharma, M., & Smith, D. 2012, "Engaging Girls in STEM: How to Plan or Revamp Your EPO Resources or Activities to be More Effective for Girls," ASP Conference Series, 457, 37
Schultz, G. R., Fraknoi, A., Smith, D., & Manning, J. 2012, "Developing Resource Guides for Astro 101 Instructors, as a Higher Education Community Collaboration from the NASA Astrophysics SEPOF," Bulletin of the American Astronomical Society, 44, in press
Sharma, M., Smith, D., Schultz, G., Bianchi, L., & Blair, W. 2011, "The Intersection of NASA Astrophysics Education and Public Outreach: A Special Interest Group Meeting," ASP Conference Series, 443, 49
Sharma, M., Peterson, K. A., Bleacher, L. V., & Smith, D. 2012, "Engaging Girls in STEM: A Discussion of Foundational and Current Research on What Works," ASP Conference Series, 457, 43
Smith, D., Eisenhamer, B., Sharma, M., Brandehoff, S., Dominiak, J., Shipp, S., & LaConte, K. 2012, "Collaborating with Public Libraries: Successes, Challenges, and Thoughts for the Future," ASP Conference Series, 457, 101

Connecting People to Science
ASP Conference Series, Vol. 457
Joseph B. Jensen, James G. Manning, Michael G. Gibbs, and Doris Daou, eds.
© *2012 Astronomical Society of the Pacific*

Teaching the Moon: A Study of Teaching Methodology Across Age Groups

Faith Tucker and Nathaniel Paust

Whitman College, 345 Boyer Ave., Walla Walla, Washington 99362, USA

Abstract.   In this study I attempted to determine the most effective teaching style for teaching elementary, middle school, and undergraduate students about lunar phases and eclipses. Within each age group, there were two sub groups, one of which was introduced to the material in a standard lecture format while the other sub-group interacted with the content through activities and demonstrations. After their respective lessons, both sub-groups were given the same post-instruction test in order to assess their comprehension of the content. The results from this experiment provided insight into effective teaching styles and common misconceptions about lunar phases and eclipses at different age levels, as well as introducing new interactive teaching activities for elementary, middle school and undergraduate students.

1.   Study Description

All three of the age groups (elementary, middle school, and undergraduate) were randomly split into two equal sub-groups, which were each separately taught the same age-appropriate material on lunar phases and eclipses. The first sub-group was taught the material entirely through lecture format instruction while the second sub-group was taught the material primarily through interactive, inquiry-based activities and demonstrations. After receiving their respective instruction, both sub-groups of each age group were given the same post-instruction test. The results of these post-instruction tests were then used to judge the relative efficacy of the teach method.

The primary learning goals for all three age groups were that student be able to do the following at an age appropriate level:

- explain why the Moon has phases;

- describe the geometry of the Sun-Earth-Moon system in general and during solar and lunar eclipses; and

- account for the approximate timing of eclipses.

2.   Activities

2.1.   Lunar Phases

- A large inflatable ball representing the Moon is entirely covered in tape, with one hemisphere white and the other blue, and then circulated around the perimeter of the classroom.

- One wall of the classroom is chosen to represent the Sun and the white hemisphere of the ball always points in that direction.

- Students in the center of the room draw a picture of what the Moon looks like at each position and tapes it on the wall behind the Moon.

- The goal of this activity is to demonstrate how the phases of the Moon are a result of the Sun continuously illuminating one hemisphere of the Moon as the Moon orbits the Earth, the visible portion of that illuminated hemisphere that is changes over the course of one month.

2.2. Eclipse Alignments

- Students hold a flashlight, large globe, and a small ball modeling the Sun-Earth-Moon system.

- The model is then arranged to produce a solar and lunar eclipse to demonstrate how and in what circumstances eclipses occur.

- The goal of this demonstration is to provide students with a three-dimensional model of how the Sun-Earth-Moon system functions and what alignments are necessary to cause eclipses.

2.3. Timing of Eclipse

- Hula-hoops with balls fastened along them are used to show the inclination of the Moon's orbit relative to the Earth's orbit around the Sun.

- The hula-hoops are interlinked and one student holds the Earth hula-hoop such that their head is in the position of the Sun while the instructor holds the Moon hula-hoop.

- The instructor models how the timing of eclipses varies if the orbits of the Moon and the Earth are coplanar or inclined relative to each other.

- The goal of this demonstration is to model the inclination of the Moon's orbit and its effect on the timing of eclipses.

2.4. Models of Phases and Eclipses

- Students use a flashlight, tennis ball, and ping pong ball to model the Sun-Earth-Moon system.

- Students are given a series of data sets or observations related to lunar phases and eclipses of increasing detail and must adjust their model to account for the increasingly complex data.

- The goal of this activity is to challenge students to formulate a model of the Sun-Earth-Moon system based on raw data and observations, which they could then use to answer questions and make predictions about the system.

2.5. Eclipse Math and Visualization

- Students work in groups to solve a number of mathematical questions regarding the geometry of the Sun-Earth-Moon system, the shadows produced during an eclipse, and the visibility of different types of eclipses.

- Students then use the Voyage 4.5 computer program to further investigate and visualize these phenomena.

- The goal of this activity is to further develop the students' conceptual understanding of how the Sun-Earth-Moon system functions and the circumstances under which phases and eclipses occur.

3. Results

In general, the data did not show a large difference between the mean scores of the lecture and interactive sub-groups in the various age groups. However, when there was some significance to the disparity it was generally in favor of the lecture sub-group. It was also found that students of all ages and across both sub-groups typically had a stronger grasp on eclipses than they did on lunar phases.

The elementary group's scores demonstrated an overall incomplete understanding of the Sun-Earth-Moon system, which made conceptualizing the geometry of the system under different circumstances beyond most of their abilities. The variation in mean scores was in favor of the lecture group in all instances except the question that asked students to arrange pictures of lunar phases into the correct order.

The middle school sub-group only involved 9 students and thus was too small of a population from which to draw any statistically significant trends, although the general trends from the elementary and undergraduate groups still held more or less true.

The undergraduate group's scores showed lass variation than those of the elementary and middle school groups. There were not any marked differences in mean scores between the lecture and interactive sub-groups for any particular question or overall.

While these data seem to imply to some extent that lecture-based teaching is preferable to inquiry-based, thorough error-analysis of this study's methodology and consideration of the majority of previous scholarship on the subject suggest that these implications are most likely a result of inconsistencies in how the study was carried out and not definite reflections on the most effective teaching style. However, such results do not preclude any useful conclusions from being made.

4. Conclusions

4.1. Conclusions from the results of this study:

- A combination of lecture and interactive teaching methods would probably be the most effective way to teach students about the lunar phases and eclipses.

- Students of all ages generally have a better conceptual understanding of eclipses than of lunar phases.

- Older students' comprehension of a the subject is less dependent on teaching style than younger students.

- Many common misconceptions about lunar phases and eclipses persist among elementary, middle school and undergraduate students.

- The use of interactive, inquiry-based teaching methods in the classroom does not in itself guarantee increased conceptual understanding on the part of the students, but it is the quality with which such inquiry-based teaching methods are implemented that matters.

- Interactive, inquiry-based teaching methods have the potential to negatively affect students' understanding of a concept by inadvertently promoting a misconception, requiring instructors to be vigilant about which activities they choose to use and the way in which they incorporate them into their lessons.

- Effective assessment is vital to producing quality astronomy education research and to astronomy education itself.

References

National Center for Mathematics and Science, Earth-Moon-Sun Dynamics Instructional Notes, http://ncisla.wceruw.org/muse/earth-moon-sun/materials/build/materials2E/inotes/index.html

Newbury, P. et al., "Exploring the Solar System with a Human Orrery," University of British Columbia: Carl Wieman Science Education Initiative

Connecting People to Science
ASP Conference Series, Vol. 457
Joseph B. Jensen, James G. Manning, Michael G. Gibbs, and Doris Daou, eds.
© *2012 Astronomical Society of the Pacific*

Astronomy Outreach Activities in Chile: IYA 2009 and Beyond

Nikolaus Vogt, Moira Evans, Jonathan Aranda, Vanessa Gotta,
Ariel Monsalves, and Evelyn Puebla

*Departamento de Física y Astronomía, Universidad de Valparaíso, Avda. Gran
Bretaña 1111, Valparaiso, Chile*

Abstract. In Chile, one of the developing countries in Latin-America, there are large social differences that persist between the richest and the poorest citizens. On the other hand, Chile has the advantage of a special and unique resource, the incomparably clear and dry skies in the desert of Atacama in the north of the country. This advantage is being exploited by the installation of large and powerful international observatories. However, the Chilean people's perception of this resource and the corresponding advantages for their country are still underdeveloped and rather poor. Therefore, we have been conducting successful outreach activities at all levels during the past few years, with special highlights during the International Year of Astronomy 2009, including participation of our undergraduate physics and astronomy students, the local media like newspapers, radio, and TV stations, talks and workshops in schools, popular talks for the general public, exhibitions, contests, and other multi-media efforts. We briefly describe these activities and outline the difference between our situation and that existing in developed countries like the USA.

1. The present situation in Chile and in Valparaiso

The advantage of the uniquely clear skies in the north of Chile is being exploited by the installation of large and powerful observatories, like Cerro Tololo, Cerro Pachón, Las Campanas, La Silla, Paranál, ALMA, and the proposed E-ELT. In spite of the enormous investment in Chile by more developed countries, the impact on the country is yet very limited: the majority of the scientific teams at the aforementioned observatories are not Chileans, and there are many foreign academics in the few universities that are active in astronomical research (currently seven), mainly due to lack of qualified Chilean scientists.

The city and Region of Valparaiso is the most populated urban nucleus in Chile, apart from the metropolitan Region of Santiago. In order to improve the perception of astronomy in Valparaiso and its surroundings, our institute has established a tradition of organizing between nine and twelve monthly astronomy outreach talks every year, all of them given by professional astronomers and invited speakers from other Chilean universities. About 50–120 persons attend regularly, and we maintain a database in order to invite them personally for each new event. The topics of these talks are quite broad, from meteorites to black holes and the Big Bang. This tradition began in 2006 and we will continue it in future.

In the following sections we describe some additional projects, mainly directed to young people at the high school level, commenced in the International Year of Astronomy 2009, but also continued since then.

2. Talks, Workshops, Exposition and Internet Blog

The Comité Mixto ESO-Chilean Government fund for the Development of Astronomy and related Technology Disciplines was used for the project *Strengthening Teaching and Astronomy Outreach in the Region V of Chile.* In the following, we describe briefly the most important activities of this project.

A total of fourteen different astronomy talks and four workshops were prepared by advanced students of the Licenciatura en Fisica Mención Astronomía, and presented on over 70 occasions in different schools of Region V, in INACAP (Instituto Nacional de Capacitación) and in summer open air activities in the Botanical Gardens of Viña del Mar. In this way, we have reached more than 4000 students to date. Talks covered topics with titles such as: Meteorites and Comets, Origins of the Solar System, Exoplanets, History of Astronomy, Black Holes, Deciphering the Universe with the LHC Particle Accelerator, Radio Astronomy and its History, Stellar Evolution, and Exploration of the Solar System. Workshops covered The Solar System, Constellations in the Southern Skies, Galaxies, and Solar Activity. This same group is carrying out observing sessions with the students.

Other activities included an exhibition in the Natural History Museum in Valparaíso, titled *Chile: Ojos Al Universo*, exhibiting posters with astronomical content and pictures of the international observatories in the north of Chile (January–May 2009). More than 30,000 people visited the exhibition.

EXPLORA-Fifth Region and the University of Valparaíso organized a contest for high school students titled *Los Relatos Del Cielo* for the writing of stories related to astronomy; the winners received astronomy books and telescopes, and their stories were published on our webpage.

Finally, in 2010 a new Astronomy Blog http://www.astronosotros.cl was created in Spanish to emphasize our relationship with the Universe and directed to amateur astronomers, students, and the general public as the main audience. The purpose of the blog is to: (1) emphasize the obvious but often overlooked fact that our home the Earth is not an isolated place in the Universe; on the contrary, it is deeply related to outer space; (2) to divulge to the general public the latest advances in astronomy and space exploration in exciting and comprehensible language; (3) to answer and clear up frequent doubts regarding many objects and exotic situations in the world of science, especially those related to physics and astronomy; (4) to familiarize primary and high school students with the methods and achievements of basic science; and (5) to maintain a collection of national and international links useful to amateur astronomers and astronomy students in Chile.[1]

---

[1]For more details on these activities funded by Comité Mixto ESO-Chile see http://www.dfa.uv.cl/divulga

3. Radio astronomy for students

Using the opportunity of the International Year of Astronomy 2009, we are conduct-ing a pioneering outreach project titled *Teaching and Outreach of Radio Astronomy at the University of Valparaiso and at High Schools in the V Region of Chile* funded by ALMA-CONICYT. It incorporates basic radio astronomy into the teaching program of the physics and astronomy courses at the University of Valparaíso and in high schools. A total of seven JOVE radio receivers were distributed to carefully selected schools in the V Region of Chile from a total of 25 applicants, those that fulfilled the neces-sary conditions regarding a suitable place to install the antenna, far away from roads with high traffic and cell phone or television antennae. Each of the schools erected an antenna in order to receive the radio signals from the Sun and Jupiter at 20.1 MHz, per-forming regular observations since about November 2010. The Sun is the only heavenly body easily observable during normal school hours, and therefore of special importance for educational purposes. It enables the students to compare the visible solar activity (sun spots and prominances) with solar emissions at radio frequencies, opening a totally new view of the Universe to the young generation.

4. Astronomy talks in all Regions of Chile

In collaboration with SOCHIAS, the Chilean Astronomical Society, we conducted the ambitious project *Taking the International Year of Astronomy to the Regions of Chile* funded by GEMINI-CONICYT. Our project included the presentation of a minimum of five talks on astronomical topics during the IYA 2009 and a similar number in 2010 in each of the 15 Regions of Chile. For this project we counted on the valuable co-operation of more than 50 colleagues from Chilean universities, each with a Ph.D. in Astronomy, who travelled to the Regions to give their talks in schools and in munici-palities, organized in collaboration with EXPLORA-CONICYT, whose regional offices worked out the details for each event. Each talk was presented twice, once for scholars in the morning and once for the general public later in the afternoon or evening. This project offered hitherto unexploited channels to inform the public and, especially the high school students, about new insights and the professional possibilities of astron-omy and related sciences in Chile. We have reached a total of ~18,000 persons, mostly young students, often at remote places which never had any previous contact with a professional astronomer.

5. Conclusions

Chile has the clearest skies on planet Earth, guaranteeing the most efficient use of ground based optical and infrared telescopes located there. About 70% of the world-wide reflecting telescope collecting area at these wavelengths is already or will soon be assembled in the north of Chile. The ALMA interferometer is also a unique radio astronomy observatory installed in Chile. Ten percent of the total observing time at all these facilities is reserved for astronomers working at Chilean research institutions.

At the present there are only 62 academic tenure track positions in astronomy dis-tributed among seven Chilean Universities with undergraduate astronomy programs and research groups. Three of these groups reside in the capital Santiago (60% of the total); the remaining four groups are at provincial universities (Antofagasta, Concepción, La Serena, and Valparaiso, together 40% of the total). All tenure track positions in astronomy are open to world-wide competition, and astronomers compose an over-all fraction of more than a third (35%) of foreign academic staff members at the Chilean universities. This fraction is very unequally distributed: in Santiago there are only 14% foreigners, while in provinces they present a majority of 68%.

These numbers reflect the urgent need for young, well prepared Chilean astronomers who can compete at international level and who are able to utilize the unique astronomical facilities in the north of Chile. They can only emerge from attractive outreach projects focused on all levels of school education, awakening interest in basic sciences which tend to be neglected by societies in countries of the third world, as in Chile. Projects such as those presented here are efforts in this direction. We are rather confident of their success in intermediate and long-term prospects.

Acknowledgments. The projects presented here were supported by the Comité Mixto ESO-Gobierno de Chile, as well as the funds GEMINI-CONICYT 32080028 (presented through SOCHIAS) and ALMA-CONICYT 31080036. We would like to thank EXPLORA-CONICYT for their excellent organization of the events in all Regions of Chile, as well as the more than 50 colleagues who collaborated in presenting the talks. Also we would like to thank all those who participated directly or indirectly to make these projects possible and to bring a new awareness of the potential and opportunities for astronomy to the people of Chile.

Connecting People to Science
ASP Conference Series, Vol. 457
Joseph B. Jensen, James G. Manning, Michael G. Gibbs, and Doris Daou, eds.
© *2012 Astronomical Society of the Pacific*

That's MY Astronaut! Could Democratic Space Tourism Contribute to Earth Stewardship?

Elizabeth Forbes Wallace

MY-Astronaut.Org, 7516 Holly Avenue, Takoma Park, Maryland 20912, USA

Abstract.   Many studies have been done on the physical and biological effects of space on the human body. The psychological effects of living in space are also being analyzed including the stressors from living in an isolated environment. But are we paying enough attention to what seems to be a positive effect on the human psyche, that is, the effect on astronauts and cosmonauts of the magnificent view of Earth from space? Does the length of time spent looking out the window affect our consciousness? Who comes back changed? And why? Such a social experiment needs more participants. Could democratic access to the view via suborbital space tourism change our Earth for the better?

1. We Did It! Now, What If. . .

"We did it!" cheered crowds worldwide as Apollo 11 crew members, the first humans to set foot on the moon, toured their home planet in 1969. WE put men on the moon. Not just Americans, but the world did. WE threw ticker tape parades. WE gave them a hero's welcome. WE the Whole Earth.

What if WE could do something as spectacularly unifying as that again?

While only four decades ago, WE were in awe of the Earthrise photograph taken by the Apollo 8 crew, and WE were inspired to start the Green Movement. But a more recent frequent traveler, Astronaut John Grunsfeld, states that some of the ecological scars Earth has suffered since then can be seen from space. Even the words of the first man in space, Cosmonaut Yuri Gagarin, who implored us to "preserve (Earth) and enhance its future, not destroy it" seem to have disappeared into a vacuum. Now our children implore us.

To succeed we need massive amounts of international cooperation. Perhaps not just the national space agency kind.

"A new type of thinking is essential if mankind is to survive and move toward higher levels," according to Albert Einstein. Perhaps WE need a "higher" education in order to survive. A physically higher education. One sporting classrooms with a view of Earth—on suborbital spaceflights.

2. A Social Experiment of the Highest Order

What happens up there? What happens to the human psyche? ISS Astronaut Leland Melvin said that the most important thing he experienced in space was the abundance of "international cooperation." He echoed Cosmonaut Alexei Leonov's claim that "every manned spaceflight has serviced to reaffirm this feeling of a unified human race."

Astronaut Gerhard Thiele from Germany told me that more ideas came to him while writing in his journal sitting next to a window on his STS 99 mission. Astronaut David Scott wonders just how the experience of sitting in the "right hand seat" on the return voyage transformed the lives of several Apollo crew members such as Alan Bean and Edgar Mitchell. Did they absorb their experience differently? What inspired them to take different or unusual career paths from art to the study of human potential?

"What if ...?" is the question that ignites all imaginations to set forth on adventures from science fiction to science fact, from social justice to social innovation.

"What if" we created the biggest social experiment to date? One with a view?

What if you had a chance to be a part of it?

2.1.   How Many of Us Need to Go?

When I asked Astronaut Grunsfeld if he agreed that if more people saw Earth from Space it could change our planet for the better. He said yes. I asked how many people he thought it would take? 535, he answered. While I was impressed that he had done the math, he continued, "That's how many representatives are on Capitol Hill, but they'll never go."

In 1970 WE numbered almost 3 billion. Now almost 7 billion of us live together.

Only .00000007 of us have seen our home from Space.

Only .00000006 of us have purchased $200,000 tickets on Virgin Galactic.

If the "overview effect" could possibly have a transformative influence on humans of any background, then perhaps WE have the responsibility to send as many people to space as possible on suborbital spaceflights.

When I asked a panel of space tourism CEO's at the Next Generation Suborbital Researchers Conference 2011 what they thought their social responsibility was regarding getting as many people to experience the overview as possible, one replied, "Send them to us and we'll fly them up!" But at what cost? $200K, $100K?

What if... you didn't have to save up for it?

2.2.   Vote For and Fund Suborbital Space Heros and Earth Stewards

"Heroes venture forth from the world of common day into a region of supernatural wonder. Fabulous forces are encountered...a decisive victory is won...The hero comes back...with the power to bestow boons on his fellow man" (Joseph Campbell).

What if...

WE, the Whole Earth population, made it our responsibility to make it possible for anyone, from any country, to declare "I want to be an astronaut!" so that they could venture forth into a region of suborbital supernatural wonder.

WE supported people from the world of common day, loggers and poets, waste management workers and musicians, fishermen and filmmakers, who then created their own international teams of six who would travel together on a Virgin Galactic flight.

WE were inspired by the way their commitment to prepare for the trip and could be a part of their hero's journey via social networking.

WE funded Earth stewards whom we felt might see our home from Space and return with a perspectives that, once shared, might incite us all to make big commitments to sustainability and peaceful personal and international relations.

WE, billions of us, watched as their spaceships were launched at Spaceport America and cheered "Those are OUR astronauts!"

What if, like so many shared adventures, their stories and visions became OURS? What if WE, the Whole Earth, could do something spectacularly unifying via democratic space tourism?

Connecting People to Science
ASP Conference Series, Vol. 457
Joseph B. Jensen, James G. Manning, Michael G. Gibbs, and Doris Daou, eds.
© *2012 Astronomical Society of the Pacific*

Astronomy Education and Public Outreach Research: A First Attempt at a Resource Guide

Andrew Fraknoi

Foothill College, 12345 El Monte Road, Los Altos Hills, California 94022, USA

Astronomical Society of the Pacific, 390 Ashton, San Francisco, California 94066, USA

Abstract.   This article contains a list of book and overview articles on astronomy education research.

Bailey, Janelle, et al., Conducting Astronomy Education Research: A Primer. 2011, W. H. Freeman. The first resource to turn to; more on classroom research than informal.

Bailey, J. & Slater, T., A Review of Astronomy Education Research, in Astronomy Education Review, 2003, vol. 2, no. 2, p. 20, http://dx.doi.org/10.3847/AER2003015

Bailey, J. & Slater, T., Finding the Forest Amid the Trees: Tools for Evaluating Astronomy Education and Public Outreach Projects, in Astronomy Education Review, 2005, vol. 3, no. 2, p. 47, http://dx.doi.org/10.3847/AER2004016

Bailey, J. & Slater, T. 2005, Resource Letter: Astronomy Education Research, in American Journal of Physics, vol. 73, no. 8, p. 677

Brogt, E., et al., Regulations and Ethical Considerations for Astronomy Education Research, (2 parts), Astronomy Education Review, 2007, vol. 6, no. 1, p. 43, http://dx.doi.org/10.3847/AER2007004; and vol. 6, no. 2, p. 99, http://dx.doi.org/10.3847/AER2007021

Lelliott, A. and Rollnick, M. 2010, Big Ideas: A Review of Astronomy Education Research 1974–2008, International Journal of Science Education, 32, 1771. Analysis of 103 peer-reviewed journal articles published between 1974–2008; most describe conceptions in astronomy, 40% investigated interventions. Has suggestions for future research directions.

Prather, E. et al. 2009, Teaching and Learning Astronomy in the 21st Century, in Physics Today, vol. 62, no. 10, p. 41. Review of research on teaching Astro 101 and results of a national study of 4000 students who took the Light and Spectroscopy Concept Inventory (Oct. 2009). http://astronomy101.jpl.nasa.gov/files/Teaching%20and%20Learning%20Astronomy%20in%20the%2021st%20Cen pdf

Review papers by Cary Sneider et al. on research on student learning about basic astronomy:

1. Gravity and Free Fall: http://dx.doi.org/10.3847/AER2006018
2. Trajectories and Orbits: http://dx.doi.org/10.3847/AER2006019
3. Phases of the Moon& Eclipses: http://dx.doi.org/10.3847/AER2005002
4. The Earth's Shape and Gravity: http://dx.doi.org/10.3847/AER2003017
5. The Seasons: http://dx.doi.org/10.3847/AER2010035

1. Evaluation of Informal Science Education (Relevant to Astronomy)

1.1. Key Documents

Friedman, Alan, ed. Framework for Evaluating Impacts of Informal Science Education Projects. 2008, National Science Foundation. Free on the web at: http://insci.org/resources/Eval_Framework.pdf (This guide is full of practical hints and specific ideas and includes chapters on evaluating exhibits, mass media, youth and community programs, learning technologies and collaborations.)

Frechtling, Joy, et al.: NSF User-Friendly Handbook for Project Evaluation (2002, National Science Foundation) http://www.nsf.gov/pubs/2002/nsf02057/start.htm (A more general introduction to evaluation from the education division of NSF.)

Elements of a Program Evaluation (from the NASA IDEAS Program): http://ideas.stsci.edu/Evaluation.shtml (A concise introduction and checklist)

1.2. Web-based Articles

Dussault, Mary, How Do Visitors Understand the Universe, at http://www.astc.org/resource/visitors/universe.htm (Some studies testing the understanding of astronomical concepts—rather than mere factoids—by museum visitors.) See also: Atlas, Toby, et al. Front-end Evaluation for Cosmic Questions Exhibit at: http://www.cfa.harvard.edu/seuforum/download/Cosmic_front_end_eval_2000.pdf and Karp, J., et al., Summative Evaluation for Cosmic Questions: http://www.cfa.harvard.edu/seuforum/download/CQ_Exec_Sum_Final_2003.pdf

Flagg, Barbara, Summative Evaluation of the Planetarium Show Black Holes: The Other Side of Infinity, at: http://www.informalscience.org/evaluation/report_view.php?id=283

Goodman Research Group, Evaluation of the Origins: NOVA Mini-series (and its Outreach), at: http://www.informalscience.org/evaluation/report_view.php?id=268

Korn, Randi, Evaluation of the MarsQuest Museum Exhibit, at: http://www.informalscience.org/evaluation/report_view.php?id=34

372   *Fraknoi*

Storksdieck, M., et al., Amateur Astronomers as Informal Science Ambassadors: Results of an Online Survey, at: `http://www.astrosociety.org/education/resources/ResultsofSurvey_FinalReport.pdf`

Stroud, Nicholas, et al., Toward a Methodology for Informal Astronomy Education Research, in Astronomy Education Review, vol. 5, issue 2, p. 146: `http://dx.doi.org/10.3847/AER2006023` (Action Evaluation and How it can be Used in Museums.)

1.3. Websites

Boddington, Andy & Coe, Trudy, So Did It Work: Evaluating Public Understanding of Science Events, at: `http://www.copus.org.uk/pubs_guides_sodiditwork.html` (A quick guide to event evaluation.)

Informal Science Web Site Evaluation Page: `http://www.informalscience.org/evaluation/` (This is a web site for informal science education professionals, funded by NSF. Their evaluation page has a number of good general resources and a number of finished evaluation reports specific to astronomy and physical science.)

On-line Evaluation Resource Library: `http://oerl.sri.com/` (Repository of documents and plans for evaluation professionals.)

Visitor Studies Association Archives: `http://www.visitorstudiesarchives.org/index.php` (Past publications of organization devoted to the systematic study of how their audiences perceive and use museums.)

Council for Advancement of Informal Science Education: `http://insci.org/` (A new NSF-funded group to bring together those doing science outside the classroom. Website has some impact studies and promises to grow into a useful resource.)

Field-Tested Learning Assessment Guide for Science Instructors: `http://www.flaguide.org/`

Concept Inventories for Teaching Astronomy (including inventories about the moon's phases, light and spectroscopy, etc): `http://astronomy101.jpl.nasa.gov/teachingstrategies/teachingdetails/?StrategyID=4`

Each year, Astronomy Education Review (`http://aer.aas.org`) lists papers in the field not published in AER.

1.4. Print Articles

Burtnyk, Kimberly, Impact of Observatory Visitor Centers on the Public's Understanding of Astronomy, in Publications of the Astronomical Society of Australia, 2000, vol. 17, p. 275. A first study of the impact of a visit to one of the observatories in Australia.

Fields, D. 2009, What Do Students Gain from a Week at Science Camp: Youth Perceptions and the Design of an Immersive, Research-Oriented Astronomy Camp, in International Journal of Science.

Education, vol. 31, p. 151. Compared goals and design of such a camp to participant's perceptions, gained through interviews.

Geller, H. and Frazier, W. 2010, Assessing Planetarium Programs for Contents and Pedagogy, The Planetarian, 39(2), 9. Two university researchers on how they did an evaluation of a K–12 planetarium facility for a county and general consideration for such evaluations. (June 2010)

Miller, Jon, The Evaluation of Adult Science Learning, in Narasimhan, L., et al., eds. NASA Office of Space Science Education and Public Outreach Conference 2002 (2004, ASP Conference Series vol. 319, p. 26.) Suggests some methods for measuring adult learning in a planetarium setting.

Author Index

ASTRONOMICAL SOCIETY OF THE PACIFIC

THE ASTRONOMICAL SOCIETY OF THE PACIFIC is an international, nonprofit, scientific, and educational organization. Some 120 years ago, on a chilly February evening in San Francisco, astronomers from Lick Observatory and members of the Pacific Coast Amateur Photographic Association—fresh from viewing the New Year's Day total solar eclipse of 1889 a little to the north of the city—met to share pictures and experiences. Edward Holden, Lick's first director, complimented the amateurs on their service to science and proposed to continue the good fellowship through the founding of a Society "to advance the Science of Astronomy, and to diffuse information concerning it." The Astronomical Society of the Pacific (ASP) was born.

The ASP's purpose is to increase the understanding and appreciation of astronomy by engaging scientists, educators, enthusiasts, and the public to advance science and science literacy. The ASP has become the largest general astronomy society in the world, with members from over 70 nations.

The ASP's professional astronomer members are a key component of the Society. Their desire to share with the public the rich rewards of their work permits the ASP to act as a bridge, explaining the mysteries of the universe. For these members, the ASP publishes the Publications of the Astronomical Society of the Pacific (PASP), a well-respected monthly scientific journal. In 1988, Dr. Harold McNamara, the PASP editor at the time, founded the ASP Conference Series at Brigham Young University. The ASP Conference Series shares recent developments in astronomy and astrophysics with the professional astronomy community.

To learn how to join the ASP or to make a donation, please visit http://www.astrosociety.org.

ASTRONOMICAL SOCIETY OF THE PACIFIC
MONOGRAPH SERIES

Published by the Astronomical Society of the Pacific

The ASP Monograph series was established in 1995 to publish select reference titles.
For electronic versions of ASP Monographs, please see
http://www.aspmonographs.org.

INFRARED ATLAS OF THE ARCTURUS SPECTRUM, 0.9-5.3μm
eds. Kenneth Hinkle, Lloyd Wallace, and William Livingston (1995)
ISBN: 1-886733-04-X, e-book ISBN: 978-1-58381-687-5

**VISIBLE AND NEAR INFRARED ATLAS
OF THE ARCTURUS SPECTRUM 3727-9300Å**
eds. Kenneth Hinkle, Lloyd Wallace, Jeff Valenti, and Dianne Harmer (2000)
ISBN: 1-58381-037-4, e-book ISBN: 978-1-58381-688-2

ULTRAVIOLET ATLAS OF THE ARCTURUS SPECTRUM 1150-3800Å
eds. Kenneth Hinkle, Lloyd Wallace, Jeff Valenti, and Thomas Ayres (2005)
ISBN: 1-58381-204-0, e-book ISBN: 978-1-58381-689-9

**HANDBOOK OF STAR FORMING REGIONS: VOLUME I
THE NORTHERN SKY**
ed. Bo Reipurth (2008)
ISBN: 978-1-58381-670-7, e-book ISBN: 978-1-58381-677-6

**HANDBOOK OF STAR FORMING REGIONS: VOLUME II
THE SOUTHERN SKY**
ed. Bo Reipurth (2008)
ISBN: 978-1-58381-671-4, e-book ISBN: 978-1-58381-678-3

A complete list and electronic versions of ASPCS volumes may be found at
http://www.aspbooks.org.

All book orders or inquiries concerning the ASP Conference Series, ASP
Monographs, or International Astronomical Union Volumes published by the ASP
should be directed to:

Astronomical Society of the Pacific
390 Ashton Avenue
San Francisco, CA 94112-1722 USA
Phone: 800-335-2624 (within the USA)
Phone: 415-337-2126
Fax: 415-337-5205
Email: service@astrosociety.org

For a complete list of ASP publications, please visit
http://www.astrosociety.org.